For Alleve,

Here at long last is my book. I thought you would like to have a copy to look through. It is a reference volume being marketed primarily to libraries.

With love,
Kate
December 31, 2008

Religion and the Physical Sciences

Recent Titles in
Greenwood Guides to Science and Religion

Religion and the Physical Sciences

Kate Grayson Boisvert

Greenwood Guides to Science and Religion
Richard Olson, Series Editor

Greenwood Press
Westport, Connecticut • London

Library of Congress Cataloging-in-Publication Data

Boisvert, Kate Grayson, 1943–
    Religion and the physical sciences / Kate Grayson Boisvert.
        p.   cm. — (Greenwood guides to science and religion)
    Includes bibliographical references and index.
    ISBN 978–0–313–33284–5 (alk. paper)
    1. Religion and science.   I. Title.
    BL240.3.B65   2008
    201′.65—dc22   2007048217

British Library Cataloguing in Publication Data is available.

Library of Congress Catalog Card Number: 2007048217
ISBN: 978–0–313–33284–5

First published in 2008

Greenwood Press, 88 Post Road West, Westport, CT 06881
An imprint of Greenwood Publishing Group, Inc.
www.greenwood.com

Printed in the United States of America

The paper used in this book complies with the
Permanent Paper Standard issued by the National
Information Standards Organization (Z39.48–1984).

10 9 8 7 6 5 4 3 2 1

To the Light
That will illuminate
The mystery of the cosmos

Contents

Contents   ix

Illustrations

Series Foreword

For nearly 2,500 years, some conservative members of societies have expressed concern about the activities of those who sought to find a naturalistic explanation for natural phenomena. In 429 BCE, for example, the comic playwright, Aristophanes parodied Socrates as someone who studied the phenomena of the atmosphere, turning the awe-inspiring thunder which had seemed to express the wrath of Zeus into nothing but the farting of the clouds. Such actions, Aristophanes argued, were blasphemous and would undermine all tradition, law, and custom. Among early Christian spokespersons there were some, such as Tertullian, who also criticized those who sought to understand the natural world on the grounds that they "persist in applying their studies to a vain purpose, since they indulge their curiosity on natural objects, which they ought rather [direct] to their Creator and Governor" (Tertullian 1896–1903, 133).

In the twentieth century, though a general distrust of science persisted among some conservative groups, the most intense opposition was reserved for the theory of evolution by natural selection. Typical of extreme antievolution comments is the following opinion offered by Judge Braswell Dean of the Georgia Court of Appeals: "This monkey mythology of Darwin is the cause of permissiveness, promiscuity, pills, prophylactics, perversions, pregnancies, abortions, pornography, pollution, poisoning, and proliferation of crimes of all types" (Toumey 1994, 94).

It can hardly be surprising that those committed to the study of natural phenomena responded to their denigrators in kind, accusing them of willful ignorance and of repressive behavior. Thus, when Galileo Galilei was warned against holding and teaching the Copernican system of

astronomy as true, he wielded his brilliantly ironic pen and threw down a
gauntlet to religious authorities in an introductory letter "To the Discerning Reader" at the beginning of his great *Dialogue Concerning the Two Chief
World Systems*:

Several years ago there was published in Rome a salutory edict which, in order to
obviate the dangerous tendencies of our age, imposed a seasonable silence upon
the Pythagorean [and Copernican] opinion that the earth moves. There were those
who impudently asserted that this decree had its origin, not in judicious inquiry,
but in passion none too well informed. Complaints were to be heard that advisors
who were totally unskilled at astronomical observations ought not to clip the wings
of reflective intellects by means of rash prohibitions.

   Upon hearing such carping insolence, my zeal could not be contained. (Galilei
1953, 5)

   No contemporary discerning reader could have missed Galileo's anger
and disdain for those he considered enemies of free scientific inquiry.
   Even more bitter than Galileo was Thomas Henry Huxley, often known
as "Darwin's bulldog." In 1860, after a famous confrontation with the
Anglican Bishop Samuel Wilberforce, Huxley bemoaned the persecution
suffered by many natural philosophers, but then he reflected that the
scientists were exacting their revenge:

Extinguished theologians lie about the cradle of every science as the strangled
snakes beside that of Hercules; and history records that whenever science and
orthodoxy have been fairly opposed, the latter has been forced to retire from the
lists, bleeding and crushed, if not annihilated; scotched if not slain. (Moore 1979,
60)

   The impression left, considering these colorful complaints from both
sides is that science and religion must continually be at war with one
another. That view of the relation between science and religion was reinforced by Andrew Dickson White's *A History of the Warfare of Science
with Theology in Christendom*, which has seldom been out of print since it
was published as a two volume work in 1896. White's views have shaped
the lay understanding of science and religion interactions for more than a
century, but recent and more careful scholarship has shown that confronta-
tional stances do not represent the views of the overwhelming majority of
ether scientific investigators or religious figures throughout history.
   One response among those who have wished to deny that conflict constitutes the most frequent relationship between science and religion is to
claim that they cannot be in conflict because they address completely different human needs and therefore have nothing to do with one another.

This was the position of Immanuel Kant who insisted that the world of natural phenomena, with its dependence on deterministic causality, is fundamentally disjoint from the noumenal world of human choice and morality, which constitutes the domain of religion. Much more recently, it was the position taken by Stephen Jay Gould in *Rocks of Ages: Science and Religion in the Fullness of Life* (1999). Gould writes:

I ... do not understand why the two enterprises should experience any conflict. Science tries to document the factual character of the natural world and to develop theories that coordinate and explain these facts. Religion, on the other hand, operates in the equally important, but utterly different realm of human purposes, meanings, and values. (Gould 1999, 4)

In order to capture the disjunction between science and religion, Gould enunciates a principle of "Non-overlapping magisterial," which he identifies as "a principle of respectful noninterference" (Gould 1999, 5)

In spite of the intense desire of those who wish to isolate science and religion from one another in order to protect the autonomy of one, the other, or both, there are many reasons to believe that theirs is ultimately an impossible task. One of the central questions addressed by many religions is what is the relationship between members of the human community and the natural world. This question is a central question addressed in "Genesis," for example. Any attempt to relate human and natural existence depends heavily on the understanding of nature that exists within a culture. So where nature is studied through scientific methods, scientific knowledge is unavoidably incorporated into religious thought. The need to understand "Genesis" in terms of the dominant understandings of nature thus gave rise to a tradition of scientifically informed commentaries on the six days of creation which constituted a major genre of Christian literature from the early days of Christianity through the Renaissance.

It is also widely understood that in relatively simple cultures—even those of early urban centers—there is a low level of cultural specialization, so economic, religious, and knowledge producing specialties are highly integrated. In Bronze Age Mesopotamia, for example, agricultural activities were governed both by knowledge of the physical conditions necessary for successful farming and by religious rituals associated with plowing, planting, irrigating, and harvesting. Thus religious practices and natural knowledge interacted in establishing the character and timing of farming activities.

Even in very complex industrial societies with high levels of specialization and division of labor, the various cultural specialties are never completely isolated from one another and they share many common values and assumptions. Given the linked nature of virtually all institutions in any culture it is the case that when either religious or scientific institutions change substantially, those changes are likely to produce pressures for change in the other. It was probably true, for example, that the attempts of Presocratic investigators of nature, with their emphasis on uniformities in the natural world and apparent examples of events systematically directed toward particular ends, made it difficult to sustain beliefs in the old Pantheon of human-like and fundamentally capricious Olympian gods. But it is equally true that the attempts to understand nature promoted a new notion of the divine—a notion that was both monotheistic and transcendent, rather than polytheistic and immanent—and a notion that focused on both justice and intellect rather than power and passion. Thus early Greek natural philosophy undoubtedly played a role not simply in challenging, but also in transforming Greek religious sensibilities.

Transforming pressures do not always run from scientific to religious domains, moreover. During the Renaissance, there was a dramatic change among Christian intellectuals from one that focused on the contemplation of God's works to one that focused on the responsibility of the Christian for caring for his fellow humans. The active life of service to humankind, rather than the contemplative life of reflection on God's character and works, now became the Christian ideal for many. As a consequence of this new focus on the active life, Renaissance intellectuals turned away from the then dominant Aristotelian view of science, which saw the inability of theoretical sciences to change the world as a positive virtue. They replaced this understanding with a new view of natural knowledge, promoted in the writings of men such as Johann Andreae in Germany and Francis Bacon in England, which viewed natural knowledge as significant only because it gave humankind the ability to manipulate the world to improve the quality of life. Natural knowledge would henceforth be prized by many because it conferred power over the natural world. Modern science thus took on a distinctly utilitarian shape at least in part in response to religious changes.

Neither the conflict model nor the claim of disjunction, then, accurately reflect the often intense and frequently supportive interactions between religious institutions, practices, ideas, and attitudes on the one hand, and scientific institutions, practices, ideas, and attitudes on the other. Without denying the existence of tensions, the primary goal of the volumes of this series is to explore the vast domain of mutually supportive and/or transformative interactions between scientific institutions, practices, and knowledge and religious institutions, practices, and beliefs. A second goal is to offer the opportunity to make comparisons across space, time, and cultural configuration. The series will cover the entire globe, most major faith traditions, hunter-gatherer societies in Africa and Oceania as well as advanced industrial societies in the West, and the span of time from classical antiquity to the present. Each volume will focus on a particular cultural tradition, a particular faith community, a particular time period, or a particular scientific domain, so that each reader can enter the fascinating story of science and religion interactions from a familiar perspective. Furthermore, each volume will include not only a substantial narrative or interpretive core, but also a set of primary documents which will allow the reader to explore relevant evidence, an extensive bibliography to lead the curious to reliable scholarship on the topic, and a chronology of events to help the reader keep track of the sequence of events involved and to relate them to major social and political occurrences.

So far I have used the words "science" and "religion" as if everyone knows and agrees about their meaning and as if they were equally appropriately applied across place and time. Neither of these assumptions is true. Science and religion are modern terms that reflect the way that we in the industrialized West organize our conceptual lives. Even in the modern West, what we mean by science and religion is likely to depend on our political orientation, our scholarly background, and the faith community that we belong to. Thus, for example, Marxists and Socialists tend to focus on the application of natural knowledge as the key element in defining science. According to the British Marxist scholar, Benjamin Farrington, "Science is the system of behavior by which man has acquired mastery of his environment. It has its origins in techniques . . . in various activities by which man keeps body and soul together. Its source is experience, its aims, practical, its *only* test, that it works"(Farrington 1953). Many of those who study natural knowledge in preindustrial societies are also primarily interested in knowledge as it is used and are relatively open regarding the kind of entities posited by the developers of culturally specific natural knowledge systems or "local sciences." Thus, in his *Zapotec Science: Farming and Food in the Northern Sierra of Oaxaca*, Roberto González insists that

Zapotec farmers . . . certainly practice science, as does any society whose members engage in subsistence activities. They hypothesize, they model problems, they experiment, they measure results, and they distribute knowledge among peers and to younger generations. But they typically proceed from markedly different premises—that is, from different conceptual bases—than their counterparts in industrialized societies. (González 2001, 3)

Among the "different premises" is the presumption of Zapotec scientists that unobservable spirit entities play a significant role in natural phenomena.

Those more committed to liberal pluralist society and to what anthropologists like González are inclined to identify as "cosmopolitan science," tend to focus on science as a source of objective or disinterested knowledge, disconnected from its uses. Moreover they generally reject the positing of unobservable entities, which they characterize as "supernatural." Thus, in an *Amicus Curiae* brief filed in connection with the 1986 Supreme Court case which tested Louisiana's law requiring the teaching of creation science along with evolution, for example, seventy-two Nobel Laureates, seventeen state academies of science, and seven other scientific organizations (Seventy-two Nobel Laureates 1986, 24) argued that

Science is devoted to formulating and testing naturalistic explanations for natural phenomena. It is a process for systematically collecting and recording data about the physical world, then categorizing and studying the collected data in an effort to infer the principles of nature that best explain the observed phenomena. Science is not equipped to evaluate supernatural explanations for our observations; without passing judgement on the truth or falsity of supernatural explanations, science leaves their consideration to the domain of religious faith.

No reference whatsoever to uses appears in this definition. And its specific unwillingness to admit speculation regarding supernatural entities into science reflects a society in which cultural specialization has proceeded much farther than in the village farming communities of southern Mexico.

In a similar way, secular anthropologists and sociologists are inclined to define the key features of religion in a very different way than members of modern Christian faith communities. Anthropologists and sociologists focus on communal rituals and practices which accompany major collective and individual events—plowing, planting, harvesting, threshing, hunting, preparation for war (or peace), birth, the achievement of manhood or womanhood, marriage (in many cultures), childbirth, and death. Moreover, they tend to see the major consequence of religious practices as the intensification of social cohesion. Many Christians, on the other hand, view the primary goal of their religion as personal salvation, viewing society as at best a supportive structure and at worst, a distraction from their own private spiritual quest.

Thus, science and religion are far from uniformly understood. Moreover, they are modern Western constructs or categories whose applicability to the temporal and spatial "other" must always be justified and must always be understood as the imposition of modern ways of structuring institutions, behaviors, and beliefs on a context in which they could not have been categories understood by the actors involved. Nonetheless it does seem to us not simply permissible, but probably necessary to use these categories at the start of any attempt to understand how actors from other times and places interacted with the natural world and with their fellow humans. It may ultimately be possible for historians and anthropologists to understand the practices of persons distant in time and/or space in terms that those persons might use. But that process must begin by likening the actions of others to those that we understand from our own experience, even if the likenesses are inexact and in need of qualification.

The editors of this series have not imposed any particular definition of science or of religion on the authors, expecting that each author will develop either explicit or implicit definitions that are appropriate to their own scholarly approaches and to the topics that they have been assigned to cover.

Richard Olson

Acknowledgments

My special thanks go to a number of people who have made this work possible. First, I am most grateful to Richard Olson, the series editor, for proposing the project and for reviewing many chapters, and I am extremely indebted to Kevin Downing, the Greenwood Press Editor, for his support and tremendous patience. For their readings of specific chapters, I also wish to thank colleagues Mark Lewis and Kurt Crowder. The help of my husband, Nick, has been inestimable, not only for his extremely insightful critiques of most chapters but also for his artistic insights about the creative process and daily support of all kinds. This book could not have been done without him. Cherie Plumlee was also most helpful in securing photography and permissions. I am also very grateful to Sufism Reoriented for allowing a rendering of Meher Baba's chart of creation from their publication *God Speaks* and to artist Norman Remer for his beautiful execution of the drawing. Finally, for critiquing an important chapter on Eastern cosmology I wish to thank Sufism's President Ira Deitrick, and for being able to review several chapters and providing much encouragement and support, I am most deeply grateful to Murshida Carol Weyland Conner.

Chronology of Events

1859	Publication of *On the Origin of the Species* by Charles Darwin
1869–1870	First Vatican Council of the Roman Catholic Church
1879	Birth of Albert Einstein
1888	Initiation of Gifford Lectures, on-going yearly series of invited talks at four Scottish universities on natural theology, religion, science, and philosophy
1888	Pope Leo XII reopens the Vatican Observatory, which continues to operate today, mostly directed by Jesuit astronomers
1893	First World Parliament of Religions in Chicago; Hindu leader Vivekenanda delivers inspiring message of world and religious unity
1894	Birth of Meher Baba and Georges Lemaitre
1896	Henri Becquerel discovers radioactivity of uranium
1897	J. J. Thompson discovers the electron
1898	Pierre and Marie Curie observe emission of smaller particles from new element radium
1900	Max Planck proposes quantum hypothesis for radiant energy
1905	Einstein's "miracle year"—publishes three seminal papers on special relativity, the photoelectric effect, and Brownian motion (supporting existence of atoms)

1906	J. J. Thompson proposes "plum pudding model" for the atom
1911	Ernest Rutherford proposes nuclear model of the atom
1913	Niels Bohr explains light spectrum of hydrogen with quantum model of the atom
1914–1918	World War I
1916	Einstein presents general theory of relativity
1916	Karl Schwarzschild predicts existence of black holes
1918	Harlow Shapley determines that Earth is not the center of the Milky Way Galaxy
1919	General relativity confirmed during solar eclipse
1919	Rutherford experiment achieves transmutation of elements: nitrogen converted to oxygen
1922	Alexander Friedmann demonstrates that general relativity predicts an expanding universe
1922	Hypothesis for chemical origin of life suggested by Alexander Oparin
1922	Vesto Slipher summarizes findings that almost all fuzzy spiral nebulae have redshifts—are receding from Earth
1923	Edwin Hubble determines that fuzzy spiral nebulae are distant galaxies beyond the Milky Way
1923	Louis de Broglie suggests that matter particles have wave-like properties
1925–1927	Werner Heisenberg and Erwin Schrodinger develop theory of quantum mechanics, including Heisenberg uncertainty principle and Schrodinger wave equation
1925	Meher Baba begins life-long silence and Scopes Monkey trial begins on July 10
1925	Arthur S. Eddington launches study of stellar structure
1927	Georges Lemaitre proposes Big Bang model of expanding universe and connects it to galactic redshifts; suggests "primeval atom" event later
1928	Paul Dirac predicts antimatter particles while combining quantum theory with special relativity

1929   Theory of thermonuclear fusion as the source of energy of the sun and stars first investigated by Fritz Houtermans and Robert Atkinson

1929   Hubble demonstrates expansion of the universe by discovering the linear redshift-distance relation for galaxies.

1929   Publication of Alfred North Whitehead's *Process and Reality*— major work of process philosophy

1932   Subrahmanyan Chandrasekhar predicts existence of collapsed white dwarf and neutron stars

1932   Discovery of neutron by James Chadwick; first antimatter particle discovered

1933   Fritz Zwicky discovers unseen mass in cluster of galaxies

1935   Einstein and associates present "Einstein-Podolsky-Rosen" (EPR) paradox challenging a quantum theory prediction of non-local entanglement between particles

1936   Pontifical Academy of Sciences re-established by Pope Pius XI "to honor pure science wherever it is found, assure its freedom and promote its researches;" Academy was to have worldwide membership from many faiths.

1938   Hans Bethe proposes thermonuclear fusion of hydrogen nuclei as the energy source of the sun and stars and makes first complete analysis of a specific reaction sequence, the proton-proton chain

1938–1940   Teilhard de Chardin writes *Phenomenon of Man*; published after his death in 1955

1938–1939   Nuclear fission of uranium atoms observed by chemists Otto Hahn and Fritz Strassman and explained by Lise Meitner and Otto Frisch

1939   Einstein writes letter to President Roosevelt about nuclear research in Germany and the possibility of building an atomic bomb.

1939–1945   World War II

1941   Establishment of American Scientific Affiliation, a fellowship of Christian scientists committed to religious faith and the integrity of science

1942   First controlled nuclear fission chain reaction accomplished by Enrico Fermi

1942–1945   Manhattan project to build atomic bomb, directed by J. Robert Oppenheimer; first explosion on July 16, 1945 and use in World War II against the Japanese on August 6, 1945

1948   Proposal of Steady State theory of cosmology by Fred Hoyle, Hermann Bondi, and Thomas Gold

1948   Ralph Alpher, Hans Bethe ("in absentia") and George Gamow suggest element synthesis in expanding, cooling early universe

1948   Alpher and Robert Hermann predict the remnant microwave radiation from Big Bang

1950   Roman Catholic Pope Pius XII issues encyclical *Humani generis* giving restrained acceptance of evolution with certain cautions

1950   Pope Pius XII endorses the Big Bang as evidence for creation by a transcendent God

1950   Hoyle coins term "Big Bang" on radio program, while criticizing theory

1952   Explosion of first hydrogen bomb

1953   Francis Crick and James Watson discover structure of DNA

1953   Meher Baba publishes *God Speaks*

1953   Stanley Miller experiment synthesizing amino acids from chemicals of early Earth

1954   Founding of Institute for Religion in an Age of Science (IRAS) by joint scientific-religious group to promote positive relation between the two fields

1957   Hoyle, William Fowler and the Burbidges show that cosmic abundance of heavy elements predictable by nuclear fusion in stars

1957   First orbiting spacecraft, Sputnik, launched by the Soviet Union; begins age of space exploration

1959–1962   First spacecraft reach other celestial bodies

1959   Darwin Centennial at University of Chicago; Sir Julian Huxley presents epic of cosmic evolution

1959	Dalai Lama flees Tibet and establishes headquarters in India
1960	Frank Drake conducts first radio search for extraterrestrial intelligence in Project Ozma
1961	Publication of *Genesis Flood* by Henry Morris and John Whitcomb; creationist revival begins
1962	East–West Gathering convened by Meher Baba in Pune, India
1962–1965	Second Vatican Council convened by Pope John XXIII
1964	Murray Gell-Mann and George Zweig propose existence of quarks
1964	John Bell proposes experiment to determine if particles can act in a non-local, non-mechanistic manner, according to quantum predictions; experiments in the early 1980s by Alain Aspect and collaborators verified nonlocality
1965	Arno Penzias and Robert Wilson discover cosmic microwave background radiation
1966	Ian Barbour publishes *Issues in Science and Religion*, presenting methodology for the study of science and religion
1967	Steven Weinberg, Sheldon Glashow, and Abdus Salam propose electroweak theory—unification of electromagnetic and weak nuclear forces
1967	First pulsar detected by Jocelyn Bell Burnell and Antony Hewish; later identified with neutron stars, superdense remnants of exploding stars
1968	Strong Anthropic Principle introduced by Brandon Carter
1968	U.S. Supreme Court struck down Arkansas law prohibiting teaching of evolution in schools (*Epperson v. Arkansas*)
1969	Successful U.S. manned landing on the moon
1970–1973	Standard Model of Elementary Particles developed
1973	500th anniversary celebration for Copernicus by national science organizations
1973	Center for Process Studies established
1974	Stephen Hawking predicts radiation from black holes and their eventual evaporation
1976	Viking Lander searches for biological activity on Mars

1978   John Paul II elected Pope

1978   Presence of dark matter in external galaxies inferred from rotation curves

1981   Guth proposes theory of extremely rapid "inflationary" expansion in first second of the universe

1981   Center for Theology and Natural Sciences founded at the Graduate Theological Union in Berkeley to promote creative mutual interaction between science and theology

1981   Pope John Paul II appoints a commission to reexamine the Galileo case and "rethink the whole question" of the relationship between science and religion.

1982   U.S. District Court judge William Overton ruled unconstitutional an Arkansas law requiring balanced treatment for creation science and evolution (*McLean v. Arkansas*)

1984   Vatican initiates review of Galileo case

1984–1986   First Superstring Revolution; theory of particles as vibrating strings promises unification of all forces

1985   Founding of Ian Ramsey Center at Oxford University to study religious beliefs in relation to the sciences

1987   First Mind and Life conference establishes dialogue between scientists, the Dalai Lama, and other Buddhists.

1987   Founding of John Templeton Foundation to provide financial support to research in science and religion subjects; offers yearly science and religion prize

1987   Newton Tri-Centennial Vatican Study Week convenes scientists and theologians to study: "Our Knowledge of God and Nature: Physics, Philosophy and Theology"

1987   U.S. Supreme Court ruled unconstitutional a Louisiana law requiring that creation science be taught alongside evolution (*Edwards v. Aquillard*)

1988   Papal message on Science and Religion following Vatican Study Week

1988   Founding of Zygon Center for Religion and Science (originally Chicago Center for Religion and Science)

1992–2002   CTNS/Vatican sponsorship of five research conferences on theme of divine action and its relation to natural sciences;

joint publication of essays on quantum cosmology (1993), chaos and complexity (1996), evolution and molecular biology (1998), neuroscience and the person (1998), and quantum mechanics (2002)

1992   COBE (Cosmic Background Explorer) satellite discovers very slight variations in the cosmic microwave background radiation; supports Big Bang

1993   First modern World Parliament of Religions in Chicago; essays published in *Cosmic Beginnings and Human Ends*

1995   First detection of extrasolar planets

1995   Second Superstring Revolution

1995   AAAS established Program of Dialogue between Science, Ethics and Religion, sponsors conferences "Epic of Evolution" (1997) and "Cosmic Questions" (1999)

1995–2003   Science and the Spiritual Quest program assembles leading scientists in private workshops and numerous worldwide conferences to dialogue on questions of science and religion

1996   Pope John Paul II delivers message supportive of evolution to Pontifical Academy of Science

1996   Georgetown University Center for the Study of Science and Religion established

1996   Center for Science and Culture established within political think-tank Discovery Institute to support research and education initiatives promoting Intelligent Design

1997   Mind and Life Institute conference on New Physics and Cosmology brings leading physical scientists into dialogue with Dalai Lama at his compound in Dharamsala, India

1997   Science and the Spiritual Quest Conference brings leading scientists together to discuss religion and science; initiates multi-year, project

1998   Founding of Metanexus Institute, a global interdisciplinary science and religion think tank and worldwide network with yearly conference

1998   Discovery of accelerating universe

1999   Second modern World Parliament of Religions (Cape Town, South Africa); essays published in *When Worlds Converge*

1999   Centre for Islam and Science established in Canada

2000   Conference in Islamabad, Pakistan convenes scholars from Christianity and Islam to discuss "God, Life and the Cosmos"

2003   WMAP satellite image of cosmic microwave background consistent with 13.7 billion-year age for universe composed mostly of dark matter and dark energy

2007   CTNS initiates STARS program (Science and Transcendence Advanced Research Series) to support research grants to study ways that science, in the light of theological reflection, points toward ultimate reality

2007   Pope Benedict XVI declares the creation-evolution clash "an absurdity"

Chapter 1

Introduction: Physical Science and Religion in the Twentieth Century

No time in history has ever seen more revolutionary change than the twentieth century. And nowhere was transformation more stunning than in the domain of physical science. There, old structures of thought crumbled, and in their wake new ones emerged that opened wider, deeper vistas of the physical world and humanity's place in the cosmos. A dynamic, evolving universe with possibly a beginning moment in time replaced the old static, eternal cosmos. Relativity and quantum theories revolutionized common-sense notions of the very components of physical reality—space, time, matter, energy, and causality—and transformed the old mechanistic, deterministic paradigm into a more holistic, open and relational view. New understanding of the operation of matter at the most minute levels produced the nuclear age and revealed the energy source and life cycle of the sun and stars. These realizations, together with the Darwinian evolutionary theory and the discovery of DNA, led to the grand picture of cosmic evolution, the multi-billion-year journey of the universe that began with the Big Bang and culminated in the human being and perhaps other intelligent beings. Furthermore, features of the early universe seem to have been perfectly "fine-tuned" to allow for, or perhaps even require, the development of life.

How this enormous revolution in knowledge related to religion in the past hundred years is the subject of this book. As it tells the tale of discovery and the new story of cosmic evolution, the book will examine at each step religious responses from scientists and from scholars and leaders in a variety of faith traditions. And there is no question that the modern cosmological picture evokes a strong religious response. Never before did science obtain such precise quantitative answers to age-old questions of

origin formerly in the province of theology and philosophy. In so doing, scientists brought themselves much closer to the realm of traditional
religious thought, either to supplant it, as atheists and scientific materialists would hope, to delineate its separate sphere more clearly, or to join
with religion in a new era of dialogue and joint exploration. Interestingly,
religious thinkers displayed the same spectrum of responses.

Religion in America in the twentieth century also experienced enormous
change, and movements supporting and opposing a positive relation to
science ebbed and flowed. Early liberal Protestant thought, which sought
to accommodate religious belief to contemporary world views, was challenged in the early twentieth century by more conservative forces, which
viewed science as irrelevant to religion or actively opposed its evolutionary picture. The second half of the century saw a sea change, as Roman
Catholicism underwent a new flowering of ecumenical spirit and dialogue with science, and Christian theologians on the whole embraced a
new, more hopeful orientation toward society and the future (MacQuarrie
2002, 373–75). New voices emerged from the East calling for religious unity
and interdisciplinary dialogue and presenting cosmologies that fully integrated ancient thought with modern scientific evolution. Interestingly, in
this same period conservative Christianity also grew stronger and energetically renewed anti-evolutionary campaigns. Thus the full spectrum of
relationships between science and religion has been present throughout
the century—from outright conflict to distinct separation to harmonious
dialogue and integration.

SCIENTIFIC AND RELIGIOUS THOUGHT—EARLY TWENTIETH CENTURY

An early hint of significant change to come was evident when both
scientists and philosophers declared near the turn of the twentieth century
that the end was in sight for their field of endeavor. British physicist Lord
Kelvin believed that the job of physics was nearly complete and there was
little left for physics to do, although in 1900 he did note "two clouds" on
the horizon needing resolution (Southgate et al. 1999, 99; Greene 2004, 9). A
similar comment came from English philosopher Bernard Bosanquet, who
said of philosophy, "in the main the work has been done" (MacQuarrie
2002, 21). Although such statements may seem amazing or even amusing
in hindsight, in a sense they both were right. The twentieth century would
replace classical physics and shift religious and philosophical thought
away from the traditional questions of cosmology and metaphysics.

The dominant thinking in the first decades of the twentieth century reflected the legacy from the previous century, whose most striking feature

had been the tremendous advance of the natural sciences (MacQuarrie 2002, 95). Confidence reigned supreme that the classical Newtonian mechanics coupled with the newer electromagnetic theory gave a nearly complete understanding of the physical world. The cosmos was static and infinite, with neither beginning nor end—all of it a system of bodies and solid, indestructible particles in motion, driven by fixed laws. Once set in motion, such a mechanistic, deterministic system seemed to require nothing outside of itself to operate or determine its course. It seemed compatible with a philosophy of either atheism, the belief in no God, or deism, the belief in a creator God who designed a perfect machine "at the beginning," set it to working and interfered no further.

Physical scientists were also influenced by the philosophy of Immanuel Kant, who argued that science and religion operate in separate spheres. They increasingly abandoned "natural theology," which sought to demonstrate God's existence and characteristics from studying nature. Most writings simply tried to demonstrate how the understanding of the physical world was consonant with Christian ideas (Olson 2000, 304–6). However, other belief systems, such as atheism, were compatible as well, so that perhaps a growing neutrality or independence might best describe the relation between physical science and religion at the close of the nineteenth century.

The contrast of scientific and religious ideas was especially striking in the realm of biology, where the work of naturalist Charles Darwin brought further pressure to bear on natural theology and the religious notion of special divine creation. His explanation that new species originated through natural selection acting on chance variations clearly brought into question God's role in directly designing and creating living forms. In one fell swoop, Darwin's theory not only challenged God's role but also further demoted the human being from any special status. As well as being removed from a central cosmic location, we were now no longer elevated above other creatures. Like them, humans were merely a product of natural processes and not far removed from ape-like ancestors.

Science's naturalistic explanations for what was formerly attributed to divine action thus challenged traditional religious belief more than ever during the late nineteenth century. In popular culture and two prominent publications of the time, this challenge was represented as warfare, and for some conservative Christians and scientific materialists, this was and remains the proper description. However, a large number of Christians in both the scientific and religious communities came to tolerate, accept or even embrace naturalistic explanations of the physical world and the evolutionary development of living beings. Many factors contributed: Kantian "separate spheres" thinking, liberal Protestant and Catholic modernist

adjustment of doctrines to the contemporary world, and new philosophies incorporating God and science. Despite the popular perception,
evangelical Christians for the most part also accepted both biological evolution and biblical religion. According to Protestant theologian Langdon
Gilkey, mainstream religion responded to the impact of science by gradually changing its conception of "religious" truth and the arena of divine
action. What was previously seen as factual in scripture became "symbolic
and analogical." God was no longer just the transcendent creator, who designed everything in one original event, but an immanent God who acted
within a long, slow evolutionary process. The old natural theology was
thus replaced by a "'wider teleology': a divine purpose shaping the entire
process of change over time" (Gilkey 1993, 20–22).

In light of the phenomenal progress of the natural sciences, and the
extension of the scientific method to many other fields of knowledge, it
is no wonder that religious thought at the turn of the century echoed
themes of science: optimism, evolution, the notion of substance, and a
comprehensive, metaphysical approach, according to British theologian
John MacQuarrie. In general, belief in the goodness of the world and the
power of science prevailed, he argues, coupled with confidence in positive
evolutionary progress in society as well as nature. Objects of material and
spiritual reality were substantive—they had "solid, enduring thinghood."
Finally, philosophy was "comprehensive," focusing on systematic treatment of the big questions of existence that religion also addressed. These
themes were found in most philosophies but especially those based on
science—positivism, a view that only scientific knowledge is reliable and
true, and naturalism, a belief that reality is identified with nature. Naturalism becomes materialism, if nature consists only of material particles and
energy. Much debate would ensue in the century to come about whether
these beliefs are implied by science or imposed on science by atheists who
espouse them.

Two particular events in the last two decades of the nineteenth century
bespeak of the optimism of the times and herald trends that would prevail
again toward the end of the twentieth century. One was the establishment
of the yearly Gifford Lectureships on natural theology at four universities
in Scotland. Endowed by a Scottish judge with an interest in philosophy,
Adam Gifford, the lectures have continued to the present time and represent one of the highest honors in the academic world. Several of the
lecture series have become seminal works in their field, such as William
James' *Varieties of Religious Experience,* Alfred North Whitehead's *Process
and Reality,* and Paul Tillich's *Systematic Theology* (3 *vols.*). A wide range of
viewpoints on the whole relation of science to philosophy and theology
have been expressed in these lectures, and their history tells much of the
story of science and religion in the twentieth century in mostly Protestant

British, American, and European circles. This history has been recently reviewed in an engaging way by science journalist Larry Witham in his book *The Measure of God: Our Century-Long Struggle to Reconcile Science & Religion* (2005).

A second event which signaled significant interchange to come was the formal introduction of Hinduism to the West by Swami Vivekananda in the mid-1890s. His famous first address at the World Parliament of Religions at the Chicago World's Fair on September 11, 1893 brought a three-minute standing ovation when he opened with the words "Sisters and Brothers of America." He then delivered an inspirational message about the unity of world religions, which he likened to various streams which all mingle eventually in the same sea of divinity. In the following two years he traveled around America lecturing, giving private classes, and generally stimulating interest in the Hindu subjects of Yoga and Vedanta philosophy. Vivekananda was a harbinger of a new spirit of dialogue and interchange between East and West, which would come to full expression with such eminent twentieth century figures as Meher Baba, Aurobindo, and the fourteenth Dalai Lama of Tibetan Buddhism.

SCIENTIFIC AND RELIGIOUS
THOUGHT—MID-TWENTIETH CENTURY

The century's first half brought enormous change—upheavals in world events, revolutions in physical science, and distinct shifts in the relation of religious thought to science. Two world wars, economic depression, and the rise of the nuclear age all tempered earlier optimism. Scientific advance continued unabated but revealed puzzling new conceptions and unleashed a destructive force never before imagined. In religious thought, diverse trends persisted, but conservative movements grew stronger, opposed earlier liberalization and focused on the subjective concerns of faith.

The first three decades of the twentieth century have been called by one physicist "the thirty years that shook physics" (Gamow 1985). Lord Kelvin's "two clouds"—the mysteries about light's constant speed in all directions and the impossible prediction of infinite radiation from a glowing object—broke wide open and inundated the world of physics with stunning new knowledge that mystified most of humanity, including some scientists. This was the era when Einstein's theories of relativity overturned traditional notions of space, time, and gravitation, and quantum theories sounded the death knell to determinism and common-sense notions of the nature of subatomic particles. The jam-packed 1920s marked not only the development of quantum mechanics but also the discovery of the vast realm of the galaxies and the expanding universe with a beginning at a point in time. The following decade saw the discovery of both kinds of nuclear energy—fission, which governs radioactivity, and fusion, which powers the sun and stars. The development and detonation of the first atomic, or fission bomb, in 1945 initiated the nuclear age, and the hydrogen, or thermonuclear fusion, bomb followed soon after. The nuclear arms race had begun. As physicists plumbed the depths and wielded the power of the subatomic world, astronomers were busy enlarging the known universe and using the new physics to understand the evolution of chemical elements in the early Big Bang universe and in the lives of stars. They were gradually unveiling important pieces of the grand picture of cosmic evolution.

Regarding science's relation to religion and philosophy, fascinating developments emerged from science itself. First, science was beginning to offer its own serious, quantitative answers to the question of the origin and development of the universe. This brought it face to face with religion, with mixed reaction. A Pope endorsed the Big Bang theory for its theistic implications, while an atheist rejected it on the same grounds, and the Catholic cleric who proposed the theory argued that it had nothing to do with religion! Second, quantum mechanics declared that physical observations had uncertainty, unpredictability, and observer-dependency, thus directly challenging fundamental assumptions of science itself— predictability, cause and effect, and confidence in objective reality. Vanished was the former vision of the world as composed of solid, indestructible particles operating in mechanistic, deterministic fashion. Atoms were more "complex rhythmical patterns of energy" than solid static objects— process rather than inert substance—and the universe was possibly not infinite or eternal and therefore "much less likely to be the ultimate reality" (MacQuarrie 2002, 242). The far-reaching philosophical and religious implications of these findings were explored at length by many of the physicists themselves, some of whom had a keen interest in Eastern religion and even had a mystical bent. Such an open universe offered much richer possibilities for comparison with religious ideas.

Religious thought displayed a similar radical change, and new twentieth century themes emerged that were almost the exact opposite of the earlier ideas. The shattering events of global warfare and economic instability affected humanity's self-concept and sense of where it was headed. The earlier optimism yielded to a greater pessimism, or at least a "more sober and realistic" assessment of human nature and the human condition. Belief in God was in decline; most of the prominent pre–World War I thinkers were theists, whereas afterwards, most were not, one theologian noted. While evolutionary change remained a theme, and even expanded to encompass the expanding universe and chemical evolution, it now no longer always connoted cultural progress, especially since Darwinism had been misused to exploit the weak.

Other major themes were a focus on process over substance as constituting reality, an anti-metaphysical streak in some philosophy, and a near divorce of much mainstream theology from both philosophy and science (MacQuarrie 2002, 118). Old idealistic philosophies waned, but so did science-based philosophies of naturalism and materialism, as limits to objectivity and certainty were revealed by science itself. Witham describes the time between the two World Wars as one of "great rebellion against science and reason in the West, a movement that could be called *subjectivism* for its revolt against the vaunted "objectivity' of science" (2005, 6). Retreating from comprehensive, metaphysical approaches, many theologians focused on divine revelation, faith, and personal encounter—the more subjective side of religion. Protestant fundamentalism arose to counteract earlier liberal tendencies and launch a major campaign to outlaw the teaching of evolution, while Roman Catholicism sought to stem the tide of liberal, modernist thinking.

In sum, religious thought in this transitional period saw increasing belief that science and religion belong to separate spheres, coexisting perfectly well but irrelevant to each other. Such a view was fostered also by pragmatic conceptions of truth, such as those of William James, and philosophical analysis that emphasized the very different languages and functions of science and religion (Brooke 1991, 322). Ironically, during the same period, changes in science opened the door to greater dialogue. It was the scientists of the day who seemed most interested in discussing religion.

By mid-century these themes consolidated into distinct trends and movements. With the Cold War and nuclear arms race, and the demise of Western colonial empires, world unrest and change continued, creating a general mood of anxiety, which itself became a theme for some philosophers. Religious movements continued to focus on faith and revelation and avoided metaphysical concerns. Neo-orthodoxy, associated with one of the century's most influential theologians, Karl Barth, harked back to the Reformation and the New Testament stressing that religious knowledge came not from nature but only from direct revelation by God and the subjective experience of the faithful. Existential theology focused on intense personal commitment and decision as the essential elements of faith. Some new metaphysical systems did emerge, however, such as the process philosophy of Alfred North Whitehead, which incorporated notions from the new science, and neo-Thomism, which sought to revive the medieval philosophy of St. Thomas Aquinas. Another was the epic vision of man's spiritual evolution conceived by French Jesuit paleontologist Pierre Teilhard de Chardin, whose major work, *The Phenomenon of Man,* was written in this period but published and disseminated later. All of these continue to be influential schools of thought today (MacQuarrie 2002, 253–57).

Comprehensive metaphysical systems were also offered by spiritual leaders in the Eastern world—in particular, by Meher Baba and Aurobindo Ghose in India, who sought to reformulate ancient Indian thought for the modern world. They both developed and began to publish evolutionary perspectives which incorporated the physical and biological evolution of the universe and life into a larger spiritual whole. Meher Baba's Hindu-based cosmology also encompassed elements from other world religions, such as the important role of the Christ, and during this period he first announced his goal of working to bring greater harmony among different faiths.

SCIENTIFIC AND RELIGIOUS THOUGHT—THE LATTER TWENTIETH CENTURY

The last half century has seen the growth of a vital movement of greater dialogue between science and religion but also an intensification of both religious and scientific voices of opposition. Science delivered a much more detailed and well-supported picture of cosmic evolution, and physicists continued to probe the depths of matter, seeking new theories of unified forces. Though stunning scientific advancement continued, the social and cultural revolution of the 1960s brought ambiguous reactions and even a distrust of science. Philosophical study of science further questioned scientific certainty and objectivity, thereby opening it to more productive dialogue with religion. The same revolutionary period also ushered in new interest in nontraditional and non-Western religious thought, while traditional Roman Catholic and Protestant thought underwent a renaissance (MacQuarrie 2002, 373–76). A new movement arose seeking to establish "science and religion" as its own field of academic research. World religious figures keenly interested in science assumed leadership roles in their faiths and energetically supported the dialogue, helping to create what is today a thriving, interreligious global interchange. Interestingly, the same half century has also seen the growth of creationist reaction against both science and religious accommodation to it.

Developments in Science

The spirit of the Darwin Centennial at the University of Chicago in 1959 expressed well the tenor of the times in science (Witham 2003, 4–7). There, biologists celebrated the new epic of cosmic evolution emerging from major discoveries in biology and astrophysics—the discovery of the structure and reproductive mechanism of DNA, experimental evidence for the chemical origin of life, and a well-developed theory for the origin of chemical elements in stars. The scenario of cosmic evolution suggested that life might be plentiful in the cosmos, and in the following year Project Ozma made the first radio search for intelligent life beyond Earth. The following decade brought the Big Bang's most solid evidence, the cosmic microwave background radiation. It also brought the realization that many features of the early universe were perfectly "fine-tuned" to allow for or even require the development of life—the so-called "anthropic principle." Its design implications opened new possibilities for natural theology, and an energetic debate over its significance began.

Cosmology became a joint venture of astrophysicists and particle physicists, as they worked together to unveil the realm of the very small, the very dense, and the very hot that existed only in the earliest moments of the universe. Physicists proposed and in some cases demonstrated the unification of forces at high energies and embarked with fervor on a quest to join all four known forces into a single "theory of everything," describing matter at the very first instant of time. The most promising candidate suggested that all material particles are strings of inconceivably tiny size vibrating in ten or eleven dimensions. However, many string theories are now thought to exist, suggesting the possible existence of an immense number of universes. Physicists and astrophysicists have also been working together to solve two enormous puzzles: the unknown dark matter which comprises perhaps a quarter of the universe, and the even more mysterious "dark energy" fueling the acceleration of cosmic expansion and comprising perhaps two-thirds to three-quarters of the universe.

For many scientists the continued triumphs of science confirm that science will ultimately explain everything by natural causes and show that the physical matter and energy is all that exists (philosophical naturalism and materialism). At the same time, other developments have questioned science's power and the objectivity, certainty and beneficence of the knowledge it delivers. Witham notes that by the time of the next major science celebration, the 500th anniversary of Copernicus' birth in 1973, a "robust debate over the authority of science" was well underway that "revealed the vulnerability of science both as a tool of knowledge and as a social force" (2003, 28–29).

The questioning of science's authority came from several directions: scientific results themselves, philosophy of science, and larger social and cultural forces. One hint of doubt was the failure to develop and validate a theory of life's chemical origin and to find any life beyond Earth. Another factor was the growing awareness of extra-scientific influences on the scientific process. Philosophy of science questioned the view that the scientific method was a purely logical process yielding objective, value-neutral truth and explored the influence of subjective, cultural factors. Thomas Kuhn in his well-known work *The Structure of Scientific Revolution* (1996) focused on how cultural and historical assumptions—the frameworks of thinking, or "paradigms," of an era—influenced scientific thinking and results. Thus, the image of the universe as a clock or machine emerged during the Industrial Revolution, and a focus on "information" in modern biology appears now in the computer age. Other thinkers, including Michael Polanyi, showed how the conclusions drawn in scientific inquiry and the doing of science itself are affected by the "experiencing and knowing subject." The "doer" of science brings into play a host of subjective elements such as imagination, intuition, and commitment to the mores and values of a community of like-minded scientists (Gilkey 1993, 26–29).

In the larger society, public awe at technological advancement and the power of science was tempered by growing ambiguity and even distrust over its role in weapons development and environmental degradation. The ambiguities were easily seen in the counterculture values of the 1960s where the promotion of "God in a pill," a materialistic approach to spiritual experience, coexisted with rebellion against science's role in the "military industrial complex." The same era saw a growing commitment to human rights for all people, and a new and expanded pluralism was abroad in the land, which reached right into the heart of both science and religion. The new spirit expressed itself not only in the appreciation of multiple influences on science but also in a new religious pluralism that included growing interest in nontraditional and non-Western religions and in an ecumenical science-faith dialogue.

Christian Thought in the Latter Twentieth Century

The decade of the 1960s brought a watershed in religious thought, ushering in a new era of far-reaching change. In Christianity the shift was evident across the board in Catholic and Protestant circles. For the first time in a hundred years the Roman Catholic Pope, John XXIII, convened the Vatican Council to "promote 'peace and unity of all humankind.' " Meeting for three years beginning in 1962, it was the "largest and first truly ecumenical council in the history of the Catholic Church," bringing together 3,000 people from all over the world. One Catholic religious scholar describes it as follows:

[Vatican II] opened the church to the modern world and radically changed the traditional official attitudes toward non-Catholic Christianity, non-Christian religions, and Catholics who called for freedom of thought and conscience. Self-segregation, condemnation, and proselytizing gave way to constructive dialogue with the secular world and other denominations of religions. (Shafer 2002, 351)

The Council also declared "the legitimate autonomy of human culture and especially the sciences" (Peters 1988, 274).

Without question the new stance of the Catholic hierarchy represented a sea change, although it should be noted that the Church had almost never been as adamantly opposed to science as is often assumed from the Galileo affair. Its longtime stance has been one of "autonomy" and "separation" of the "two truths" of science and religion, which should never contradict each other, given that they both come from God (Harris 2002, 256–57). Throughout the modern era the Church supported scientific endeavor through the work of individual scientist-priests and the papal establishment of the Vatican Observatory and Pontifical Academy of Sciences. With respect to evolution, Catholicism has maintained a "direct, if sometimes strained, engagement" (Harris 2002, 252). Some of its most illustrious clerics made seminal discoveries key to cosmic evolution: Gregor Mendel with modern genetics and Georges Lemaitre with Big Bang cosmology. The evolutionary Christology of Teilhard de Chardin was far more controversial, engaging as it did with theological speculation and violating the traditional separation of science and theology, and was banned by the church during his lifetime (Harris 2002, 252–55). Nevertheless, the tide was already changing, and exactly at mid-century then Pope Pius XII issued an encyclical giving restrained support to evolution, with certain cautions, and broke with traditional independence by strongly endorsing the Big Bang as evidence for divine creation.

Without question Vatican II exerted a "liberating influence" within the church, causing Catholic theology to "burst out of the rigid categories to which it had long been confined, almost like a butterfly out of its chrysalis" (MacQuarrie 2002, 374). A new group of eminent theologians were able both to honor Catholic tradition and incorporate the thought of the modern world. Chief among them was Karl Rahner, a strong contributor at Vatican II who successfully integrated evolution into a Christology, or theology of the incarnation. In a few years, John Paul II ascended to the papacy, bringing a keen interest in cosmology, philosophy, and metaphysics. In 1981 he initiated a working group to reexamine the case of Galileo and announced at its conclusion in 1992 that the church had made an error in condemning the Copernican view of the solar system and Galileo's support of it. In 1982 he declared to the Pontifical Academy of Science, "There no longer exists the ancient opposition between true science and authentic faith," and to the community of scientists, "the church is your ally" (Peters 1988, 273). At the 300th anniversary of Newton's publication of the *Principia* he convened a Vatican study week, where physicists, philosophers and theologians met to explore common ground, and which was followed by five more conferences on divine action in relation to various scientific subjects. A statement he published shortly after the initial conference was the strongest message yet from a pope urging harmonious working together of science and religion, and in 1996 he delivered a message about evolution that was considerably more positive in acknowledging scientific truth than earlier papal statements (Harris 2002, 257–58).

The decade of the 1960s was also a watershed in Protestant thought. A new spirit emerged following decades of dominance by neo-orthodoxy and existentialism and the "two separate spheres" model. The enormous influence of such theological "giants" as Karl Barth, Paul Tillich, Rudolf Bultmann, the Niebuhrs and Langdon Gilkey began to wane. As rapid changes in the world influenced theology, religious thought again reached outwards to relate to societal, practical and future concerns. Optimism revived, as continued technological advance brought increasingly greater mastery of nature and drew humanity closer together in the new information and space age. The earlier emphasis on man's "finitude and even sinfulness" following the world wars yielded to a "new humanism" which spoke of "hope, aspiration and human transcendence." Liberal Protestant theology and other "new theologies" once again showed the "immanentist, humanist trend" of the previous century. Two influential scholars, Jurgen Moltmann and Wolfhart Pannenberg, both emphasized a future or eschatological orientation and sought a "viable concept of resurrection" as a meaningful future event. The new spirit of pluralism also brought many new voices to the theological discussion, including liberation theologians of South America, black theologians of North America, and feminist theologians (MacQuarrie 2002, 374–78).

The latter part of the twentieth century also saw a flowering of conservative Christianity. The post-World War I antievolution campaign of populist-led fundamentalists was followed later by a revival among scholarly evangelicals of their traditional quest for harmony between science and the Bible. Such a re-engagement was the hallmark of the American Scientific Affiliation, formed by a group of evangelical scientists in 1941 and still highly active today, and was also the chief aim of a new postwar group of evangelical theologians, such as Bernard Ramm. Many conservative Christians, such as scientist Hugh Ross, accept most or all of science's findings, and even celebrate how science proves the Bible and how scientific concepts such as the anthropic principle are proof of design by God.

Strict creationism also experienced a revival, starting with the 1961 publication of *Genesis Flood*, which argued for a young Earth. Led by Henry Morris, the new "scientific creationism" or "creation science" movement presented physical evidence for interpreting biblical creation literally and sought equal time for teaching creation science in schools, a goal which met defeat in significant court cases in Arkansas and Louisiana in the 1980s. Nevertheless, Morris' organization of scientists, the Institute for Creation Research, has continued to support education and research and has become influential worldwide. In recent years, the cause of fighting court battles and challenging the teaching of Darwinism has been taken up by a newer movement called Intelligent Design (ID), which avoids reference to a religious deity, speaking only of an "intelligent designer." The hub of the movement is a group of Fellows at the Discovery Institute, who have also founded the International Society for Complexity, Information, and Design (ISCID) to promote their views. Despite the lack of reference to the Christian God, many evangelicals who remain conflicted about evolution are also attracted to the ID movement.

Religious Pluralism in the Latter Twentieth Century

The growth of religious diversity brought an increase of interest in non Christian religions in the Western world. Although Hindu, Islamic, and Buddhist teachers had toured and established organizations in the West previously, the mid to latter twentieth century saw a great growth in contact. This was due in part to the influx of new immigrants from Asia, Africa, and the Middle East following passage of the Immigration and Nationality Act of 1965 (Witham 2005, 259), but also to the restlessness and spiritual searching that characterized the times. Thus the latter twentieth century saw the arrival of many more teachers from the East and the establishment of many more centers and institutes to disseminate knowledge and encourage the practice of Eastern religions. By the 1980s every major world religion had established, in the United States, headquarters, centers, and orders where seekers received spiritual training. As this pluralistic trend has continued, the Christian community of the Western world has had to rethink its belief that following Christ is the only or perhaps best means to salvation. For many Christians, this original stance gradually transformed into a more inclusive position and, for some, a hearty acceptance of full religious pluralism, the idea that all religions are true paths to divinity (Witham 2005, chapter 9).

Two events and figures exemplify the worldwide transformation of religious life and growth of East-West contact in the mid-century decades. One was the enormous change that happened with Tibetan Buddhism with the flight from Tibet of its spiritual and temporal leader, the fourteenth Dalai Lama, in 1959, and his subsequent establishment of headquarters in India. From there he has undertaken a worldwide mission of peace and understanding, spreading and embodying the ideals of Buddhism far and wide. His own long-standing interest in science has made him an enthusiastic participant and leader in establishing and guiding the operation of the Mind and Life Institutes, begun in 1987, which explore the relation of science and Buddhism in understanding the mind and consciousness and also the physical universe. The tragic loss of the Tibetan homeland ironically had at least one positive result. It brought the Tibetans' cherished and charismatic leader, and his religion, into very full contact with the larger world, and his universal message of unity and compassion that inspires audiences everywhere has certainly seemed to benefit the whole of humanity. The very events of his life and his mission convey a message about the true spirituality of our times—to engage fully in serving our fellow human beings out in the world.

A similar message about inclusiveness came from Meher Baba, whose mission culminated in this mid-century era. Following an extraordinary life of contacting and helping spiritual seekers worldwide, a signal event in his work occurred in 1962. Just as the Roman Catholic Pope was convening thousands to a worldwide ecumenical dialogue at Vatican II, half way around the world Meher Baba brought together many thousands of his Eastern and Western followers in a meeting he called "The East-West Gathering." There he worked toward his goal of bringing together the different peoples of the world into more harmonious understanding and cooperative action, so that the spiritual wisdom of the East could join with the material, scientific progress of the West and with Western religions in improving the lot of all of humanity, both spiritually and materially. Part of this goal was to work towards bringing world religions into closer harmony, so that people would honor a common core at the heart of all religions. In his unique work, *God Speaks,* published in 1953, he described levels of spiritual consciousness from the perspectives of several religions, indicating that they were all referring to the same states (see Chapter 7).

THE GROWTH OF THE SCIENCE AND RELIGION DIALOGUE IN THE LATTER TWENTIETH CENTURY

In the inclusive and open climate of the 1960s was born the contemporary science and religion dialogue. One arm of the movement was begun and nurtured by a group of scientists in the United States and Britain, who also received divinity degrees and professional training in theology—"hybrids," one theologian calls them (Peters 1998, 2). They and others in related fields sought to establish "science and religion" as a rigorous academic discipline with its own methodology. Fifty years later, the movement has grown to have a significant presence in many eminent universities, seminaries, church organizations and independent centers. It has also broadened to include participants from virtually all the world's major religions, as well as some small-scale indigenous faiths.

Ways of Knowing and a New Methodology for Science and Religion

One name stands above others as "the pioneer" who forged a new path in science and religion research—Ian Barbour. His training in both physics and Christian theology spanned the science-religion divide, and he made groundbreaking contributions by developing a new conceptual framework for comparison called "critical realism," by presenting comprehensive surveys of issues in the field, and by creating the most widely used typology for analyzing different ways science and religion relate. His ideas were presented first in 1966 in a seminal text, *Issues in Science and Religion*, further developed in *Myths, Models and Paradigms* (1974) and summarized in his comprehensive review of the field, *Religion and Science: Historical and Contemporary Issues* (1997). In 2006, a conference and publication, *Fifty Years in Science and Religion: Ian G. Barbour and His Legacy*, honored and reviewed his work and developments based on it. Articles by its organizer, physicist-theologian Robert John Russell, form the basis of much of the present discussion (Russell 2004).

To explore joint methodology, Barbour first examined possible common ground between scientific and religious conceptions of truth and methods of knowing. Traditionally, these have been regarded as very different, if not diametrically opposed. Science answers the "what" and "how" questions by studying objective, verifiable, repeatable evidence from the physical world. Based on this evidence, it develops theories, which are always open to revision and are often expressed in symbolic, mathematical language. Only "natural" causes, verifiable by physical data, are acceptable as scientific explanations (hence science has "methodological naturalism"). Philosophical views differ as to what scientific theories really are: do they correspond exactly to the world (classical or "naive" realism), are they mere ideas in the human mind (idealism), or are they simply useful calculating devices (instrumentalism)? Barbour would draw on each of these ideas in formulating his conception.

Religion, on the other hand, answers "why" questions of meaning and purpose with timeless truths that do not change. Religious beliefs are supported by subjective, nonverifiable, nonrepeatable inner experiences of human beings. One "knows" a religious truth or holds a belief based on faith and intuition. For some, there appears to be direct experience through inner vision, as with great world religious figures such as the Buddha or Jesus, great saints such as Francis of Assisi and Theresa of Avila, and modern spiritual leaders considered in this volume—Meher Baba and Aurobindo Ghose. In addition, the language of religious truths is very different from scientific language, although it is also symbolic and metaphorical.

Barbour's concept of "critical realism" sought to bridge this gap. To do so, he first redefined scientific theory and criteria for judging them by blending elements from traditionally competing philosophies of science. Scientific theories, he asserted, were neither exact representations of reality nor mental constructs but something in between—incomplete, abstract, changeable understanding that partially corresponds to the real world. The criteria for theory evaluation were, first and foremost, "agreement with data" (as in realism), but also "coherence" and "scope" (as in idealism) and "fertility" (as in pragmatism). In studying religion Barbour concluded that its truths could be conceived in a similar way and were assessable by the same four criteria.

Barbour thus identified parallels that allowed him to propose his groundbreaking "methodological" bridge: the process of knowing in science has a similar structure to the process of knowing in religion, even though there are certainly notable differences. What observable data is to theory in science, religious experience is to belief in religion, as shown below:

<u>Structure of Science:</u>
Observation/Data >>> Imagination, Analogies, Models >>> Concepts, Theories
>>> Theories Influence Observations

<u>Structure of Religion:</u>
Religious Experience, Story and Ritual >>> Imagination, Analogies, Models >>>
Concepts, Beliefs >>> Beliefs influence experience and interpretation
(Barbour, 1997: 107, 111)

Each process is cyclical: a concrete experience leads the human imagination and intuition to a more abstract idea (theory or belief) which then provides a way to interpret or understand further experiences or observations, which either support or challenge it. An example from science is the following: the observed motion of galaxies led to the Big Bang theory, which predicted leftover radiation from the original fireball. Later observation discovered it, but a challenge arose when no structure in this radiation was found that could have produced galaxies. Finally, a more powerful tool successfully detected the structure and gave the theory stronger support. For the religious process an example might be the following: imagine that one has a strong feeling of a divine spirit in nature or a sacred place or has a conversion experience and develops a strong belief in God. This belief could then be supported by other experiences of love and community among fellow believers, or it could be challenged by experiencing the loss of a loved one. With prayer for divine help, one might find inner courage to face such suffering, which could then strengthen belief in God. These are obviously very different phenomena, but the process by which human beings move through experiencing and understanding them has structural similarity, Barbour contended.

Other similarities are that both processes are influenced by their historical, human context, both use metaphors in expressing their concepts, and both have subjective and objective elements—although the former predominate in religion. As noted above, both can be assessed by the same four criteria. Thus theories or beliefs must first and foremost match experiences (agreement with data), must hold together as a consistent and meaningful whole (coherence), and must apply to a wide variety of phenomena and lead to new ideas and experiences (scope and fertility). Clearly, however, there are major differences. The religious process has different and more varied types of experience as "data" and has different purposes relating to inner transformation. It also differs from science in lacking "lower-level laws" and consensus-seeking ability.

Exploring similarities and differences in religious and scientific "knowing" has been an on-going aspect of the recent dialogue, and researchers have both challenged and expanded on these ideas, as examples throughout the present study will illustrate. For example, Eastern spiritual systems suggest that "lower-level laws" do exist in the religious sphere—the law of karma being an example. Other examples will show how scientific knowledge can have an inner transformative effect, just as much as religious beliefs. Many scientists, for instance, were religiously inspired to discover the immensity and grandeur of the universe, and Einstein's "cosmic religious feeling" developed from his study of natural laws. Sometimes religious or philosophical belief affected theory choice. When Einstein's general theory of relativity suggested cosmic expansion, his own deep belief in a static universe led him to reject the idea, and he changed his equations to eliminate it. Such examples reveal significant cross-overs or bridges between the two processes of knowing

In expanding on Barbour's work, Russell has explored these bridges and identified eight different pathways by which aspects of the religious process of knowing can influence parts of the scientific process and vice versa (Russell 2001, 113). Bringing such relationships to light and using them is part of the "creative mutual interaction" that he envisions as an exciting and fruitful pathway for future collaboration. The fact that so much cross-over exists also points to a deeper truth. There may be one basic process that all of human understanding and creativity follows, manifesting differently in the various areas of life. Humans have experiences, and then mull them over—interpreting them, meditating on their meaning—and develop ideas, beliefs, theories or creative works, which then influence future experience. If this is true, it is not surprising that scientific and religious knowing are similar, deriving as they do from the same deep wellspring of the search for truth within the same human being.

Another way of expressing these bridges is to note that spiritual elements come into play in the actual practice of science, as many scientists have affirmed. Nobel-prize winning physicist Charles Townes describes a revelatory experience he had when discovering the laser and reminds us of the famous discovery of the benzene ring by German chemist Kekule while dreaming of a snake biting its tail (Townes 1998, 47). Physicist Joel Primack describes "religious" commitment to honesty and long-term effort, inner attunement with the "real" universe, and faith that the next steps will emerge from the "bottomless well of ideas and enthusiasm" (Primack 1997, 14–15). Molecular biologist Pauline Rudd speaks of deep intuition that "eclipses" the distance between observer and observed and gives rise to feelings of empathy with the object of study. Scientists thus are on a quest or journey involving sacrifice and contemplative union with the subject of inquiry—the search can be long and hard, but they love what they are studying! These are all elements of a spiritual path. Different as science and religion are, many of the same character traits, the need for faith and devotion and the faculty of intuition operate in both processes. Hindu physicist George Sudarshan goes further to affirm that "any spiritual search, whether academic or not, is bound to lead to God.... there is nothing which is not sacred" (Richardson et al. 2002, 8–9).

Ways of Relating Science and Religion: The Typologies of Barbour, Haught and Others

Perhaps the best known and most widely used of Barbour's contributions is his four-fold typology for categorizing different ways that science and religion relate: conflict, independence, dialogue, and integration (1997, chapter 4; 2000, chapter 1). Catholic theologian John Haught has a similar four-fold typology–conflict, contrast, contact, and confirmation—in which the first two are virtually identical to Barbour's (Haught 1995, chapter 1).

Conflict   In this mode, science and religion are diametrically opposed—"enemies," if you will—where each side regards the other as simply wrong and untrue. The biblical literalism of strict creationists and the scientific materialism of atheistic scientists are examples. The writings of the physicist Steven Weinberg, astronomer Carl Sagan (Chapter 4) and biologist Jacques Monod (Chapter 5) often express this view. Barbour and Haught contend that scientific materialists confuse science's method, which only accepts physical evidence, with philosophical belief that only physical things exist. Science, they argue, simply cannot find anything beyond the physical and therefore can never disprove higher levels of reality. Nevertheless, the warfare model has often been most dominant in the public mind. In the late nineteenth century two prominent publications promulgated the notion, even though it did not reflect the true spirit of the times, and today, the popular media often emphasize the exciting drama of conflict over more harmonious points of view.

Concordism: A Form of Conflict?   Both Barbour and Haught discuss a view called "concordism" in connection with conflict. In this approach, scriptural text is interpreted to correspond closely, often literally, with modern science, as if it were presenting scientific knowledge centuries before science discovered it. Examples can be found across the board in many world religions, usually among conservatives in the faith. Evangelical Christian astronomer Ross uses the extra dimensions proposed by modern string theory to explain miracles, and orthodox Jewish physicist Gerald Schroeder uses time dilation in relativity theory to reconcile the six biblical creation days with cosmology's fifteen billion years. In Islam interpreting the Qur'an to reveal exact scientific knowledge was first done to draw Muslims to modern science but has since become a widespread popular movement, often done by those not trained as either scientists or Qur'anic scholars (Iqbal 2002, chapter 10). A highly popular example of this kind of literature is *The Bible, the Qur'an and Science: The Holy Scriptures Examined in the Light of Modern Knowledge* (1976) by surgeon Maurice Bucaille. An example from Hinduism is the statement that the "conclusions of Modern Physics are exactly the conclusions of Vedanta philosophy enshrined in the Upanishads and countless other Vedantic texts" by Swami Jitatmananda in *Modern Physics and Vedanta* (2006). A number of scholars have also viewed Buddhism as "identical" to science. The Buddhist representative at the World Parliament of Religions in 1893 claimed that "in Buddhism there was no need for explanations that went beyond that of science," and much later Gerald Du Pre, identified "many 'amazing' instances of modern scientific findings being prefigured in the Buddhist . . . texts" (Cabezon 2003, 45–48).

It may seem strange to lump "concordists" with those who see science and religion as in conflict, yet they have similarities. Barbour and Haught note that they both mix up or "conflate" scientific and religious knowledge, considering them comparable on a literal level without acknowledging major differences. Most trained religious scholars try to avoid this kind of literal comparison, although they do not deny that there could be different levels of consonance at work.

Independence   Beyond conflict is the more neutral view of independence, where science and religion are like "strangers." They exist in their own separate compartments, ask different questions, refer to different domains and speak in different languages. A legacy from Kant, this view was dominant in much of the modern era and in the last century in conservative, neo-orthodox and existential Christian thought. The domains of religion are the higher realms where "God works in mysterious ways" and the inner, subjective world wherein he reveals himself to humans and the individual makes a personal commitment through faith. What happens in nature is unrelated; there science works to derive truths about the physical world quite independently from religion, using observation and reason. The independence view also characterizes the modern Catholic theology of neo-Thomism (based on the thought of medieval theologian Thomas Aquinas), which separates God's realm of "primary causality" from the realm of "secondary causes," such as the laws of nature, which can be studied scientifically. Strong voices for independence are heard among both religious scholars and scientists. Notable among them is the late eminent paleontologist Stephen Jay Gould, who argued that science and religion belong in "non-overlapping magesteria (NOMA), or separate "domains of teaching authority."

The different languages of science and religion reflect their very different functions in people's lives. The language of Genesis I is meant to reveal and inspire believers about their relation to God and creation and not give detailed information about the physical universe. It can in fact be detrimental to religious belief to equate it too closely with a scientific explanation, because next year the scientific theory may change!

Barbour also includes within "independence" the relation of complementarity, wherein science and religion are seen to coexist as separate, seemingly contradictory aspects of a larger whole, much as in quantum mechanics both particle and wave behavior coexists in the same electron. A version of this relation is exemplified by the "categorical complementarity" of evangelical Christian astronomer Howard van Till (see Chapter 6). Some participants in the Buddhism-science dialogue also suggest a relationship of complementarity, although they mean something stronger than independence—the idea that scientific and religious inquiry can be mutually supportive aspects of a whole understanding (Cabezon 2003, 49–56).

Dialogue   This relationship builds a bridge between science and religion, as in "friendship." Here similarities and parallels in both methodology and content are noted and discussed, as well as differences. The above description of Barbour's methodological bridge is a perfect example. Numerous examples will be discussed later on: the possible relevance of Big Bang cosmology to the Judeo-Christian creation account (Chapter 2), similarities between conceptions of the material world in quantum mechanics and Eastern religions (Chapter 3), and the relation of scientific theories of the future to Christian eschatology (Chapter 8).

Dialogue also arises because science has limits and depends on realms of inquiry beyond its own. For one thing, its historical roots and underlying assumptions come from beyond science itself. Much has been written about the influence of Greek and Judeo-Christian thought on the rise of science, and, while the extent of the role is arguable, the ideas of those traditions that the cosmos is orderly, understandable, and law-abiding are core presuppositions of science. Science cannot be done, especially in cosmology, without initial assumptions, taken on faith, that the same laws work throughout all space and time. Influence of religious ideas occurred in early Islam, where belief in the oneness of knowledge and the "religious obligation" of studying nature to learn about the Creator led to a great flowering of scientific activity (Guiderdoni 2003, 465).

Science is limited in raising questions that it cannot itself answer. Why does the universe have the particular orderly laws and intelligibility that it does? What gave rise to the framework of space and time and physical laws within which the cosmos operates, and what lies beyond it? Such questions transcend the power of science to answer, confined as it is to physical data and natural causes, and require dialogue with religion and philosophy.

Integration   Barbour's final relationship is the closest of all and resembles a close partnership or even marriage. In this view a theological system or religious belief can be synthesized with modern scientific understanding into a unified vision of reality. He identifies three approaches: natural theology, theology of nature, and systematic synthesis.

In natural theology the orderly workings, design, beauty and/or complexity of nature imply or at least support the existence of a Grand Designer behind it all. In the modern era, as science's naturalistic explanations grew, this kind of teleological argument waned. A more general form of the argument sees God's hand in the initial design of physical laws, as some physicists propose today (Chapter 4). An even more potent form resurfaced to explain how "anthropic" features of the early universe were perfectly attuned for life's development (Chapters 4 and 6). Intelligent Design advocates also use natural theology, although they do not infer a particular designer.

In "theology of nature" religious doctrines are reformulated to incorporate well-established scientific understanding, such as the idea that nature is a "dynamic evolutionary process" operating through both law and chance. Teilhard de Chardin's grand synthesis of evolution and Christian theology exemplifies this approach, as does the theology of Anglican scientist-theologian Arthur Peacocke, for whom evolution is God's own creativity at work "making things make themselves" (Chapter 6). New models of God arise in theologies of nature, and Peacocke and others suggest a rich array of new conceptions of God—composer, artist, mother birthing the world, communicator of information. Feminist theologians urge the adoption of new metaphors that transcend the old male, monarchical model and hold nature more sacred. Barbour contends that such an attitude must inform a new theology of nature that draws on scientific understanding of ecology to develop "environmental ethics."

Finally, in systematic syntheses, whole metaphysical schemes incorporate both scientific and religious elements into a seamless whole. Examples are the process philosophy of Whitehead and its further elaboration in process theology, the still influential philosophy of medieval scholar Thomas Aquinas (Chapter 6) and the comprehensive spiritual evolutionary cosmology of Meher Baba (Chapter 7).

Haught's "Contact" and "Confirmation" Modes   Haught's last two relationships feature the same elements as Barbour's dialogue and integration but rearrange them for different emphasis. In "contact," mutual support and open-ended dialogue broaden both sides' view while still honoring differences. In the spirit of critical realism, they both connect the real world with unknown, ultimate reality in an open, evolving way, hoping to gain a joint understanding "more illuminating than either can provide on its own." Haught's "confirmation" stresses the deep support that religion gives to science, as in Barbour's dialogue. The very motivation to do science, the "humble desire to know," springs from an inner spiritual place and rests on essentially religious assumptions that the universe is a rational, coherent, understandable whole. Religion thus supports the whole scientific endeavor in a very close relationship.

Other Typologies   Other scholars have presented different categorizations, among them Lutheran theologian Ted Peters (1998, chapter 1), religious scholar Mikael Stenmark (2004), and Episcopal theologian W. Mark Richardson (2002). Peters' eight-fold typology further divides conflict into scientism, scientific imperialism, ecclesiastical authoritarianism, and scientific creationism. His "two-language theory" resembles independence, while "hypothetical consonance" spans a range from finding common areas of inquiry to "accord" and "harmony." The word "hypothetical" conveys the need for "humility" and openness to new knowledge that comes through joint inquiry. Peters also includes "ethical overlap," where theologians contribute to developing an ethical vision for Earth's future, and "New Age Spirituality," where a holistic vision of reality involves insights from quantum mechanics, the power of human imagination, and a committed ecological awareness. New Age spirituality is exemplified by physicists Fritjof Capra and David Bohm (Chapter 3) and Brian Swimme and Thomas Berry (Chapter 4).

Stenmark and Richardson both offer three-fold schemes. In *How to Relate Science and Religion: A Multidimensional Model* (2004), Stenmark suggests basic categories of "independence, contact and monism" (the belief that all is one) in an insightful philosophical analysis that incorporates perspectives from many disciplines. Richardson's categories emerged from his study of scientists' essays relating their religious belief to scientific work in the Science and the Spiritual Quest project. Some focused on universal principles of reasoning about ultimate reality (rationalist-speculative), others on the spiritual search and dimension of feeling in understanding one's relation to the cosmos and purpose (affective-holistic), and a third group on cognitive philosophical questions about method, language, content, etc. (critical-historical).

The Present Status of the Science and Religion Dialogue

In the past twenty-five years, the dialogue has expanded and intensified at all levels, from the corridors of academia to the halls of churches and temples to public institutions and the media. It has also become a truly global conversation.

As the academic effort gained momentum, a number of major universities and seminaries established research centers devoted to this work. Two prominent ones are the Center for Theology and the Natural Sciences (CTNS) in Berkeley, California, and the Chicago (now Zygon) Center for Religion and Science. Both publish scholarly journals, *Theology and Science* from CTNS, and *Zygon*, copublished by the Zygon Center and the Institute on Religion in an Age of Science. Other centers exist at Georgetown, Princeton, Columbia and Boston Universities and at the Boston Theological Institute, and in Canada there is the Centre for Islam and Science, which publishes a scholarly journal. In this same period the Vatican and CTNS sponsored the science-theology study groups described above. Continuing strong in this period also has been the American Scientific Affiliation, which publishes *Perspectives on Science and Christian Faith*.

An influential source of support for academic initiatives during this expansionary period has been the John Templeton Foundation, founded in 1987. It offers a one million dollar yearly prize for outstanding contribution in the field of science and religion, funds grants for cutting-edge research, and publishes books. Motivated by the personal religious interest of financier John Templeton, the Foundation has funded and cosponsored many projects with the above organizations and others. Three at CTNS are exemplary: a course award program; the Science and the Spiritual Quest project (SSQ), where scientists related their religious beliefs to science in workshops and conferences worldwide; and the newest STARS program (Science and Transcendence Advanced Research Series) offering grants to study how science, in the light of theology, points toward ultimate reality. Templeton funding has certainly enabled the growth of many programs and initiatives but is not without controversy. Some academics strongly oppose what they consider undue influence of a wealthy, religious-minded individual on the direction of institutional scholarly work.

The academic discussion has passed through three overlapping stages in the last fifty years, according to Peters. The first "methodological" stage forged a joint framework for dialogue, as described above. The second "physics" phase took on the big, metaphysical questions that arose from cosmology and subatomic physics—the God question and the increasing mystery of the cosmos. Phase three has addressed questions arising in biology—how divine action and religious future promises relate to evolution and whether humanity should use genetic planning to guide the future (Peters 1998, 3–7). Until recently, the issues addressed have mostly arisen through science. Theology has been in the position of reacting to the findings of science, sometimes accommodating too much, according to Peters and Hewlett (2003). Recently, theologians began to reverse this trend, making theological questions set the agenda, as with the focus on divine action in the Vatican-CTNS studies, or at least considering them on an equal footing at the outset, as in a recent project studying the "end times" (Polkinghorne 2007).

Beyond academia, other organizations focus on broader areas like education and popular understanding, often in addition to academic research. An example is the Metanexus Institute, a worldwide think tank which includes the humanities in addressing humankind's deep questions, and offers conferences, an online forum and initiatives for local societies worldwide. Another example is the Dialogue on Science, Ethics and Religion Program (DoSER) of the American Association for the Advancement of Science (AAAS), which has sponsored conferences on evolution and cosmology and provides teaching resources. Some science-faith organizations are church-based, such as ones in the Episcopal and Lutheran churches, while others are ecumenical. In one grass-roots initiative, the Clergy Letter Project, 10,000 clergy signed a proevolution statement and continue to organize "evolution Sundays" for church discussion. Also aimed at a more popular level is the independent journal *Science and Spirit*, which considers ways of bringing science and religion together in a way that is meaningful to our everyday lives.

In recent decades the dialogue has broadened to include most of the world's religions, which, like Christianity, have all had to accommodate one way or another to the rise of modern science. The history and present status of these interactions are told in a number of excellent sources, among them other volumes in this series, as well as *Buddhism and Science* (Wallace 2003), *Islam and Science* (Iqbal 2002), *Jewish Tradition and the Challenge of Darwinism* (Cantor and Swetlitz 2006), *Science and Religion in India* (Gosling1976), and *God, Humanity and the Cosmos* (Southgate et al. 1999, chapters 9 and 12). Helpful summaries can also be found in the *Encyclopedia of Science and Religion* (van Huyssteen 2003).

A full spectrum of views about how science and religion relate exists within each of the major religions—from conflict to integration and confirmation. The main targets for conflict with science are evolution and the primacy of material reality in Western science. Some elements within each major religion oppose evolution as a purely materialistic process which cannot accommodate divine action, this view being strongest perhaps in Islam, and weakest in Hinduism and Buddhism. Scholars of other religions seem to share with Christians the same confusion about whether science actually dictates a materialistic philosophy. Another source of conflict for Hinduism and Buddhism is the scientific primacy of the material world over consciousness, because for them consciousness or divinity are more often regarded as the primary reality. For Islam, what is questionable is the separation of material reality from a traditional metaphysical framework that includes matter and spirit in the same hierarchical "ladder of being." Christian theologians grapple with the same issues, and today, leading scholars and spiritual leaders from all religions seek understanding that can heal these rifts. Eminent Islamic scholars speak of the need to integrate physical science into the framework of a universal "perennial philosophy," whose principles transcend cultures and separate faiths but lie at the heart of all religions (Iqbal 2002, 307; Nasr 1993, 53–54). A similar approach was followed by Meher Baba and Aurobindo in integrating consciousness and cosmic purpose with material evolution in an overarching universal spiritual system.

Such ideas are today explored in a large and ever growing number of centers, institutes, workshops, programs and conferences worldwide that bring single-faith or interfaith groups into dialogue with science. Examples of permanent research organizations are the Centre for Islam and Science, the Buddhist Mind and Life Institutes, and the Project of History of Indian Science, Philosophy and Culture. Individual meetings and projects abound, and the following are merely representative. Two symposia were held at modern meetings of the World's Parliament of Religions—in Chicago in the Parliament centenary year 1993 and in Capetown, South Africa in 1999. *Cosmic Beginnings and Human Ends* (Matthews and Varghese 1995) and *When Worlds Converge* (Matthews et al. 2002) presented essays from the Chicago and Capetown meetings, respectively. Taoist, Confucian, Hindu, Buddhist, Jewish, Christian, and feminist perspectives were all represented. Themes addressed were cosmic evolution, religion and ecology, technological responsibility, and the promises and challenges of joint collaboration. A similar topic, "Cosmology and Teleology," was the subject of an interfaith, multiple-perspective conference sponsored by the AAAS and Georgetown University's center in 1997 that led to the publication *Science and Religion: In Search of Cosmic Purpose* (Haught 2000b). CTNS' course award and SSQ programs also drew participants from many faiths, and their essays can be found in *Bridging Science and Religion* (Peters and Bennett 2003) and *Science and the Spiritual Quest* (Richardson et al. 2002). A first ever publication of Christian and Muslim perspectives in the same volume, *God, Life and the Cosmos* (Peters et al. 2002) followed a joint conference in Islamabad in 2000.

This short review of the current situation hardly does justice to the enormous scope of the science and religion discussion which has intensified and blossomed in recent years. It fills the literature, airwaves and cyberspace with a multitude of voices affirming every possible science–religion relationship: antiscience fundamentalist proselytizers, science-praising evangelicals from all faiths, antievolution ID advocates, personal testimonials for harmony, diatribes against religion from atheistic scientists, praise of religion from believing scientists, independence arguments from both academics and religious folk, praise of cosmic evolution as the new religion, scientist-theologians endeavoring to forge integrated perspectives, and perhaps enlightened leaders who envision integration clearly. It is hard to sort out all these voices. The dialogue is underway, and the story is being written at the present time, making it impossible to predict its future course. One thing is certain from surveying the past century, however. Tremendous change has occurred, both within science and within religion and in the nebulous area that joins them. Religion is now far more willing to engage with science, and many scientists are eager to reciprocate. Astronomers write such books as *God's Universe* (Gingerich 2006), and a leading DNA researcher pens *The Language of God* (Collins 2006) describing the basic molecule of life. Even as voices of opposition ring loud and clear, many scholars and laymen are forging a relationship between science and religion that is meaningful and productive. There is every reason to assume this will continue.

Themes in Science and Religion

Three simple themes, among many, have characterized the history of science in the modern era and will clearly emerge in the chapters to follow that trace physical science during the last century. They apply as well to the science and religion dialogue.

The first is that "things are not what they seem." Not until the sixteenth to seventeenth centuries did humanity accept that everything in the cosmos did not revolve around us, although it certainly looks that way. The stars, sun and moon appear to revolve around us every day, but now we know that Earth's rotation produces this effect. Matter looks very solid and smooth, but now science tells us that atoms are mostly empty space—dense, tiny, hard nuclei surrounded by a "cloud" of undulating, wave-like "electron-stuff," whose rapid vibrations give the appearance of solidity.

Science and religion both deal with an "unseen world." Who has "seen" the nucleus of the atom, or the force of gravity, or the dark matter of the physical universe? And who has "seen" God, in a way easily portrayed to others? Leaps of faith are required in both instances. The appearances of the world belie both of these realities. The way things "seem" is not the way they are. From "nothingness" emerged the physical universe, according to science, the Kabbalah of Judaism, and Meher Baba. Deep metaphysical questions about what is truly real and eternal and what is passing underlie this theme.

A second theme is the "Copernican principle"—the modern era's relegation of Earth and humanity to a less central, important or elevated place in the cosmos. First, Copernicus declared that Earth no longer was the center of things. It rapidly spun on its axis and circled the sun, but at least humans were special and elevated above all other creatures—we had minds! Darwin took care of that, showing but a small separation between us and the apes. Then the solar system was no longer the center of the Milky Way Galaxy, and the Galaxy was one of hundreds of billions of other galaxies in the whole universe! Smaller still became Earth and man. Somewhere along the way, the Copernican principle became the "principle of mediocrity," that assumes there is nothing special about us in any way. Copernicanism has now gone wild: our own universe may be just one among an infinitude of other universes, where an infinitude of other planets with intelligent beings may exist. We are one among trillions upon trillions, with no special status of any kind. Our relegation seems complete.

What does religion say about "Copernicanism?" The vastness and plenitude of life may celebrate the infinite God, some say, while for others they make us seem truly insignificant. Others focus on the many factors that converge to make Earth just perfect for life and the virtually nil chance that they would all occur together. More than one religious perspective contends in fact that Earth is truly special, that humans are indeed the pinnacle of creation, and that only here can they realize their full spiritual potential.

A third theme is the "search for unity." The whole history of science can be seen as one long story of finding unity in phenomena formerly perceived as separate. Heaven and Earth became one in Newtonian mechanics and gravitational theory. Acoustics and heat came under the umbrella of mechanics, just as optics and light were subsumed into electromagnetic theory. Two forces unite at high enough energies and act as one, and a fervent search is underway to unite all four forces into one grand "theory of everything," the Holy Grail of physics that Einstein labored unsuccessfully for years to find. Such conditions may have prevailed at the first instant of the universe and hold the key to cosmology and the nature of space and time, matter and energy.

The drive for unity is at the very heart of the science and religion dialogue. The search for everything from common ground to a deeply integrated vision motivates the whole exchange. Few have expressed the urgency of the task more eloquently than Alfred North Whitehead:

When we consider what religion is for mankind, and what science is, it is no exaggeration to say that the future course of history depends upon the decision ... as to the relations between them. We have here the two strongest general forces ... which influence men and they seem to be set one against the other—the force of our religious intuitions and the force of our impulse to accurate observation and logical deduction ... The clash is a sign that there are wider truths and finer perspectives within which a reconciliation of a deeper religion and a more subtle science will be found. (Whitehead 1962, 162)

The interesting suggestion here is that both science and religion will need to change and yield—become "deeper" and more "subtle"—to find a pathway to unification. Perhaps the whole process is a spiritual one—a path of seeking unity and becoming whole. What is a spiritual journey if not the search for unity? And yet another level of unification seems possible in this dialogue. Science might provide a neutral reference point for interfaith conversation. As one scholar, Thomas Berry, remarked at a World Parliament, "The story of the universe is the only thing that can bring the religions of the world together" (Matthews and Varghese 1995, vii).

A fourth theme must join the others in the merged landscape of science and religion—the search for a Supreme Being. Although science has abandoned addressing this question directly, its own limits bring up the question. Does the Big Bang imply a creation moment and a creator? What lies beyond? Certainly, science cannot say. How can evolution's mechanistic processes allow for divine action? Theologians can only guess. If God draws us to him out of the future, how do scientists' wild scenarios of the future universe relate to this? This issue is front and center of the debate, and especially prominent in the public consciousness.

These themes threaded through all of twentieth century science and religion, as the following history of scientific discovery and story of cosmic evolution will reveal. These ideas characterized much of the thinking of scientists and religious scholars alike as they absorbed the new knowledge, reflected on its deeper meaning, and sought for understanding that merged the two great enterprises of the human spirit.

Chapter 2

—◆—

The Beginning of the Universe: Scientific and Religious Perspectives

The early twentieth century witnessed a dramatic change in the conception of man's place in the cosmos and the nature of the universe itself. Physical science joined biology in presenting to the world an evolutionary picture of the cosmos—a dynamic, expanding universe that may have had its origin at a point in time. By the early 1930s both observation and theory agreed that the space of universe was growing and carrying the galaxies along with it, and a law describing this expansion was well established. In the same period astronomers discovered other stunning features of the physical world—that the Earth and solar system have no central location in our Galaxy and that the Galaxy itself is but one of many other vast stellar systems in the cosmos.

In physics the transformation went even deeper, revolutionizing basic conceptions of space, time, and matter itself. Einstein's theories revealed the relativity of space, time, and mass to an observer's motion, the equivalence of mass and energy, and the shaping of space–time by the matter and energy occupying it. General relativity and later hypotheses yielded models of an expanding universe that had burst forth from a very condensed unstable state, enlarged and cooled over eons of time. Prediction and later observation of remnant radiation from the initial event confirmed the validity of what came to be called the "Big Bang" theory and caused the demise of rival theories. Special relativity unveiled the energy–mass relation and, along with quantum theory, laid the groundwork for comprehending nuclear reactions and element production in the early universe and in stars. Understanding dawned that stars themselves do not live forever but process matter and build heavy elements in long life cycles stretching over millions or billions of years.

The religious implications of these discoveries were many and varied, as the wide-ranging views of scientists, theologians and church leaders revealed. Earth was once more relegated to a secondary position, this time within the much larger system of the Galaxy, which itself was but one among legions of similar immense systems. Implications of the grandeur of the picture but also of Earth's insignificance in the midst of this enormity were obvious. The growth and evolution of the universe and of the stars within it, while presenting an exciting picture, also suggested impermanence. More profound perhaps was the invitation to reflect metaphysically on the relative, but also unified nature of the most basic structural elements of the physical world, such as time, space, matter and energy. Most striking of all were the strong religious implications of a temporal origin—a "creation point"—and its possible consonance with the biblical creation event, with the Christian concept of "creation from nothing" and with the idea of cosmic expansion from a point presented in Kabbalistic writings of Judaism and in the cosmology of Meher Baba.

DISCOVERING THE BEGINNING AND OUR PLACE IN THE UNIVERSE

The first hint of the expanding universe came in 1912 when Vesto Slipher determined that several fuzzy, spiral-shaped nebulae, or gaseous clouds, were receding from Earth at rapid speeds. Motion is easily detectable in the spectrum of light received from an astronomical object. By the Doppler Effect, the recessional motion of an object stretches its emitted light waves to a longer or "redder" wavelength, and they appear at a "redshifted" position in the spectrum (see Figure 2.1). Within a decade Slipher had measured forty such objects and found redshifts for thirty-six of them.

At the time, it was understood that these nebulae were part of the Milky Way Galaxy, in which our solar system occupied the central position. Stars, it was believed, filled an infinite space; they had to, in fact, to prevent the universe from collapsing. Apart from random motions of the stars themselves and orbital motions of planets, the whole system was static, eternal, and infinite. A major breakthrough in understanding the Milky Way's enormity and our cosmic location came in 1918 when Harlow Shapley used variable stars to determine that the Earth is nowhere near the center and that our Galaxy has a diameter of 300,000 light years (a figure later revised downward by a factor of three). Surely, the receding nebulae must be within such an enormous system, he reasoned, and eminent astronomer Arthur Eddington declared in 1923: "One of the most perplexing problems of cosmogony is the great speed of the spiral nebulae" (Lightman 2005, 235).

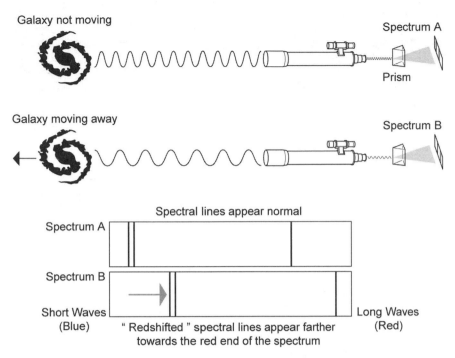

Figure 2.1 The Doppler Effect for Galaxies. When a galaxy recedes from us, its emitted light waves are stretched to a longer or "redder" wavelength and thus show a "redshift." All but a few nearby galaxies show such redshifts in their spectral lines, indicating that they are almost all moving away from us. Later understanding revealed that such motion was due to the stretching or expansion of space itself. (Illustration by Jeff Dixon)

Another breakthrough occurred when astronomers realized that the fuzzy nebulae were, in fact, extragalactic. In the mid 1920s astronomer Edwin Hubble used variable stars to determine the distance to the nearest of these objects, the Great Spiral Nebula in Andromeda (see Figure 2.2). At 900,000 light years (now revised to 2.5 million) it was well beyond our Galaxy, as were many more he measured—all immense "island universes" like the Milky Way itself. These objects were clearly important in the "big picture of the universe," and Hubble set about determining distances to the receding nebulae. His first results, published in 1929, showed a definite relationship: with a few exceptions the redshifts, or velocities, grew larger for increasingly distant galaxies (see Figure 2.3). Though aware of the "scanty material, so poorly distributed," he boldly asserted that the correlation was linear, suggesting a fundamental law at work: the velocities of the receding galaxies were directly proportional to their distances. This "leap of faith" was subsequently borne out when he and Milton Humason extended the study to much greater distances two years

Figure 2.2 The Great Galaxy in Andromeda. In the 1920s astronomers discovered this beautiful "spiral nebula" to be a separate "island universe" well beyond our own Milky Way Galaxy. They now know that it lies at a distance of 2.5 million light years and contains 1 trillion stars. (Credit: NASA/JPL-Caltech/P. Barmby; Harvard-Smithsonian CfA)

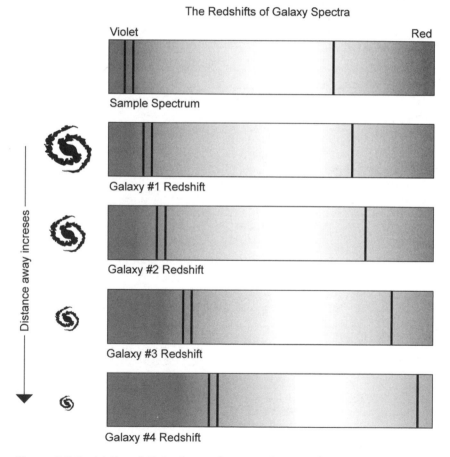

Figure 2.3 Redshifts of Light from Distant Galaxies. After determining the distance and photographing spectral line shifts for many galaxies, astronomers found a distinct correlation. The more distant the galaxy, the greater was its redshift and therefore the greater was its speed of recession. (Illustration by Jeff Dixon)

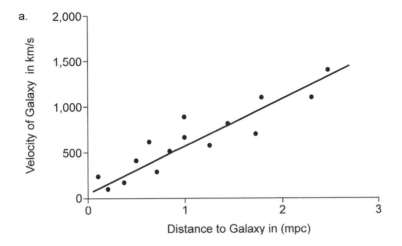

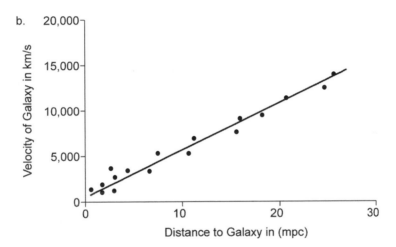

Figure 2.4 Hubble Law of the Expanding Universe. In 1929 Edwin Hubble plotted distances and velocities of two dozen galaxies out to a distance of 2 mpc (6.5 million light years) (top diagram). Although the data were quite scattered, Hubble asserted there was a linear relationship—a direct proportionality between distance and speed. His claim was borne out two years later when he and Milton Humason measured galaxies out to 30 mpc (98 million light years) (bottom diagram). (Illustration by Jeff Dixon)

later (see Figure 2.4). From his diagram, which expresses the rate at which the universal speed increases with distance, one can estimate the time over which the galaxies have been separating, or the "age" of the universe. Such estimates over time have ranged between 10 and 20 billion years.

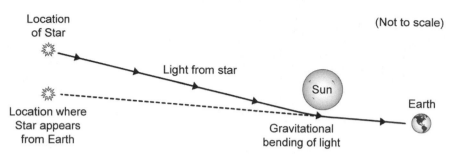

Figure 2.5 Bending of Starlight by Gravity of Sun. Einstein's general theory of relativity predicted that light should bend near a strong gravitational field. This prediction received clear confirmation when the light of distant stars directly behind the sun bent around the sun to become visible to us during a total solar eclipse in 1919. By showing how matter and energy thus shape the structure of space–time, Einstein's general theory revealed how all the basic elements of the physical world are interdependent and interwoven. (Illustration by Jeff Dixon)

Sometime before Hubble's first paper, theoretical calculations had also suggested an expanding universe. Such work derived from Einstein's 1915 general theory of relativity, which linked matter and energy to the unified space–time of special relativity. In this theory gravity is viewed differently from other forces; it results from the curvature of space–time in the vicinity of mass and energy. One popular description emphasized how radically the theory changed our conceptions of the cosmos:

The universe is [no longer] a rigid and immutable edifice where independent matter is housed in independent space and time; it is on the contrary an amorphous continuum, without any fixed architecture, plastic and variable constantly subject to change and distortion. Wherever there is matter and motion, the continuum is disturbed. Just as a fish swimming in the sea agitates the water around it, so a star . . . or galaxy distorts the geometry of the space-time through which it moves. (Barnett 1957, 85)

The theory's prediction that light waves would bend near a massive object received triumphal confirmation in 1919 when light from distant stars was observed to curve around the sun during total eclipse (see Figure 2.5). Much later astronomers observed the same "gravitational lensing" of the light of very distant galaxies passing through more nearby clusters of galaxies (see Figure 2.6).

From the equations of general relativity Einstein and others developed cosmological models. Although he saw the possibility for an expanding model, his belief in a static universe led him to subtract out such cosmic repulsion or "antigravity," using a term he called the "cosmological constant" or "Lambda." He strongly resisted models for a changing universe,

Figure 2.6 Gravitational Lensing of Light from Distant Galaxies. According to general relativity, just as light bends around the sun, so light from distant galaxies is bent when passing through massive groups of galaxies. Rich clusters of thousands of galaxies, like Abell cluster 2218 above, thus act as gravitational lenses—giant "telescopes in space"—collecting and focusing the light of more distant invisible galaxies directly behind them. Such "lensed" images are seen as arcs of light in this Hubble Space Telescope photograph. One extremely faint "double image" galaxy in this photograph turned out to be one of the most distant objects known—a "baby galaxy" at a distance of 13.4 billion light years. (W. N. Colley and E. Turner, Princeton University; J. A. Tyson, Bell Labs, Lucent Technologies; and NASA)

which were soon suggested by a young Russian mathematician Alexander Friedmann and a few years later by a young Belgian physicist Georges Lemaitre. Most significantly, Lemaitre connected the theory with recent discoveries of redshifts and extragalactic positions of the spiral nebulae. In his major but not well publicized 1927 paper he stated, "The receding velocities of extragalactic nebulae are a cosmical effect of the expansion of the Universe," and he predicted they should show a linear correlation with distance (Lightman 2005, 230–45).

Unbeknownst to Hubble, the theory was thus in place to support his epochal discovery. In 1931 in the library of the Mount Wilson Observatory Einstein honored Hubble's achievement and announced his own abandonment of the static model with its famous lambda term and his full acceptance of an expanding cosmos (Lightman 2005, 244). In the ensuing decade three major versions of the expanding universe would become

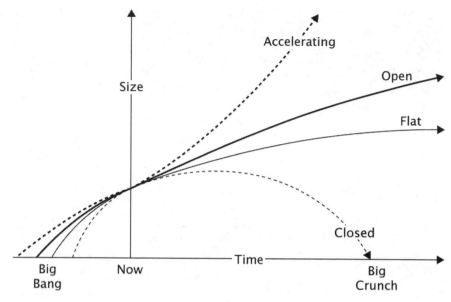

Figure 2.7 Models of the Expanding Universe. Einstein's general theory of relativity led to three "standard" models for the expansion and eventual fate of the universe: the closed model that collapses into a "big crunch," the "flat" or boundary model that expands forever but almost stops, and the open model that expands forever at a faster speed. In all three cases, the universe slows down over time. Astronomers were immensely surprised to determine in 1998 that in the past 4 to 5 billion years, the universe has actually been accelerating in its expansion (see Chapter 8). (Ilustration by Jeff Dixon)

standard "Big Bang" models—two "open" models that expanded forever and one that eventually "closed" back on itself in a universal contraction (see Figure 2.7). Much of the significant work of cosmology in the twentieth century would focus on attempts to distinguish which of these models was correct.

In the early 1930s Lemaitre hypothesized that the expanding universe began in a superdense state that he called the "primeval atom." Cosmologist George Gamow says that "primeval nucleus" would be a more apt term for his conception—a giant, dense body of nuclear fluid, much larger than ordinary nucleus, which became unstable with the onset of expansion and fragmented into many pieces (Gamow 2004, 50–51). In creating his hypothesis of the moment "before" time began and the "atom" that existed, Lemaitre connected the known existence of radioactive substances and the quantum understanding that the behavior of a single "particle" can never be known exactly—even its position in space and time (Kragh 1999, 48). He described the process as follows:

If the world had begun with a simple quantum, the notions of space and time would altogether fail to have any meaning at the beginning; they would only begin to have a sensible meaning when the original quantum had been divided into sufficient number of quanta. If this suggestion is correct, the beginning of the world happened a little before the beginning of space and time . . . we could conceive the beginning of the universe in the form of a unique atom, the atomic weight of which is the total mass of the universe. This highly unstable atom would divide in smaller and smaller atoms by a kind of super-radioactive process. (Kragh 1999, 47)

He acknowledges the speculative nature of his idea, including the need for more thorough understanding of nuclear physics to support it (Lemaitre 1965, 342) and the puzzle of how a single "undifferentiated quantum" could possibly result in the enormous variety of the world today. But he believes that quantum theory allows for such a possibility and that "the whole matter of the world must have been present at the beginning, but the story it has to tell may be written step by step" (Kragh 1999, 48). In a later description he portrays this great cosmic event as it looks from our vantage point:

The evolution of the world can be compared to a display of fireworks that has just ended; some few red wisps, ashes, and smoke. Standing on a cooled cinder, we see the slow fading of the suns, and we try to recall the vanished brilliance of the origin of the worlds. (Gamow 2004, 51)

By the late 1940s several developments raised new challenges but also promised new successes with Big Bang model. British astrophysicist Fred Hoyle joined cosmologist Herman Bondi and astrophysicist Thomas Gold in presenting a rival theory to the Big Bang—the steady state theory, which proposed that, although the universe is expanding, new matter is created constantly to maintain a constant density over time. At the same time, Ukrainian born cosmologist George Gamow's interest in the origin of chemical elements led him and physicist Herman Alpher to connect the cosmic abundance of light elements observed today with nucleosynthesis from thermonuclear fusion reactions in the very early, hot, dense, expanding universe. This extension of the Big Bang hypothesis to include nuclear events in the early universe marked a major advancement and at mid century established Gamow as the theory's strongest proponent.

In the same year Alpher and Robert Hermann also predicted that the hot early universe should have left a tell-tale signature radiation, once a field of high energy light waves but now stretched out by the expansion of space and dissipated into a very cold, low energy field of randomly oriented microwaves (see Figure 2.8). The accidental discovery of this "cosmic microwave background" radiation by physicists Arno Penzias and Robert Wilson in 1965 offered the strongest evidence yet for the validity

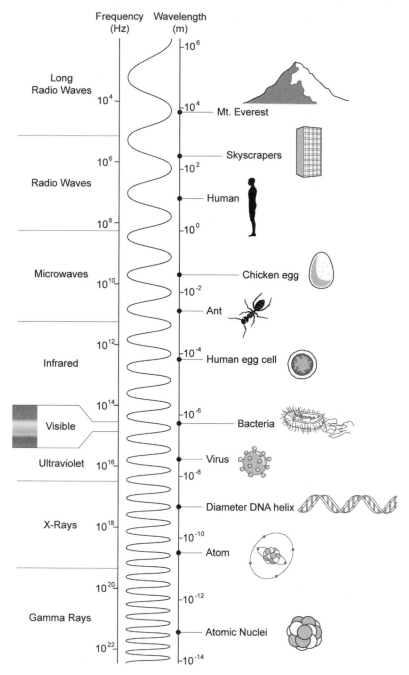

Frequency (Hz)

Wavelength (m)

- 10^6
- Long Radio Waves — 10^4
- 10^4 — Mt. Everest
- 10^6 — 10^2 — Skyscrapers
- Radio Waves — Human
- 10^8 — 10^0
- Microwaves — 10^{10} — Chicken egg
- 10^{-2} — Ant
- 10^{12} — 10^{-4} — Human egg cell
- Infrared
- 10^{14} — 10^{-6} — Bacteria
- Visible
- Ultraviolet — 10^{16} — Virus
- 10^{-8}
- Diameter DNA helix
- X-Rays — 10^{18} — 10^{-10} — Atom
- 10^{20} — 10^{-12}
- Gamma Rays — Atomic Nuclei
- 10^{22} — 10^{-14}

Figure 2.8 The Electromagnetic Spectrum. Astronomers now study the full range of electromagnetic radiation—from the least energetic radio waves with wavelengths the size of people and mountains up to the most energetic gamma rays with wavelengths the size of the atomic nucleus. As the universe evolved, the gamma rays of the primeval fireball gradually stretched and weakened with the expansion of space, becoming today a field of extremely cold microwaves. (Ilustration by Jeff Dixon)

Figure 2.9 Arno Penzias and Robert Wilson at their Microwave Antenna. In 1965 Arno Penzias and Robert Wilson were puzzled by an annoying hiss in the new sensitive microwave receiver they had built to improve telephone communication. This nuisance turned out to be the cosmic background radiation—the afterglow of the primordial fireball—one of the century's most significant discoveries and a strong support for the Big Bang theory. (Lucent Technologies' Bell Laboratoties. Courtesy AIP Segrè Visual Archives, Physics Today Collection)

of the Big Bang (see Figure 2.9), and caused astrophysicist Fred Hoyle to abandon his rival steady state theory. This remarkable radiation represents the earliest "snapshot" of the universe scientists have, for it was released at the time atoms formed a few hundred thousand years after the Big Bang. Its remarkable smoothness—that is, its lack of temperature variation across space of more than one part in 10,000, reveals how homogeneous the universe was at the time but also raises the question of how clumpy structures like galaxies could ever have formed. Eventually in 1992 satellite measurements of tiny variations showed that very small density fluctuations did indeed exist at that early time and could have produced galaxies, thereby vindicating once again the Big Bang paradigm. Hailed as one of the great discoveries of the century, the observation was described in religious terms even by skeptics. Project Director George Smoot said, "If you're religious, it's like seeing God," and astrophysicist Michael Turner declared, "They have found the Holy Grail of cosmology" (Barbour 1997, 198; Ross 1995, 19).

Calculations of element production in the early universe were possible because of the enormous strides made during the first half of the twentieth century in understanding nuclear reactions, which operated according to quantum principles and also the principle of the interchangeability of mass and energy first elucidated in Einstein's theory of special relativity. Einstein had been led to develop this theory to explain the puzzling observation of light's constant velocity when viewed from all moving frames of reference. This simple observation, Einstein showed, had bizarre consequences: moving observers would measure such seemingly fixed properties as distance and time and even the simultaneity of events differently from those who were stationary. Even mass was relative, becoming larger relative to other frames of reference as its velocity neared light speed, where it would be infinite! Therefore, nothing with mass could ever reach that speed. Mass could also be annihilated and converted into pure energy, by the well-known equivalence formula $E = mc^2$. This principle has been a cornerstone in understanding the changes in mass which always accompany two kinds of nuclear transmutations—nuclear fission, in which radioactive nuclei decay, and nuclear fusion, in which nuclei merge to become a heavier nucleus. This knowledge, under the pressure of global war, unleashed both the atomic or fission bomb in the 1940s and the hydrogen or fusion bomb in the 1950s but also led to a deepened understanding of the nature of matter and processes in stars. German physicist Hans Bethe worked out details of thermonuclear reactions that power the sun, creating helium from hydrogen, in the late 1930s, and Eddington and Indian astrophysicist Subrahmanyan Chandrasekhar also studied interior stellar structure and processes. In the late 1950s Hoyle, in a seminal joint paper with physicist William Fowler and astronomers Geoffrey and Margaret Burbidge, demonstrated that almost all the chemical elements in nature could have been created through nucleosynthesis in the interiors of stars. A major cornerstone of science's picture of cosmic evolution thus fell firmly into place.

PHILOSOPHICAL AND RELIGIOUS REFLECTIONS OF SCIENTISTS

The scientists who made these discoveries were well aware of the immense philosophical and religious implications they carried, and a number of them were moved to write about the relation between science and religion. Influence of personal belief upon scientific work was also evident, and this period holds some of the most striking examples of such extrascientific influence on theory.

Responses to Cosmic Immensity and Earth's Significance

One philosophical response to the immensity of the cosmos and mankind's further demotion emphasized our increasingly peripheral status. Shapley became "the evangelist for the philosophical implications of the human displacement," according to journalist Larry Witham, who quotes him as saying, "The solar system is off center and consequently man is too... man is not such a big chicken. He is incidental" (Witham 2003, 39). In his popular book, *Of Stars and Men: Human Response to an Expanding Universe* (1964), Shapley argued not only that we are peripheral but that an ever more enormous and expanding cosmos made it almost certain that other beings pervaded the universe. This was because of the sheer number of possible suns and because of the greater rate of sun and planet formation in a younger, more condensed universe (Dick 1998, 46). Clearly, we were not the greatest among them.

Others, on the other hand, celebrated our continued unique importance and praised God's effulgent creativity. Eddington, a Quaker who wrote of mysticism and religion in popular books such as *Science and the Unseen World* (1929) and *Why I Believe in God: Science and Religion, as a Scientist Sees It* (1930), acknowledged that "the contemplation of the galaxy impresses us with the insignificance of our own little world; but we have to go still lower in the valley of humiliation" (Eddington 1928, 165). By this he referred to the realms of distant galaxies each with their millions upon millions of stars. Yet he himself regarded the plethora of stars as evidence of nature's bounty to produce on one planet her culminating achievement, man, in the same way that many acorns are scattered so that one might sprout into a great oak tree (Dick 1998, 77–78). British cosmologist E. A. Milne echoed these sentiments in his *Modern Cosmology and the Christian Idea of God* (1952), where he went further to propose that the infinite number of galaxies and planets are sites where an infinitely creative God has introduced biological evolution (Milne 1952, 152–54).

Scientists' Religious Reflections on the New Cosmological Theories

For some Christian theists, the discovery of a beginning to the cosmos was a strong argument for divine creation. In the 1940s such views were voiced by Milne and also by British physicist Edmund Whittaker, who wrote *The Beginning and End of the World* in 1943. Whittaker suggested that science's discovery of an absolute beginning would allow for a revived argument for God's existence as the ultimate cause. Science thus supported a transcendent, deistic God who created all at the beginning, but not an immanent "pantheistic" God who would originate and grow with the universe. The moment of creation itself was beyond science. Milne concurred on this point: "We can make no propositions about the state of affairs *at t = 0*; in the divine act of creation, God is unobserved and unwitnessed, even in principle" (Kragh 1999, 252). This recognition underlies a basic point often missed: the Big Bang is really not a theory about the very beginning. Its realm of relevance is immediately following time zero—it describes only the expansion and cooling that has transpired ever since. As for the exact moment of origination it can say nothing; there, according to general relativity, the curvature of space–time and thus density become infinite and the known laws break down. Such a "point" is called a singularity. Einstein and Lemaitre both recognized that a singularity was unavoidable but did not regard it as having physical reality (Kragh 1999, 54–55), and Lemaitre's Big Bang model only envisioned expansion from a preexisting "primeval atom." Late twentieth century quantum cosmologies have been able to penetrate further back but never exactly to time zero; in fact some suggest that one might avoid a "creation" moment (see Chapter 3).

For other Christian theists, the idea of connecting the "beginning of the expansion" to the "creation of the world" was not initially important or even desirable philosophically, and Lemaitre himself as well as Eddington at first explored cosmologies that did not include the "beginning of the world in a realist sense" (Kragh 1999, 45). Eddington, although a theist, found the notion of a sudden beginning distasteful, for it represented the end of space-time. In a talk in 1931 he stated that "philosophically, the notion of a beginning of the present order of Nature is repugnant to me . . . it leaves me cold," and if a beginning had to exist, then it should be *"not too unaesthetically abrupt"* (Eddington 1933, 56). What he preferred was a "world . . . beginning to evolve infinitely slowly from a primitive uniform distribution in unstable equilibrium" and gradually approaching a static state (Kragh 1999, 44–45). Elsewhere he gives the idea poetic expression:

The corridor of time stretches back through the past. We can have no conception how it all began. But at some stage we imagine the void to have been filled with matter rarified beyond the most tenuous nebula. The atoms sparsely strewn move hither and thither in formless disorder . . . Then slowly the power of gravitation is felt. Centres of condensation begin to establish themselves and draw in other matter. The first partitions are the star-systems such as our galactic system; sub-condensations separate the star-clouds or clusters; these divide again to give the stars. (Eddington 1928, 167)

Despite acknowledging his philosophical preferences, Eddington never went so far as to draw religious conclusions from science, arguing instead: "I repudiate the idea of proving the distinctive beliefs of religion either from the data of physical science or by the methods of physical science" (1928, 333).

No one could agree more than Lemaitre, who throughout his career maintained that science and religion should be kept strictly separate. As a Catholic priest, however, he was religious himself and recognized the connection that so many people naturally made between scientific ideas of the "beginning of the world" and the religious notion of "God's creation of the world." He seemed to have a great appreciation for God's hidden nature and the satisfaction of discovery. In a paragraph he subsequently deleted from his original presentation of the Big Bang hypothesis, he wrote, "I think that everyone who believes in a supreme being supporting every being and every acting, believes also that God is essentially hidden and may be glad to see how present physics provides a veil hiding the creation." In an exchange with astronomer James Jeans about answers they would like from an all-knowing oracle, Lemaitre concluded that he would most like for the oracle to say nothing "in order that a subsequent generation would not be deprived of the pleasure of searching for and of finding the solution" (Kragh 1999, 48–49).

Our mental capacities were a gift from God to encourage us to explore and discover the universe on our own. This "epistemic optimism," as Kragh labels it, sees that "God would hide nothing from the human mind, and consequently there could be no real contradiction between Christian belief and scientific cosmology." As Lemaitre expressed it to a Catholic audience, "The universe . . . is like Eden, the garden which had been placed at the disposal of man so that he could cultivate it and explore it" (Kragh 1999, 59–60). This metascientific reflection is very different from what he was often accused of, creating a cosmology to suit or support theology. He expressed this conviction clearly at a physics conference:

As far as I can see, such a theory [of the primeval atom] remains entirely outside any metaphysical or religious question. It leaves the materialist free to deny any transcendental Being . . . For the believer, it removes any attempt to familiarity with god . . . It is consonant with the wording of Isaias [sic] speaking of the "hidden God' hidden even in the beginning of the universe . . . Science has not to surrender in face of the Universe and when Pascal tried to infer the existence of God from the supposed infinitude of nature, we may think that he is looking in the wrong direction. (Kragh 1999, 60)

Lemaitre spoke forcefully against interpreting the Bible to support science, and using science to judge the Bible. In 1933 he argued that "[believing] the Bible pretends to teach science . . . is like assuming that there must be authentic religious dogma in the binomial theorem . . . Should a priest reject relativity because it contains no authoritative exposition of the doctrine of the Trinity?" To believe that biblical writers understood all things, including immortality and salvation, is to misunderstand the purpose of the Bible. He himself took care never to conflate science's world beginning with God's creation, as many cosmologists do, and to resist invoking religion even if a scientific resolution of the "moment of creation" remains always beyond reach. Such thinking fails to honor both science and theology. God is neither a scientifically testable idea nor can he be limited to being a scientific theory. Cosmology can be relevant for religion, but only if one respects that they are in different spheres of truth (Kragh 1999, 59–60).

Despite his strong professions of an independence view, because Lemaitre was a priest, others continued to speculate that beliefs had motivated his theory of the world's beginning. One of the most vocal was Hans Alfven, who said that he believed "the motivation for his theory was Lemaitre's need to reconcile his physics with the Church's doctrine of creation *ex nihilo*." He and fellow Swedish physicist Oskar Klein found the Big Bang theory "unscientific and mythical" but likewise rejected the steady state theory as unnecessarily invoking new physical laws in proposing the continuous creation of matter. Their own "plasma" cosmology presumed a universe of equal parts matter and antimatter held apart by electromagnetic fields in a very thin, cold environment. In a contraction period, radiation pressure from the increasing annihilation of protons and their antiparticles caused the universe to begin expanding in a kind of "cosmic explosion" in one region of the universe. More recently, in *The Big Bang Never Happened* (1991), science writer Eric Lerner argues for the positive features he sees in this theory: a universe that operates according to everyday principles and that is infinite in space and time.

Perhaps the most vocal opponent of the Big Bang theory and one with strong anti-religious views was Hoyle, who, with Gold and Bondi, developed its strongest competitor, the steady state theory. Their hypothesis proposed a universe which obeys a perfect cosmological principle, that the universe looks the same, not only from all points in space, but also from all points in time. It is infinite and unending. Since expansion is accepted as an observed fact, the theory postulates the continuous creation of matter "out of nothing" at the small rate required to maintain a constant density—approximately one atom per century in a volume equal to the Empire State Building. In popularizing the theory on a series of BBC radio broadcasts in 1949, Hoyle argued for it as the new creation theory and criticized the Big Bang. It was in this context that he was the first to give the Big Bang its name:

On philosophical ground this big bang assumption is much the less palatable of the two. For it is an irrational process that cannot be described in scientific terms . . . On philosophical grounds too I cannot see any good reason for preferring the big bang idea. Indeed it seems to me in the philosophical sense to be a distinctly unsatisfactory notion, since it puts the basic assumption out of sight where it can never be challenged by a direct appeal to observation. (Kragh 1999, 192)

In the print version of his talks, *The Nature of the Universe* (1950), he mounted a strong criticism of Christian belief and its "nonsensical" promise of eternal life for an unembodied immortal soul. Religion for him was merely "a desperate attempt to find an escape from the truly dreadful situation in which we find ourselves." This strong attack produced backlash among the public and, he claims, among scientific colleagues.

Reactions against religion thus played some part in the development of the scientific theory. According to science historian Helge Kragh "there can be little doubt that the discussions among Hoyle, Gold and Bondi, which led to the tentative formulation of the steady-state theory in 1947, were colored negatively by the views expounded by ... religious scientists." While antireligious sentiment did not necessarily motivate the creation of the new theory, as atheists they may have taken some delight in positing a universe with no need for a Creator. The idea of a moment of origin Hoyle found both philosophically and scientifically distasteful, reminiscent of primitive mythologies and begging a religious explanation. While Hoyle himself aligned atheism with the steady state and theism with the Big Bang, this simplistic categorization did not always hold. The version of the Big Bang he most targeted was Gamow's, but few of Hoyle's criticisms applied to him. Gamow was not religious, had no desire to "have his science drawn into the fuzzy realm of theology," and, in fact, preferred a Big Bang version that avoids a beginning—the oscillating model (see Figure 2.7), wherein the universe eventually collapses into a Big Crunch leading to another Big Bang, ad infinitum (Kragh 1999, 252–56). Additional exceptions to Hoyle's simplistic categorization include theologian William Inge, who felt that the perpetual Steady-State universe accorded better with our conception of God's will than the Big Bang, and radio astronomer Bernard Lovell, a devout Christian influenced by process thought, for whom any creation of matter, all at once or in little pieces, found its source in divinity. Even the strong theist Whittaker expressed "warm admiration for the work as a whole." Milne, however, was true to Hoyle's conception in holding that "a cosmology consistent with Christian belief had to begin with a point singularity created by God" (Kragh 1999, 190, 243, 254).

Influence of belief on theory has no better example than Einstein's decision to revise his cosmological model to eliminate expansion, even though he felt the added term seemed "gravely detrimental to the formal beauty of the theory" (Kragh 1999, 10). Although he claimed there was no relation between his theory of general relativity and theology, such a step seems to indicate at least a philosophical bias toward an unchanging, eternal universe. It has been suggested that he was influenced by the philosophy of Spinoza, who interpreted the biblical phase "the Heavens endure from everlasting to everlasting" as confirmation of "immutable existence," and who stated in his *Ethics* that "God is immutable [and] . . . all his attributes [including space and time] are immutable." Later, however, when observations of the expansion began to roll in, Einstein called the cosmological constant "the greatest blunder of my life" and praised the achievements of both Hubble and Lemaitre. To Hubble's wife, he once said, "Your husband's work is beautiful," and upon hearing Lemaitre present the Big Bang theory reportedly commented, "This is the most beautiful and satisfactory explanation of creation to which I have ever listened" (Jammer 1999, 62–63; Lightman 2005, 244).

Einstein's Religious Beliefs and Views on Science and Religion   A recent study by physicist Max Jammer has focused solely on Einstein's philosophical and religious beliefs, their possible influence on his work, and the theological implications of his scientific work. In *Einstein and Religion: Physics and Theology* (1999), he describes Einstein's early abandonment of traditional Judaism as his scientific education drew him into a much larger world, which Einstein later described as follows:

Out yonder there was this huge world . . . which stands before us as a great, eternal riddle . . . and beckoned like liberation . . . I soon noticed that many a man whom I had learned to esteem and to admire had found inner freedom and security in devoted occupation with it. The mental grasp of this extrapersonal world . . . swam as [the] highest aim half consciously and half unconsciously before my mind's eye . . . The road to this paradise was not as comfortable and alluring as the road to the religious paradise; but it has proved itself as trustworthy, and I have never regretted having chosen it. (Jammer 1999, 28)

Einstein continued over his life to describe to various audiences his own definition and philosophy of religion. In 1927 when a fellow dinner guest heard him labeled "deeply religious" and incredulously asked him if it were true, Einstein replied,

Yes, you can call it that. Try and penetrate with our limited means the secrets of nature and you will find that, behind all the discernible concatenations, there remains something subtle, intangible and inexplicable. Veneration for this force beyond anything that we can comprehend is my religion. To that extent I am, in point of fact, religious. (Jammer 1999, 40)

In looking for philosophical influences on Einstein's ideas, many have noted that the religious thought most consonant with his beliefs was that of Baruch Spinoza, a seventeenth-century Jewish philosopher excommunicated for atheism and free thinking. Sometimes called the "God-intoxicated man," Spinoza believed, not in a traditional Judeo-Christian personal God, but an infinite God who was the great "cosmic order" itself working according to unchangeable laws of cause and effect. Like Spinoza, Einstein believed in "unrestricted determinism" and in the "existence of a superior intelligence that reveals itself in the harmony and beauty of nature" but not in a personal God, who rewarded and punished, answered prayers, and was some kind of "superman." For both, the perfect workings of the cosmos and the rationality of the universe alone deserved religious adulation, and humans must accept inevitable laws of cause and effect determining the course of our existence (Jammer 1999, 43–46).

Religious beliefs can affect scientific work in two ways, by providing inner strength to endure grueling work required and by influencing the content of the results. Both seem to be the case with Einstein. While Einstein never directly acknowledged religious feelings as a source of strength, he did think that profound belief in the rationality of the universe allowed Kepler and Newton to persevere through years of work to understand celestial motions. He himself was driven to remain in solitary confinement for two weeks while developing general relativity. Perhaps his deep love of music, which for him was an "expression of religious feeling," and his use of it while developing the theory, indirectly speak to the point. Most notably, his deep lifelong commitment and tireless persistence in seeking a unified theory of gravity and electromagnetism in spite of repeated failures must be connected to a profound conviction about the oneness of the natural world. This belief he shared with Spinoza, who believed in the divinity of nature and the oneness of God (Jammer 1999, 55–58).

Belief in oneness thus motivated both Einstein's drive and the content of his sought after unified theory. This belief lies at the heart of his earlier theories as well, since both relativity theories propose hidden unities, or at least equivalences, of basic elements of the universe. Space and time are put on the same footing in special relativity, which spatializes time and temporalizes space, and mass is shown to have an equivalence in energy. In general relativity, an additional unification shows how mass–energy is linked to the curvature of space–time. Further evidence for influence of beliefs was evident in Einstein's resistance to quantum mechanics, whose indeterminacy went against the grain of his uncompromising determinism. Ironically, it was his own beliefs in unity that led him to see light as similar to matter and propose one of the foundational ideas of quantum mechanics—the particle nature of light.

Starting in the 1930s Einstein began to write and speak about the relation between science and religion, perhaps stimulated by two interviews in 1930. In the first, he was asked to pronounce upon whether science could deliver a new conception of God or is completely unrelated to religion. Both ideas, he replied, "disclose a very superficial concept of science and also of religion. As to whether science inspires and aids humanity spiritually more than traditional religion, he spoke rather of the "spirit that informs modern scientific investigations":

> I am of the opinion that all the finer speculations in the realm of science spring from a deep religious feeling, and that without such feeling they would not be fruitful. I also believe that, this kind of religiousness, which makes itself felt today in scientific investigations, is the only creative religious activity of our time.

Further, our morality and aesthetic sensibilities and spiritual yearnings are "tributary forms in helping the reasoning faculty toward its highest achievements." Science has moral foundations but cannot provide them, and "every attempt to reduce ethics to scientific formulae must fail." Can science save Western society from its decline? No, he said, science does not give us something to worship; "Mankind must exalt itself. Every cultural striving, whether it be religious or scientific, touches the core of the inner psyche and aims at freedom from the Ego–not the individual Ego alone, but also the mass Ego of humanity."

Einstein's second interview that year was with the Indian poet and visionary, Rabindranath Tagore, with whom he shared not only a love of music but religious beliefs that transcended organized religion (see Figure 2.10). Both saw that divinity was not separate from the world, an idea that Tagore expressed poetically: "With every breath we draw we must always feel this truth that we are living in God . . . [in a] great universe full of infinite mystery . . . an eternal symphony." They differed, however, on a crucial philosophical point about the dependence of truth and beauty on man. While Tagore believed that both were linked to human beings, Einstein believed that truth and reality had independent existence and declared himself more religious than Tagore (Jammer 1999, 68–72).

Einstein clarified his beliefs in an essay entitled, "What I Believe," where he wrote of the "purpose of life [and] the ideals of kindness, beauty, and truth without which his life would be empty." He also defined his religious belief:

> The most beautiful experience we can have is the mysterious. It is the funda-
> mental emotion which stands at the cradle of true art and true science . . . It was
> the experience of mystery—even if mixed with fear—that engendered religion. A
> knowledge of the existence of something we cannot penetrate, our perceptions
> of the profoundest reason and the most radiant beauty, which only in their most
> primitive forms are accessible to our minds—it is this knowledge and this emotion
> that constitute true religiosity; in this sense, and in this sense alone, I am a deeply
> religious man. I cannot conceive of a God who rewards and punishes his creatures,
> or has a will of the kind that we experience in ourselves.

Figure 2.10 Albert Einstein and Rabindranath Tagore. In 1930 these two great luminaries of Eastern and Western culture met to discuss their common beliefs and interests, among them a mutual love of music and their belief in a divinity that was not separate from the world. Both men were committed to inner ideals of beauty and truth transcending traditional religion, but disagreed as to whether truth existed independent of man. Einstein, who believed that it was, declared himself the more religious of the two. (AIP Emilio Segrè Visual Archives)

Einstein's disbelief in a personal God, Jammer contends, is a view that is actually found in traditional Judaism—in the second commandment against "graven images." It is also reflected in the writings of medieval philosopher Maimonides, who wrote in his *Guide for the Perplexed* "our knowledge of [God] consists in knowing that we are unable to compre- hend Him," and in Spinoza's view that God is "beyond description and imagination" (Jammer 1999, 72–75).

Einstein expanded on all these ideas in two major essays that bracketed the decade of the 1930s (see Primary Sources). In the first, entitled "Religion and Science," which he wrote for the *New York Times*, he suggested that religion developed from a primitive basis in fear, to a "social or moral conception of God" and finally to a "cosmic religious feeling," which expresses an indescribable yearning for experiencing the deep unity of cosmos, transcending any notion of a personal God. Religious "geniuses" of past ages, including saints and heretics, had this feeling, the "strongest and noblest driving force behind scientific research." Reaction to his ideas ranged from praise to condemnation. Strongest ridicule came from then Catholic professor and priest Fulton Sheen, who argued that the *New York Times* had "degraded itself" with this "sheerest kind of stupidity and nonsense." Wondering who would ever sacrifice his life for the Galaxy, Sheen closed with the comment, "There is only one fault with his cosmical religion: he put an extra letter in the word—the letter 's.' " Positive reaction generally came from liberal Jews, who praised his reverential attitude toward the mysterious, the "magnitude of [his] soul," and his bringing to light a "new perspective of life." Reaction, however, did not always fall along religious lines. Orthodox Jews condemned the essay, while Catholic theologian Hans Kung wrote sympathetically of it. The main objection that most clergy had was Einstein's strict determinism, which argued against man's freedom of will and ethical duty. Einstein maintained this belief but noted that it engendered compassion for fellow humans and did not change practically the need to be moral. Man must still act as if he were free (Jammer 1999, 75–87).

In 1940 Einstein wrote an essay, "Science and Religion," for presentation at a religious conference in New York (see Primary Source No. 5). Defining religious sensibility as the freedom from selfishness and affiliation with higher "superpersonal" values, whether connected to a Deity or not, he went on to argue for strong "reciprocal relations" between science and religion, summarizing his view in the well-known phrase, "Science without religion is lame, religion without science is blind." His final qualifying comments that the harmonious relation between science and religion was most threatened by the concept of a personal God sparked enormous controversy and were interpreted by many religious people and theologians as a denial of God. But Einstein had made very clear in earlier writings to distinguish between his denial of a personal God and atheism, and he felt that belief in a personal God was "preferable to the lack of any transcendental outlook of life." He wondered "whether one can ever successfully render to the majority of mankind a more sublime means in order to satisfy its metaphysical needs" (Jammer 1999, 51). Following criticism of his 1940 essay, he expressed dismay at being quoted in support of views that God does not exist (Jammer 1999, 93–97).

Jammer summarizes a wealth of varied responses from fundamentalist ministers to Kansas housewives, attorneys, historians, rabbis and bishops and closes with the response of two eminent theologians of the day: the Protestant Paul Tillich and the Catholic Kung. Tillich argued that the more "mature" ideas of a personal God need not conflict with science. Einstein's own concept of God—the "grandeur of reason incarnate in existence, which in its inexhaustible depth, is inaccessible to man"—was not dissimilar to "developed" conceptions of God that have prevailed from ancient Greece to the present time. This idea was of a "common ground" of being, which both "manifests" in the realms of nature and human meaning and is also "hidden in its inexhaustible depth." We sense this inaccessible "divine depth of our being" but can only express it in symbols, one of which is the Personal God. The symbol of the "personal" must be used, argues Tillich, to elevate the conception beyond the "It-ness" of the subhuman world. Kung also noted more similarity between Einsteinian and Christian views of God than is first apparent. He identi-fied Einstein's "cosmic reason . . . as an expression of reverence before the mystery of the Absolute" and noted that the term "person" is merely a "cipher for God," a symbol of a dimension of reality ungraspable in ordi-nary words and transcendent of ordinary personhood. Jammer concludes, "The religious philosophy of Einstein and that of these leading theologians are, after all, not as disparate as may have been expected" (Jammer 1999, 97–114)

The question arises as to whether certain statements of Einstein actu-ally did reflect belief in a personal God and whether Einstein's belief in the mysterious rendered him a mystic. In a well-known statement of his scientific quest, he once said to a student, "I want to know how God cre-ated this world. I am not interested in this or that phenomenon . . . I want to know His thoughts, the rest are details." And on another occasion, he stated, "What I am really interested in is knowing whether God could have created the world in a different way; in other words, whether the requirements of logical simplicity admits a margin of freedom." In eval-uating whether Einstein showed a degree of inconsistency or was only using a turn of speech, Jammer concludes from the latter statement that the "reference to God . . . was merely a manner of speaking."

The label of mystic was soundly denied by Einstein, who distanced himself from that mode of experience on more than one occasion. In the last year of his life he wrote, "What I see in nature is a magnificent structure that we can comprehend only very imperfectly, and that must fill a thinking person with a feeling of 'humility.' This is a genuinely religious feeling that has nothing to do with mysticism."

RELIGIOUS VIEWS OF THE ORIGIN OF THE UNIVERSE

Religious perspectives on scientific origin tend to fall into three cate-gories: those that regard a Big Bang origin as directly relevant to or even confirming a religious account, those that regard them as irrelevant or in conflict, and those that see an indirect relevance. For Christians, a ma-jor point to evaluate is the significance of science's "time zero" and its relevance to the doctrine of *creatio ex nihilo*—"creation from nothing." A spectrum of different views on these issues is reviewed briefly by Robert Russell (1996, 2001) and Willem Drees (1990, chapter 1), among others.

The Judeo-Christian Account of Creation

The most familiar creation account in Christianity and Judaism is the opening verses of Genesis I (King James version): "In the beginning God created the heavens and the earth. And the Earth was without form, and void; and darkness was upon the face of the deep. And the Spirit of God moved upon the face of the waters. And God said. Let there be light: and there was light." In six successive days, God created by his spoken word all the aspects of creation from the firmament, sun and moon, to land, vegetation, animals and finally, as his culminating achievement, man and woman. While the first verse shows some resemblance to other near Eastern creation myths of the time, such as the Babylonian, in which the world began in a chaotic watery state, the Hebrew account differs significantly in affirming one omnipotent, sovereign, and transcendent creator God. In this, the Hebrews wished to distinguish their monotheism from the polytheism of neighboring cultures. Likewise, Christians later interpreted this account as a creation from nothing, or *creatio ex nihilo,* even though there appears to be ambiguity as to the pre-existence of some kind of substance from which God fashioned the universe. By affirming the doctrine of creation from nothing, Christians wished to establish God's transcendence and the "otherness" of the world from him. The intent was to distinguish their views from the gnostic belief that material substance is evil and from pantheistic beliefs that God is immanent in the world and hence the world itself is divine. Thus *creatio ex nihilo* became a central doctrine of the Christian faith, not so much as to define the creation moment as to express our complete dependence on God (Barbour 1997, 200–201).

The Judeo-Christian Creation Account—Confirmation by Science

When the Big Bang theory was proposed—with its explosion of radiation, a possible comparison to God's sudden creation of light in Genesis seemed clear, and the new theory greatly interested the Roman Catholic Pope of the time. In a 1951 address to the Pontifical Academy, Pope Pius XII made a statement praising the theory and arguing that no discord exists between the church and the astronomers and that scientific evidence points to a "transcendent Creator." His message strongly supported the Big Bang: "Everything seems to indicate that the material content of the universe had a mighty beginning in time, being endowed at birth with vast reserves of energy, in virtue of which, at first rapidly, and then ever more slowly, it evolved into its present state." He described the research, pointed out that the figures obtained for the size and age in no way contradicted the opening words of Genesis, and then drew his conclusions:

Clearly and critically, as when it [the enlightened mind] examines facts and passes judgment on them, it perceives the world of creative omnipotence and recognizes that its power, set in motion by the mighty *Fiat* of the Creating spirit billions of years ago, called into existence with a gesture of generous love and spread over the universe matter bursting with energy. Indeed, it would seem that present-day science, with one sweep back across the centuries, has succeeded in bearing witness to the August instant of the primordial *Fiat Lux,* when along with matter, there burst forth from nothing a sea of light and radiation, and the elements split and churned and formed into millions of galaxies. . . .

. . . Thus, with that concreteness which is characteristic of physical proofs, it has confirmed the contingency of the universe and also the well-founded deduction as to the epoch when the world came forth from the hands of the Creator. Hence, creation took place. We say: therefore, there is a Creator. Therefore, God exists!

This statement disturbed the Catholic scientist most associated with the theory, Lemaitre, who was present at the address. Together with a Vatican astronomer, he convinced the pope that such direct linkages between science and theology were detrimental to both, and "never again did Pope Pius XII try to make cosmology suit Christian dogma" (Kragh 1999, 256–58).

Some conservative Christians and orthodox Jews share the views of Pope Pius XII that science's discovery of a temporal origin gives strong, even confirming evidence for God's existence and the truth of the biblical account. Astronomer turned evangelist Hugh Ross describes in *The Creator and the Cosmos: How the Greatest Scientific Discoveries of the Century Reveal God* (1995) how he became convinced of science's confirmation of the truth of God's word through avid study of the Bible and science. He found that God has given in his holy book all the truths that science is finding. Ross focused especially on the COBE discovery of density ripples in the early universe as confirmation of the Big Bang and creation according to Christianity. In a similar vein, orthodox Jewish physicist Gerald Schroeder argues in *Genesis and the Big Bang: The Discovery of Harmony between Modern Science and the Bible* that the events following the Big Bang and stretching over billions of years and the six days of creation in Genesis are "identical realities that have been described in vastly different terms" (Schroeder 1992, 26). He quotes medieval Kabbalistic theory, Nahmanides' *Commentary on Genesis,* about God's initial contraction that caused the universe to expand and God's choice of ten aspects in forming the cosmos. Schroeder believes this process is hinted at in the ten repetitions of the phrase "And God said" in Genesis I. He further connects them to the ten dimensions of space-time postulated by modern string theory (Schroeder 1992, 59). He claims that much hidden meaning in the Bible is very subtly expressed, by "variations in the grammar, by unusual patterns of letters, and even by nuances in the calligraphy" (Schroeder 1992, 181–82).

Finding hidden meaning in scripture has also led fundamentalists in Islam to "claim that every major discovery of modern science was long anticipated in the holy scriptures of their faith," according to Pakistani physicist Pervez Hoodboy in *Islam and Science* (1991). Both he and Kurt Wood give examples of how the Qu'ran explains, among many other things, the universal expansion, atomic structure and special relativity. This is an example of the "concordist" approach, wherein sacred texts are thought to convey scientific knowledge directly (Southgate et al. 1999, 317).

Priest and science historian Stanley Jaki provides an additional example of using cosmology to argue for God's existence (Jaki 1989). For him, science can find no final explanation and is doomed to regressing into an infinite series of questions. This shows the "inadequacy of scientific answers" and the ultimate "contingency" or dependence of the universe and everything in it, which can only find an explanation in a Creator. He declares, "The singularity of the universe is a gigantic springboard which can propel upward anyone ready to exploit its metaphysical resilience and catch thereby a glimpse of the absolute" (Southgate et al. 1999, 18–19).

The Judeo-Christian Account and Science—Conflict and Independence

Another viewpoint about the relation of the biblical account to Big Bang cosmology is that the two are actually opposed and incompatible, or perhaps just irrelevant. Firm believers in incompatibility are the scientific creationists. While they certainly believe that there was a moment of creation by God, they are firmly opposed to science's contention that this event occurred some billions of years ago and led to a long, slow evolution of matter and then living forms. Their fervent belief is that creation occurred exactly as it is written in Genesis, in the course of one week and no longer ago than 10,000 to possibly 15,000 years ago. Within the space of six, 24-hour days, God created the elements of our world and its living beings exactly as it is written, one by one in a special creation for each. They also make thermodynamic arguments against the possibility of the Big Bang. Since most of their opposition is to the notion of evolution, strict creationists are discussed more fully in Chapter 6.

In contrast to the perspective of total incompatibility, another set of views respects both Big Bang cosmology and Christian theology about creation for the truth they offer in their own domains but regards them as simply independent and irrelevant to each other's concerns. Thus, Protestant theologian Langdon Gilkey questioned whether Christian theology could know facts about "the finite extent of time," or science could know about "a first moment of the universe" in the sense of an absolute beginning (Gilkey 1959, 258). Both Ian Barbour (1997, 202–4) and John Haught (1995, 109–14) give clear descriptions of this position. Barbour stresses that Genesis I is above all about "theological affirmations" about the goodness and order of the world, its utter dependency on God and God's omnipotence, freedom, transcendence, and purposiveness. It defines our relation to the Creator and the created order. These affirmations should be seen as independent of the assumed cosmological features of the time: a "three-decker" cosmos with waters above and below the Earth that was believed by a prescientific society. We should have no expectation that such an account would yield knowledge comparable to today's science. At the same time we should not expect science to yield truths about ultimate questions and understanding of the religious meaning of creation.

Some theologians thus suggest that an origin of physical time is thus not of crucial importance to the religious notion of creation. A one-time event at a temporal beginning is not really relevant to our deeper feelings of gratitude for the great gift of existence. Even if the world had no beginning, from a religious point of view, it would still need a "transcendent grounding" to maintain its being and could still be "the expression of that primordial love that we call God." Theologian Keith Ward expresses the thought well:

This popular misconception, that "the creation" is the first moment of the space-time universe, and that the universe continues by its own inherent power, wholly misconstrues every classical theistic tradition. It is irrelevant to a doctrine of creation *ex nihilo* whether the universe began or not; that the universe began was usually accepted because of a particular reading of Genesis I. The doctrine of creation *ex nihilo* simply maintains that there is nothing other than God from which the universe is made, and that the universe is other than God and wholly dependent upon God for its existence. (Haught 1995, 111)

Two British Anglican scientist-theologians, Arthur Peacocke and John Polkinghorne, echo these ideas. Peacocke asserts that science can never do anything that would contradict a religious doctrine of creation, whose very essence is the notion of owing one's very being to God and not a temporal beginning. The findings of modern relativity theory have shown that time itself is bound up with the natural world and hence is created by God. Science may end up finding there was an origin point in space–time, and even going beyond it, but such probing would have no effect on the doctrine of creation. However, science has contributed an important perspective in revealing an evolving cosmos where creation is still occurring. This calls for including in Christian belief the important notion of *creatio continua,* or ongoing creation (Peacocke 2004, 78–79). Polkinghorne stresses that the essential element of creation doctrine is to show a clear distinction between the Creator and the created world he brought about by his complete free choice. To make a world separate from himself, he had to withdraw himself and make space for his own creative activity to work. This religious notion differs completely from a scientific version of creation *ex nihilo* that one might envision where "God started things off by manipulating a curious kind of stuff called 'nothing' (1996, 73–75). Vatican astronomer William Stoeger voices one of the strongest positions for complete independence. He believes that "even establishing a rough parallel, or consonance, between the 'beginning of time' in the Big Bang and the 'beginning of time' in the doctrine of creation (I insist on distinguishing this latter concept from the radical meaning of *creatio ex nihilo*) is very questionable." He does not think that science will ever probe an "*absolute* beginning—before which *nothing* existed, before which time of any sort was not—which would require the direct influence of God" (Stoeger 1988, 240).

The Judeo-Christian Account and Science—Dialogue and Contact

Another perspective views the scientific notion of origin at a time zero as indirectly relevant to or somewhat consonant with the Christian doctrine of creation; a fruitful dialogue or contact can be established between them. Haught gives several arguments for this view (1995, 115–18). The fact that science has shown the universe to have a temporal beginning implies that it is finite, and therefore contingent, in that it must have depended on something else to bring it into existence—it did not have to be. But to answer why the universe exists is to go beyond science. It is thus not so easy to separate scientific and religious concerns about beginnings. As one Christian scientist expresses it, "People need to understand that *science continually raises philosophical questions that go beyond the competence or purview of science*" (Hearn 1986, 28). A second way in which science has touched theology on this subject is its revelation of a cosmos that "is still in the making." To think of creation as just an instant in the distant past ignores the full spiritual significance of "creation." Such a limitation would be "unbearable" to French Jesuit paleontologist and visionary Pierre Teilhard de Chardin, who wrote: "The fact is that creation has never stopped. The creative act is one huge continual gesture, drawn out over the totality of time. It is still going on; and incessantly even if imperceptibly, the world is constantly emerging a little farther above nothingness." An additional connection Haught notes is the fact that both scientific and religious searches for origins arise from the same deep inner quest and that "much of the energy motivating science's look backward into our ultimate cosmic roots stems from the ineradicably mythic orientation of human consciousness."

Australian theologian Mark Worthing also sees relevance in Big Bang cosmology for the creation doctrine in that present research is not inconsistent with an original creation out of nothing. The intriguing similarities he notes arose especially after he considered the new quantum cosmologies of the late twentieth century (see Chapter 3), but he cautions that one can attribute only an "interesting correspondence . . . nothing more." Current cosmologies stimulate dialogue, especially for those favorably inclined to both scientific and religious aspects, but one can not use them to build on each other, without violating the integrity of both fields. In this view he concurs with the following statement of science historian and philosopher Ernan McMullin, which is an excellent example of what might be called "weak consonance":

What one *could* readily say . . . is that if the universe began in time through the act of a Creator, from our vantage point it would look something like the Big Bang that cosmologists are now talking about. What one cannot say is, first, that the Christian doctrine of creation "supports" the big Bang model, or second, that the Big Bang model "supports" the Christian doctrine of creation. (McMullin 1981, 39)

In *Cosmos as Creation: Theology and Science in Consonance* (1989) Lutheran theologian Ted Peters has collected a number of essays on this issue by leading scholars. In his own article he acknowledges the temptation for theologians to "embrace Big Bang thinking wholeheartedly, perhaps even to baptize it theologically," but he urges three cautions. First, science changes, and alternate theories could conceivably replace the now favored Big Bang. Second, the unknowable first micro-instant of time in current theory may become less mysterious with theoretical advances and become less of an "absolute beginning." But such a step would just enlarge the cosmos that science would know, and not necessarily bring us any closer to the real creation event that is beyond the cosmos. Third, the real power of God, Peters affirms, is not in a past original creation but in a new reality he will initiate in the future. This realization limits the importance of the past, to some extent, and challenges us to a much deeper faith, for science's predictions of the future seem at first glance to be much less amenable to such religious hope, as the last chapter will discuss.

Understanding of Creation in the Kabbalah

The Kabbalah, meaning "the Received," is a 2,000-year-old mystical Jewish tradition which seeks to comprehend the inner workings of God as Creator and our relation to him both in his hidden or transcendent and manifest or immanent forms. A number of modern scholars have noted striking parallels between conceptions of creation in modern cosmology and the Kabbalah, among them Daniel Matt, a scholar of Jewish spirituality, and Harvard astrophysicist Howard Smith. Their approach is one of establishing dialogue—viewing each explanation of creation in the light of the other. Rather than attempting to demonstrate that early Kabbalists possessed pre-knowledge of modern cosmology or that one could synthesize the two, the views are seen as complementary.

In presenting Kabbalistic ideas in the light of the Big Bang creation story, Matt identifies God as the unnamable oneness of all things–he can only be "named" by what he is not. Thus "Ein Sof," literally "there is no end," signifies both the infinite, transcendent being of God and the notion that there is nothing that is not divine. Another concept is "Ayin," or "nothingness"—a seemingly strange term for God, but one also used by such Christian mystics as St. John of the Cross, who spoke of "nada" in connection with God. For Kabbalists, the term is not meant to negate God's existence but only to stress his "no-thing-ness," the fact that he cannot be named or limited by being associated with any quality but is the oneness of all things. This mystical nothingness is not a total void but is rich in potential to become all things. Thus Kabbalists invert the traditional Jewish, Christian, and Islamic doctrine of creation "out of nothing" to conceive of creation as the emergence of all that manifests out of divine nothingness. Matt likens this idea to the primordial vacuum of modern cosmology, which still has a reservoir of hidden energy even at absolute zero temperature. The primordial nothingness of Ein Sof is likewise "coated with a trace of divine light" (Matt 1996, 36–41).

In both Kabbalistic and scientific conceptions, the universe emerged out of a single point. The singularity of science is a black hole, an "infinitely dense point in spacetime," which absorbs anything that goes in but from which anything can emerge. Thirteenth-century Kabbalist Moses de Leon expresses it as follows:

The beginning of existence is the secret concealed point. This is the beginning of all the hidden things, which spread out from there and emanate, according to their species ... when the concealed arouses itself to exist, at first it brings into being something the size of the point of a needle; from there is generates everything.

The central text or "masterpiece of the Kabbalah," the Zohar, which appeared in 13th-century Spain but was attributed to a Galilean spiritual leader a millennium before, describes the origin of this point, which is often equated with the wisdom of God:

> A blinding spark flashed
> within the concealed of the concealed,
> from the mystery of the Infinite,
> a cluster of vapor in formlessness.
> Under the impact of breaking through,
> one high and hidden point shone.
> Beyond that point nothing is known.
> So it is called Beginning.

As creation continues, "as God begins to unfold, the point expands into a circle," just as in science the emergence of matter and energy from the singularity begin a cosmic expansion. Kabbalists have many images for this process, one being the breathing out of God causing space to expand. As with closed versus oscillating scientific models, there are different interpretations within the Kabbalah as to whether the expansion will stop or is part of a "secret rhythm of creation."

With the emanation of light, time and space came into existence but material substance was a homogeneous mixture of unseparated energy and matter. The process by which matter came to be also is also conceived in parallel ways in the two systems. In science, some of the initial energy emanating from the singularity transformed into particles, following Einstein's law of the equivalence of matter and energy. The light "congealed" into matter. In the Kabbalah the process is described as one of "concealment" of the light, which paradoxically is necessary in order for it to be revealed and become visible, in the same way that the exact shape and size of the brilliant sun can only be revealed when its light is viewed through a screen or filter (Matt 1996, 41–45).

One may wonder how such conceptions of points and expansions can be read into the familiar biblical verses about creation. Smith gives us a hint in describing the richness of the Hebrew text, with its many layers of meaning and subtle analogy hidden in particular choices of grammar, words, letters, and even punctuation. Kabbalists excelled at the practice of "midrash," which he described as "the creative explication and amplification of scriptural texts by means of legends, parables, and homilies that build on word or name associations and on ellipses in the narrative." He gives an example with the opening phrase of Genesis I, "B'Reishet" or "In the Beginning." When this phrase is unpacked by such interpretations, its meaning becomes: "At the beginning of time, the universe was created from *nothing* in a series of stages initiated by an intentional act of desire or will, with a burst of covenantal fire expanding dramatically outward from a microscopic point called the *Resheit*" (Smith 1996, 21–23).

Both books go on to describe a major element in Kabbalistic thought: the ten Sefirot or states of God, which depict all the states and qualities of life as aspects of God's being as he works in the world, but also represent a map of a soul's spiritual journey as he draws closer to God. Throughout his book Smith draws parallels between each of these states and various processes in the physical world.

The Origin of the Universe in the Cosmology of Meher Baba

Similar parallels with science appear in the modern spiritual cosmology of Indian spiritual leader Meher Baba, who likewise asserts the unity of all things, the emergence and expansion of creation from a single point, and the creative role of "nothingness." In consonance with Kabbalism, he affirms the existence of ten states of God, of which one is radical transcendence. His cosmology is considered at length in Chapter 7, and only a few summary points are included here.

Baba's system, most fully expounded in his major work, *God Speaks*, affirms a particular ancient Hindu belief that everything is identical with God and hence all of creation is one great being. The apparent separateness is only an illusion perceived by a limited consciousness that is ignorant of the real Self within all, and the goal of all creation is for created beings to outgrow this limitation and realize themselves to be God. Creation began when God once had an urge to know himself and experience fully his own divine nature. This sudden whim disturbed his tranquil state and manifested itself as a most finite "creation" or "Om" point. There the still, unconscious aspect of his being surged forth and expanded outwards to explore the infinite possibilities of its own being.

The whim brought forth all the opposites latent within God's total being, both the infinite God who is Everything (the infinite consciousness) and the infinite shadow of that self, which is Nothing (the infinite unconsciousness). The first aspect knew himself immediately to be God, while the second began a process of coming to know itself as God through a long, slow evolution of consciousness experiencing the many different forms of created life. Thus it was the "Nothing," the unconscious aspect of God that manifested as creation. As one source describes it, "What we call creation is the manifestation of countless forms out of nothing" (Abdulla 1954, 15). In *God Speaks*, Baba explains the emanation as follows: "The most finite point from where the Nothing projects out as Nothingness is called the *Creation Point* or the *Om Point* . . . [and] the projection of the most finite Nothingness . . . gradually expands *ad infinitum* and manifests apparently as infinite Nothingness or as infinite Creation (Abdulla 1954, 77–78).

Thus both Baba's thought and Kabbalism have rich conceptions of "nothingness," just as does modern science, with its initial fertile state of primordial vacuum. Nothingness in both systems is seen as a meaningful way to think about that aspect of God's existence which stands directly between the most transcendent "beyond everything" state and the state of God as manifested existence. In both cases, also, God's being is seen as having ten complex dimensions or states, and although the specific states are quite different, each set corresponds to different modes in which God exists and/or relates to the created world, and each maps a ladder by which the soul draws closer to God.

Chapter 3

The Nature of Matter and Energy and the Quest for a Final Unified Theory

The first three decades of the last century swept the world of physics with profound new insights about the physical world. By the time they ended, scientists knew the basic structure of the atom but had also discovered mysterious limits to what they could know and predict about matter and raised questions about what the "objective world" really is. The next two decades brought nuclear energy to the fore to produce a deeper understanding of these forces but also unleashed the most destructive power ever conceived by man. The understanding of nuclear energy also revealed the secrets of the sun and stars, the birthplace of the chemical elements essential to life. The century's second half found deeper structure still in nuclear particles and experimental evidence that two of nature's forces act as one at high energy. Theory suggests that all subatomic "quantum" forces—nuclear and electromagnetic—might act as one, but can gravity join them, too? Einstein's dream of such a final unified theory is now the driving force of one of the most fervent quests in modern science. A promising candidate, string theory, proposes a beautiful idea but so far lacks both uniqueness and experimental verification.

Finding a unified theory is the key to understanding the first instant of time in the Big Bang universe, for it was then that all forces merged. In that remote epoch the entire universe was the size of an atom, so theories that govern the realms of the very large and the very small must be applied together—general relativity and quantum theories—in a merged theory of "quantum cosmology." Also developed in the last thirty years was the theory of superfast, "inflationary" expansion of the universe during the first tiniest fraction of a second, an idea developed to solve certain problems with the standard Big Bang model. Inflation and quantum cosmology have both led to the possibility of many universes—perhaps a grand "megaverse" in which our own universe is but one among a multitude.

Philosophical and religious implications abound in all these discoveries and theories of modern physics. Quantum theory, in particular, brings forth deep metaphysical and epistemological questions about the relation of the mind and the observer to physical reality and about causation and the nature of physical reality itself. If a particle is defined by a set of probabilities and not by predictable behaviors and properties—that is, if chance plays such a heavy role in "determining" its future–does this diminish God's all-knowing and all-powerful nature, or does it allow him more freedom to act in the world? The dual wave-particle nature of both matter particles and light seems paradoxical. How does one understand this? Three basic "schools" of thought interpret quantum theory, and each gives its own answers to these deeper questions.

Philosophical and religious implications also emerge from ideas about atomic structure, the interchangeability of matter and energy, the most fundamental "particle," and the unification of forces. At the human level solid matter appears impenetrable and stable under ordinary circumstances, yet physics tells us that the atom is mostly empty space, that energy and matter transform into each other, and that every particle of matter or energy may be a tiny string vibrating in many dimensions. The implications are that interwoven space and energy comprise physical reality, that everything is all one thing vibrating differently, and that even the forces started out in union with each other. Such messages of the underlying unity of all are also prominent in Judaism's Kabbalism and in Eastern religion, which also stresses the impermanence of the physical world. Origin accounts in many religions and mythologies describe the emergence of the world and its beings from a primordial unity. Strings and threads appear in world mythology, where they connect the physical with other realms, and in the cosmology of Meher Baba, where sanskaric "threads," form all levels of "matter" from physical to mental.

Quantum cosmology is also rich with implications. The possibility it raises of the universe arising from "nothing" has parallels with religious notions in Kabbalism and Meher Baba's system, as discussed in the last chapter. Christian theologians compare this concept with *creatio ex nihilo* to distinguish different degrees of "nothingness." The possibility of multiple universes can also be found in Eastern religion.

QUANTUM THEORY: THE REVOLUTION
THAT SHOOK PHYSICS

A cloud hung over physics in the late nineteenth century. Despite the glowing successes of classical physics in explaining a host of celestial and earthly phenomena, in one situation it predicted an impossible result. Its calculations predicted that a hot object, like an enclosed heated oven, should emit an infinite amount of radiation! German physicist Max Planck realized that one could solve the problem by assuming that the particles in the radiating object could emit and absorb energy only in discrete lumps or packets, called quanta. There must be an "elemental drop" of energy that could be divided no further, and he determined quantitatively what that finite limit was—the frequency times a tiny number, 6.55×10^{-27} erg seconds, now known as Planck's constant. This number would set the scale for all sizes in the atomic and subatomic realms—the size of the atoms, the smallest conceivable transistor or computer, the tiniest unit of time and the density of the original substance at the universe's origin. The "discovery . . . of the atomistic structure of energy . . . became the basis of all twentieth century research in physics," declared Einstein in his 1948 eulogy for Planck. It is interesting that the modest, reserved Planck, who continued to urge caution in applying the quantum idea, was the scientist who began the revolution that "changed all of physics" (Lightman 2005, chapter 1).

Einstein carried Planck's idea one step further in 1905 by suggesting that light itself exists in individual, discrete units, called quanta, and is not a continuous stream of waves filling space, as our everyday senses suggest. His own inner quest for unity led him to ask why there should be a "profound formal difference" in the physical view of light and matter, one smooth and continuous and the other grainy and particulate. Perhaps light came in elemental units also. Clues came from an experiment showing that light waves shining on a metal surface had to have a minimum energy to dislodge electrons. In his Nobel-prize-winning paper of 1905 Einstein explained this "photoelectric" effect by proposing that a single quantum of light is absorbed by an electron at any one time, and only if the quantum's energy is high enough can the electron break free of its bonds and escape. Light, in other words, behaves like a particle—a little billiard ball that imparts momentum to the electron. An experiment in 1923 confirmed Einstein's theory by showing that when electrons were bombarded by X-rays, each electron recoiled as if hit by a tiny ball. Thus began the paradoxical "wave-particle" theory of light. Einstein considered this his "most revolutionary" idea and is often regarded along with Planck as the "father of quantum physics," although as discussed later, he had severe reservations about the theory (see Figure 3.1). Einstein developed this idea in what is called his "miracle year," where, working in isolation as a lowly clerk in a Swiss patent office, he produced five outstanding papers, including ones introducing special relativity and explaining "Brownian motion" by atomic theory. Immediate recognition and new academic appointments followed (Lightman 2005, chapter 4).

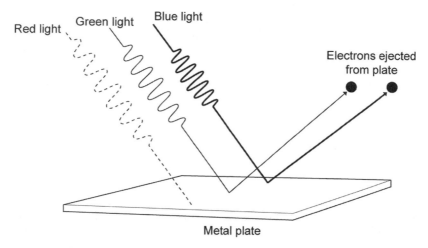

Figure 3.1 The Photoelectric Effect: How Light Behaves Like Particles. Electrons escape from a metal surface only when impinging light has less than a certain wavelength. Einstein explained this "photoelectric effect" by proposing that light is composed of discrete quantized packets of energy, later called photons, a theory confirmed by experiment in 1923. Since light also acts like a wave, it was thus thought to have "wave-particle duality." Einstein's belief in the underlying unity of disparate phenomena helped lead him to this realization. (Illustration by Jeff Dixon)

Several years later, Danish physicist Niels Bohr used the new "quantum" idea to advance knowledge of the atom. Two years before his seminal paper on the quantum atom, his mentor Ernest Rutherford had discovered the atomic nucleus. This gave rise to the "solar system model" of the atom—a central sun-like nucleus surrounded by orbiting planet-like electrons, but mysteries remained. Why didn't the negative electrons spiral into the positive nucleus, and why did light from a particular chemical element always appear only at certain specific colors, or frequencies? These conundrums would be solved, at least for one element, by Bohr's quantum atom. In this model electrons could only occupy certain orbits or energy levels— and none in between. When they "jumped" from one orbit to a lower one, they would emit a quantum of light that equaled the energy difference of the orbits. This "jumping" between orbits—without traversing the space in-between!—could not be pictured. Using this model, Bohr was able to predict exactly the observed specific frequencies in the spectrum of hydrogen. It was a hybrid classical-quantum model in which classical theory predicted the forces working in the stationary orbits and quantum theory described the mysterious jumps between them. Physicist-science writer Alan Lightman speculates that it was Bohr's "special fascination with paradoxes" that led him above others to formulate this idea and

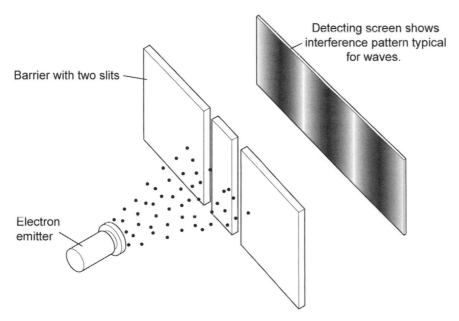

Barrier with two slits

Detecting screen shows
interference pattern typical
for waves.

Electron
emitter

Figure 3.2 The Electron Double-Slit Experiment: How Particles Behave Like Waves. In the 1920s French physicist Louis de Broglie postulated that electrons have wave-like properties. This idea was confirmed in experiments where electron beams passing through two slits created an interference pattern typical for light waves. Even if only one electron passed through the slits at a time, an interference pattern developed, as if single electrons had gone through both slits at the same time! (Illustration by Jeff Dixon)

discover the quantum atom, which earned him recognition as the "father of atomic physics" (Lightman 2005, chapter 8).

Another crucial step on the quantum journey was taken by a French duke, Louis de Broglie, who revised Bohr's quantum atomic model in such a startling way that someone called his paper "la comedie francaise." De Broglie's interest in waves and music led him to think of the atom as some sort of musical instrument, emitting a fundamental tone and overtones. Using Bohr's orbits, he envisioned that "mysterious pilot waves" moved along with electrons in their orbits, with one whole wave fitting into the first orbit, two into the second, etc. Particles outside of atoms might exhibit these waves too, which could be observed. Sure enough, two years later experiments using electron beams revealed the typical interference pattern of light and dark strips on a screen that occur with light waves (Gamow 1985, 80–85). Much later in the famous "double-slit" experiment, electrons passing through two slits showed the same pattern exhibited by light (Figure 3.2). Most mysteriously, the interference fringes build up when only one electron passes through the slits—a single electron "wave" appears to pass through two slits at once, just as a single light wave does! Not only was light a particle, but an electron had properties of a wave!

All these seminal ideas were subsumed into a full-blown theory of "quantum mechanics" by two giants in the field, German physicist Werner Heisenberg and the Austrian Erwin Schrödinger. In 1925 Heisenberg was able to predict the spectral colors of hydrogen using matrix algebra— multiplying together a matrix (array of numbers) representing many possible values of an observed property with a matrix for measurement. It was quite unclear what these abstract matrices meant physically, but Heisenberg knew he was finding something important. He later described the "transcendental creative moment" when he realized the theory would work:

… At first, I was deeply alarmed. I had the feeling that, through the surface of atomic phenomena, I was looking at a strangely beautiful interior, and felt almost giddy at the thought that I now had to probe this wealth of mathematical structures that nature had so generously spread out before me. I was far too excited to sleep. (Lightman 2005, 194–95)

In the next year Schrödinger followed a different approach, using a "wave function" derived from DeBroglie's "matter waves." In an atom he associated electron motion with three-dimensional waves filling the whole space of the atom, shaped and vibrated at different rates by electromagnetic forces. While they were much like sound or light waves, it was not at all clear "what" was vibrating. Nevertheless, his wave equation exactly predicted the frequencies of light in the spectrum of hydrogen, just as Heisenberg's matrix mechanics had. Schrödinger later proved that the two formulations were mathematically identical and could be derived from each other, a result as surprising, joked physicist George Gamow, as learning that "whales and dolphins are not fish like sharks or herring but animals like elephants or horses!" (Gamow 1985, 105). Both men received the Nobel Prize for developing quantum mechanics (Gamow 1985, chapters IV, V). Also considered a cofounder is British physicist Paul Dirac, who developed similar results in a still different way and also incorporated special relativity into the theory.

Heisenberg was also responsible for formulating the uncertainty principle, a startling idea that shook science at its foundation. Heisenberg reasoned that we cannot ascribe physical meaning to an object until we measure it—almost suggesting that it may not even exist in between measurements, or it could have multiple states. Even when we measure it, an inevitable uncertainty creeps in. Consider the trajectory of a particle in space. He claimed that it was impossible to determine its position and momentum (speed times mass) to any degree of accuracy we chose. To discern its position required shining light on it, but light imparts momentum, and disturbs the motion, making it more uncertain. More powerful light gives a more accurate position but disturbs the motion even more. Thus, the more accurately one determined position, the less accurately one could know the momentum, and vice versa. Heisenberg's "uncertainty principle" expressed this fundamental limit to the accuracy of simultaneous measurement: the product of the uncertainties in position and momentum must always equal or exceed Planck's constant. Other pairs of uncertainties, such as energy and time, or angle and angular momentum also obeyed this principle. Lightman describes the impact:

Heisenberg announced to the world that a good part of nature is permanently hidden from view. The state of the physical world, or even a single electron, hovers in a cloud of uncertainty. Consequently, and in contradiction to centuries of scientific thought, the future cannot be predicted from the past. (Lightman 2005, 191)

In quantum mechanics, all that can be known are probabilities for various outcomes, and Schrödinger's wave functions thus described not physical things but probabilities for a future state. This was often expressed by saying that before measurement, a particle possessed a set of superposed possible states—it "was" many possible things at once. In a later formulation of quantum theory, Richard Feynman obtained probabilities by assuming that a particle took all possible paths between two points, each path being a wave of different size and phase. Summing over all these "histories" canceled out many waves and yielded a set of probable paths consistent with Schrödinger's approach. Many questioned the odd idea of superposition of states and devised thought experiments to critique it. One famous example was Schrödinger's cat, placed in a box with a contraption set to release deadly gas if triggered by some event with a 50 percent probability—say the decay of a radioactive substance. Is the cat 50 percent dead and 50 percent alive before someone opens the box to find out? Such a thought seemed absurd. Another thought experiment was the Einstein–Podolsky–Rosen (EPR) paradox that Einstein and two collaborators devised. Let us say two particles can only exist in complementary spin states, +1 and −1, but until measurement, neither exists in either state. If they are then taken light years apart and measured simultaneously, they will have opposite spin states. If their spin was truly indeterminate before measurement, how did they coordinate their spins, without "communicating" faster than light? This was "spooky action at a distance," Einstein said, that "no reasonable definition of reality could be expected to permit" (Ferris 1997, 276–77).

Decades of philosophical debate ensued over the meaning of all this quantum weirdness. What did the uncertainty mean? Is it just our inability to know something before measurement, or is it intrinsic to the physical system itself? Are there still unknown variables that truly

Figure 3.3 The Fifth Solvay International Conference of Physicists in 1927. Probably the most famous of the conferences founded by Belgian industrialist Ernest Solvay was the one in October of 1927 on the topic of "Electrons and Photons," where the world's most eminent physicists met to discuss the newly proposed quantum theory. Two of the principal figures, Albert Einstein and Niels Bohr, differed on the interpretation of the theory and began a years-long exchange or "debate" over its meaning. At one point Einstein famously quipped, "God does not play dice," to which Bohr replied, "Einstein, stop telling God what to do." (Photograph by Benjamin Couprie, Institut International de Physique Solvay. Courtesy AIP Emilio Segrè Visual Archives)

determine things, as Einstein believed? Each of these in turn has religious implications. Discussions of all these issues occurred at some of the many conferences attended by the quantum luminaries, such as the Solvay conferences in the 1920s and 1930s (see Figure 3.3).

Many good sources give readable accounts of quantum theory, among them Heinz Pagels' *The Cosmic Code* (1982), John Polkinghorne's *The Quantum World* (1986), and Nick Herbert's *Quantum Reality* (1985). Accessible accounts of quantum theory can also be found in books with wider scope, from the brief description in Stephen Hawking's *Brief History of Time* (1998, chapter 4) to longer chapters in many books on string theory, such as Brian Greene's *The Elegant Universe* (1999, chapter 4).

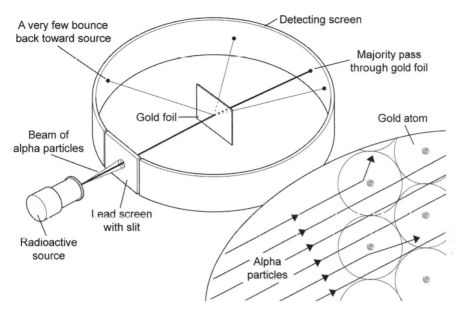

Figure 3.4 The Gold Foil Experiment: How the Atomic Nucleus Was Discovered. In 1911, New-Zealand born physicist Ernest Rutherford probed the atom by passing heavy positive alpha particles through a very thin gold foil. He assumed that the particles would easily pass through soft spongy "atoms," veering slightly when passing near an electron. Much to his surprise, particles very occasionally hit something extremely tiny and hard that bounced them straight backwards. He had discovered the nucleus of the atom, over 10,000 times smaller than the atom itself. (Illustration by Jeff Dixon)

THE JOURNEY INTO THE ATOM AND
THE FORCES OF NATURE

In the same period that quantum theory developed, the interior structure of the atom came to be known. The word "a-tom" became a misnomer, for it means "that which is indivisible," and the atom indeed has smaller pieces. In the last years of the nineteenth century, several discoveries revealed the emission of particles and radiation from inside atoms. Henri Becquerel and the Curies discovered spontaneous emissions from uranium, polonium and radium. Study of cathode ray beams led J. J. Thompson to discover the electron, and in 1906 to propose his "plum pudding" model for the atom, where electrons, like raisins, swarmed in a pudding-like ball of positively charged material.

New Zealander Ernest Rutherford, regarded as the greatest experimental physicist of his day, tested this model using fast-moving "alpha particles" shooting out of radium as atomic probes. In experiments devised with Hans Geiger (see Figure 3.4), he observed deflections of these particles as they passed through atoms in thin metallic sheets and found, much to his enormous surprise, that occasionally the particles bounced back in the same direction—a truly impossible feat for a plum pudding atom! Learning of these results, Rutherford famously commented: "It was almost as incredible as if you fired a fifteen-inch shell at a piece of tissue paper and it came back and hit you." But only one in 20,000 alpha particles bounced back, so whatever they hit was very dense, very hard and incredibly small. Rutherford had discovered the "nucleus," and it was 10,000 times smaller than the atom itself. If an atom could be blown up to the size of a baseball stadium, the nucleus would be the size of a pea. Most of the atom was empty space! (Lightman 2005, chapter 5) His "solar-system-like" model of a nucleus and orbiting electrons is still with us today, although somewhat changed by quantum theory. Rutherford, often called the "father of nuclear physics," was also the first to observe nuclear fusion—a different type of nuclear reaction than radioactive decay—when he saw cloud chamber tracks revealing the merging of nitrogen and helium nuclei to produce oxygen. Such transmutation, the age-old dream of the alchemists, would soon be understood as the process creating chemical elements and energy in stars (Jastrow 1967, 36–37). Rutherford also predicted the neutron, which his former student James Chadwick found in 1932.

By the early 1930s the basic parts and structure of the atom were known: a nucleus composed of tightly bound protons and neutrons (or "nucleons") with a cloud of much lighter "orbiting" electrons filling a volume 10,000 times larger (see Figure 3.5). These basic constituents of matter were governed by four forces. Gravity, the weakest but always attractive force, affects all objects with mass and acts over an infinite distance. A trillion trillion trillion times stronger is the electromagnetic force, which governs all particles with electric charge, binds electrons to the nucleus, and also operates over an infinite range, although it cancels out over short distances. The strong force, binding nucleons in the nucleus, is a hundredfold more powerful than electromagnetism, but operates only at very short range, a trillionth of a centimeter or so. The discovery of the neutron and its process of decay into other particles revealed a fourth force—the weak nuclear force—weaker than electromagnetism by several factors of ten.

The next three decades saw enormous growth in understanding and development of nuclear energy. In 1929, thermonuclear fusion was first proposed as the sun's energy source, and a decade later detailed reaction sequences were understood. New particles were involved: the elusive neutrino and antimatter in the form of positively charged electrons, or positrons. "Pair creation" of positrons and electrons is often seen in bubble chambers as tracks spiraling in opposite directions (see Figure 3.6). It provides clear evidence of the interchangeability of pure energy and particles with mass, a key process in all nuclear reactions. In the 1930s nuclear fission of uranium was both observed and explained, leading to

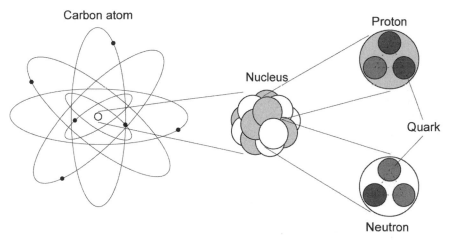

Figure 3.5 The Structure of the Atom and Nucleus. By the early 1930s the three basic components of the nuclear atom were known—protons and neutrons bound together by nuclear forces in a tiny, dense, compact nucleus and a surrounding cloud of wave-like electrons bound to the nucleus by electrical forces. In the 1960s physicists explored deeper structure in the proton and determined the existence of smaller particles of fractional charge—"up" and "down" quarks—that comprise both the neutron and proton. (Illustration by Jeff Dixon)

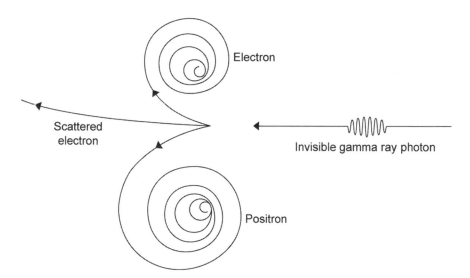

Figure 3.6 Bubble Chamber Tracks: Conversion of Energy into Matter. One of the clearest examples of the interchangeability of matter and energy occurs in bubble chambers when an incoming gamma ray nears an atomic nucleus and its energy is converted into an electron-positron pair. The tracks of such pairs can be seen spiraling in opposite directions in the chamber's electrical field. In the reverse process, an electron and a positron can come together and "annihilate" into a gamma ray of pure energy. (Illustration by Jeff Dixon)

the first chain reaction in 1942, the first atomic bomb in 1945 and much later, to nuclear power plants. By 1952 the thermonuclear hydrogen bomb was developed and at decade's end, fusion was understood as the process by which all the chemical elements had been synthesized in stars.

The next steps in probing the depth of matter and the forces of nature came from new generations of particle accelerators, where high-speed collisions produced a frenzied multiplicity of new particles, some of which lived the barest fraction of a second before decay. Over a hundred have now been seen. Physicists tried to classify and order these particles to make some sense of them. There were "hadrons" affected by the strong force, and "leptons," the lighter electrons and neutrinos affected by the weak force. "Fermions" were matter particles while "bosons," generally speaking, were force-carrying particles, such as the photon or light which "carried" or "mediated" the electromagnetic force. Breakthroughs came in the 1960s and 1970s when deeper structure was found in the proton and neutron. In 1964 Murray Gell-Mann and George Zweig suggested that three smaller entities, quarks, might compose each of these particles. Envisioned more as mathematical abstractions than real particles, they had each fractional charge of $\pm 2/3$ or $\pm 1/3$, a radical notion doubted by most, since the elemental unit of charge on the electron and proton was thought to be the smallest possible. In later accelerator experiments, deflections of near light-speed electrons whizzing through protons revealed inner structure and confirmed that protons and neutrons were each made of three quarks, bound together by strong force-carrying particles called gluons (see Figure 3.5). Within a few years, a theory was in place that described their interactions (Lightman 2005, chapter 21).

By the mid-1970s there were theories describing all the forces: general relativity for gravity, quantum electrodynamics for electromagnetism (QED), quantum chromodynamics (QCD) for the strong force and, most intriguingly, the electroweak theory that predicted that at high enough energies the electromagnetic and weak forces operated as a single unified force. (The latter three are subsumed under the name of the Standard Model.) One of the developers of this "electroweak" theory, Steven Weinberg, was led to this realization by considering an all-important principle, symmetry. In physics, this term refers not to visual balance of form, as in art, but to physical laws—that they should act in the same way from different viewpoints. An example is the invariance principle Einstein introduced with special relativity—the speed of light should be the same in all constantly moving frames of reference. The underlying symmetry Weinberg considered was that the weak force worked identically on different particle–antiparticle pairs of electrons and neutrinos. Quantum blurriness could even make particles share identities, being part one and part the other! In working out the math of the theory in a quantum framework,

Weinberg realized that the electromagnetic force was automatically included. His new theory actually described a unified electroweak force, carried by four force-carrying particles, one the photon and the other three heavy "bosons" detectable only at very high energies. Within a few years, all had been observed in accelerators which mimicked the heat of the early universe when these forces actually acted as one force. Today they act separately—the symmetry is "broken" (Lightman 2005, chapter 20).

Symmetry principles led to other quests for unification. The Grand Unified Theory joined the strong and electroweak forces at even higher energies, but its prediction that the proton will decay has not been observed. Also awaiting verification are supersymmetry theories which suggest an underlying unity between matter particles and force particles, such that each particle has a heavy "supersymmetric" partner of the other stripe— a "sparticle." Quarks have partners called "squarks," and leptons have "sleptons," etc. Driving the quest for deeper unification are several factors, among them aesthetic dissatisfactions and mathematical problems with the quantum field theory of the Standard Model. For one thing, it is too complicated. Almost thirty different parameters are arbitrary and unexplainable—how were they chosen? Physicists would like them to emerge naturally out of the theory. Also, infinite properties appear when quantum effects are calculated near the point particles. They can be eliminated mathematically, but the situation is dissatisfying and begs for a better solution. More importantly, the theory is simply incomplete. Not only does it not include gravity but it is also incompatible with gravity theory in realms where they both must apply. Beyond all this, however, is the fervent quest for a single elegant "theory of everything" that unites all physical phenomena—Einstein's dream. It is above all the quest for unity, beauty, and simplicity. Some think they may have found this Holy Grail in string theory.

UNIFIED THEORIES

String theory is a radical idea which appeared unexpectedly in the latter half of the twentieth century. It arose neither from first principles, as with relativity theories, nor from any experimental result but in serendipitous fashion, by accident. As one preeminent string theorist, Ed Witten, expressed it, "string theory is a part of twenty-first-century physics that fell by chance into the twentieth century" (Greene 1999, 19). In 1968 a young physicist, Gabriele Veneziano, found a formula that described many strong force interactions, which turned out to be an equation for vibrating strings. Even when the theory conflicted with observation, inspired researchers pressed on; one, John Schwarz, believed that "the mathematical structure of string theory was so beautiful and had so many

miraculous properties that it had to be pointing toward something deep"
(Greene 1999, 137).

In the mid-1970s excitement rose when string theory predicted the gravi-
ton, the messenger particle of gravity, promising to unify gravity with
other forces. However, it was not until the mid-1980s, when conflicts with
quantum theory were resolved, that the physics community embraced the
theory. During the "first superstring revolution" from 1984 to 1986 over
a thousand papers showed how string theory naturally predicted many
aspects of the Standard Model. As leading researcher Michael Green said,
"The moment . . . you realize that almost all of the major developments
in physics over the last hundred years emerge—and emerge with such
elegance—from such a simple starting point, you realize that this incredibly compelling theory is in a class of its own" (Greene 1999, 139). The road
forward was challenging. Strings vibrated in ten dimensions, and new
mathematics had to be developed to deal with such a multidimensional
geometry. Equations were difficult, and approximations used failed to deliver essential answers. And there appeared to be five different types of
string theory! In 1995 Witten made a breakthrough by suggesting that all
five theories were but versions of a more encompassing "M-Theory." The
"second superstring revolution" had begun (Greene 1999, chapters 6, 12).

What is a string, why is it so radical, and why is it so promising as a
theory of everything? The basic idea is that every particle in existence is
not a point-like entity but a tiny one-dimension string or loop, a "vibrating, oscillating, dancing filament" as much smaller than an atom as an
atom is smaller than the solar system, or 10^{-33} cm., a unit called the Planck
length. Like a violin string, a fundamental particle string can vibrate at
many different resonant frequencies. In the same way that different vibrational modes of a violin string produce different musical notes, different
vibrational patterns of a fundamental string produce the various masses or
force charges of particles. The more energetic the vibration, the greater the
energy or mass of the particle (see Figure 3.7). Thus, all elementary strings
are exactly the same thing—all matter is made of the same stuff—and all
particles and their interactions are unified in a single theory (Greene 1999,
chapter 6).

String theory also radically changes our conception of space and time
by requiring that strings oscillate in nine or ten dimensions of space. Only
three dimensions have become large in our macroscopic universe, while
the others remain curled up and invisible. This extradimensional geometry
is highly significant: the intricate shape and size of these tiny dimensions
determines how a string vibrates, which determines what the particle
is. Thus geometry is the basis of particle physics—space determines the
nature of matter! As science writer Timothy Ferris summarizes it, "it's a
way to make everything out of nothing [meaning space]" (Ferris 1997, 222).

a.

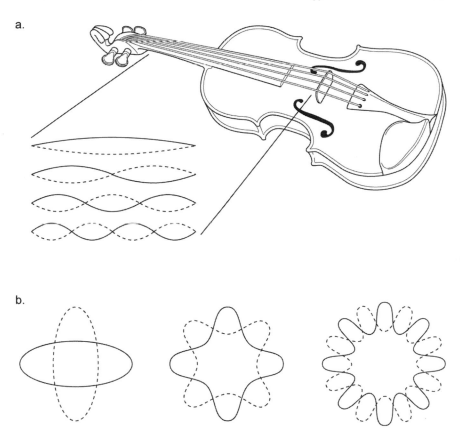

b.

Figure 3.7 Particles as Vibrating Strings. In string theory, first proposed in the 1970s, all particles are seen not as point-like entities but as extremely tiny one-dimensional strings or loops. Just as a single violin string can sound many different musical notes by vibrating at different rates, so an elemental "string," vibrating in different modes, can become all the different matter and energy particles in existence. This highly unifying theory has the great advantage of combining the two previously incompatible major theories of physics— quantum theory and general relativity. (Illustration by Jeff Dixon)

This linkage of spatial geometry with subatomic matter is another level of unification the theory suggests (Greene 1999, chapter 8).

One of the greatest promises of string theory is the unification of gravity and quantum theory. While the two theories work perfectly in their own domains of the very large and massive or the very small, they conflict where things are both very massive *and* very small, such as black holes or the first instant of the universe. Relativity's smooth slightly curving geometry of space cannot work within the quantum framework of microscopic frenzy and roiling fluctuations of space that the uncertainty principle dictates. Point-particles are jostled wildly by quantum fluctuations, but strings, which are extended in space, are too big to be affected and thus smooth out its devastating effects. Another promising feature, at least at first, was the theory's simplicity. It seemed that only one parameter was needed to determine everything else, the tension on the strings, whose enormous value—a thousand trillion trillion trillion (10^{39}) tons—caused the loops to shrink down to their tiny size (Greene 1999, chapter 6).

Beautiful and promising as the theory is, it is unfinished and beset by many challenges and unanswered questions. Why are there six or seven dimensions and what determined their shape? How did they curl up? Will a solution to the exact equations, once found, predict the properties of our universe? (Greene 1999, chapter 12) The discovery of a broader M-theory, which may incorporate separate earlier versions, has suggested new features–an additional spatial dimension and the possibility that the elemental entity has more dimensions than a string. There might be two-dimensional sheets or membranes, or three-dimensional blobs or higher dimensional "p-branes." Some dimensions of branes may be infinite and have significant roles in cosmology. The great enormity of the universe may, in fact, be a colossal brane; ours may be a "brane-world" (Greene 2004, 386). In recent years, it has dawned that there may not be a unique final theory—that very many possible "string theories" exist, which may or may not correspond to real "universes" or domains within a "megaverse." Furthermore, string theory has not been the only route to a final theory. In his book *Three Roads to Quantum Gravity*, physicist Lee Smolin describes the additional approaches of "loop quantum gravity" and "black hole thermodynamics," which differ in their respective pictures of quantum space–time (like different "windows" onto the same world) but share the view that space and time exist in discrete bits. He conjectures that they will all contribute to the final theory and possibly be integrated in a new "holographic principle," by which the universe is constructed from only the flow of information (Smolin 2001, chapter 12).

Many excellent books explain the features of string theory to the layman. Among them are John Gribbin's *The Search for Superstrings, Symmetry, and the Theory of Everything* (1998), an early work by P. C. W. Davies and J. Brown called *Superstrings: A Theory of Everything* (1988) which includes interviews with major theorists, and probably the best and most thorough, *The Elegant Universe* (1999) by string theorist Brian Greene. Volumes which focus primarily on multiple dimensions are Paul Halpern's *The Great Beyond* (2004) and leading theorist Lisa Randall's *Warped Passages* (2005). A very readable review of all of fundamental physics is Stephen Webb's *Out of This World* (2004).

INFLATION, QUANTUM COSMOLOGY,
AND MULTIPLE UNIVERSES

Around the time that string theory became prominent, a startling new idea entered the world of cosmology—inflation—proclaimed by cosmologist Joseph Silk as the "only new idea in cosmology since Einstein" (Ferris 1997, 242). It arose to account for unexplained cosmic features: why the microwave background has an even temperature everywhere, why the universe is so flat, why its expansion rate is critically fine-tuned for long-term growth, and how density clumps arose to make galaxies. According to inflationary theory, when the universe was the tiniest fraction of a second old (10^{-36} seconds), its expansion rapidly accelerated, and cosmic space itself grew faster than the speed of light, inflating from something nuclear or atomic sized to something very large. Inflation thus took a tiny region lumpy with quantum fluctuations but even in temperature throughout and made it big and smooth and flat but still slightly lumpy and equally hot everywhere—just what was needed to make our universe. After the briefest flash the expansion slowed to a more normal rate.

Why inflation happened has to do with what physicists call the "vacuum" energy of empty space. Quantum theory tells us that the "vacuum" of space is never empty but is a roiling sea of fluctuating energy fields with particles and energy surges popping in and out of existence. Usually all the fluctuating energy cancels out to zero, but sometimes it doesn't, and a residual energy lurks in a "false" vacuum. When released, it surges forth in an explosive way stretching space wildly and transforming itself into a maelstrom of particles and energy. This theory was first proposed by Alan Guth, who gave an account of it in *The Inflationary Universe*.

There have been a number of different ways that theorists have explored the creation moment itself (Trefil 1983, chapter 13), but all share the hypothesis that a "quantum fluctuation" in the vacuum, a random event, gave rise to the whole universe in what is called "vacuum genesis." Recall that according to the uncertainty principle, energy and time are two quantities that "trade-off:" a huge energy fluctuation could last only a short time, while a small one could last a long time. It turns out that the total energy of the universe may be zero—the energy of its movement exactly offset by the energy expended to expand it against gravity. Such a "zero energy" cosmos, emerging by chance, could last forever! As one researcher puts it, "our universe is simply one of those things that happens from time to time." Of course, nothing in this account explains why there should be physical laws that allow fluctuations to make universes or why there should be a universe at all.

Both inflation and quantum cosmology can allow for the possibility of multiple universes. In Andre Linde's version of inflation—chaotic inflation—many different domains with different vacuum energies are constantly inflating but only some of them do so in a way that can turn into a stable universe. Thus there was a preexistent infinite and eternal "mother" universe which continually produces regions that inflate and evolve into separate universes. Alex Vilenkin has also proposed eternal cosmic inflation and the idea of quantum creation of the universe from nothing to provide support for the possible existence of other universes. He has written an excellent review of this subject in *Many Worlds in One* (2006).

Most versions of quantum cosmology hold that our universe is just one single domain in a much larger megaverse or multiverse, which is eternally expanding and producing new universes. In each of these there could be a different vacuum energy and different laws of physics. When put together with the possibility of many different string theories being valid, there may in fact be a colossal number of possible domains and/or other universes. In *Cosmic Landscape* (2006) Leonard Susskind mentions 10^{500} possible "valleys" to explore in the landscape of string theory!

One unusual multiple universe idea is that there could actually be a "cosmological natural selection" by which a certain kind of universe is "selected" according to its ability to produce other universes. This reproduction process occurs through black holes, which can produce new big bangs and new universes. Universes with the largest number of black holes will eventually dominate the population. Lee Smolin describes his biological paradigm for cosmology in *The Life of the Cosmos*.

String theory itself has given rise to a whole different idea of multiple universes and "big bang" events, described in a number of sources examining cosmological implications of branes in M-theory. Most notable among them is a recent volume, *Endless Universe: Beyond the Big Bang* (2007), by the proponents of a new "cyclical model" for the universe, Paul Steinhardt and Neil Turok. Their hypothesis, evolved from their earlier "ekpyrotic" idea, is that the big bang was simply one incident in an infinite cycle of colossal collisions between our universe and a parallel world, which are two three-dimensional universes floating in higher dimensional space. Over the next trillion years dark energy (the accelerating mechanism in our present universe) will continue to expand and stretch both universes to become ever flatter, emptier and more parallel to each other. Finally, a spring-like interbrane force draws them together, and they collide and rapidly rebound in a tiny fraction of a second, while some of their energy is converted to particles and radiation.

It is evident that the riches of joining quantum theory and string theory with cosmology are just beginning to be mined. One thing is clear: with this merger the canvas has become much broader. Although some object to the unnecessary complication of hypothesizing many universes, or to the seeming impossibility of verifying them, the multiverse seems here to stay. And, as the next section and later chapters will show, it provides rich fodder for religious and philosophical reflection.

PHILOSOPHICAL AND RELIGIOUS REFLECTIONS ON QUANTUM THEORY

Physicists themselves have generally given three different philosophical interpretations of the status of quantum theories and the uncertainty they describe. In a sense, they represent varying degrees of acceptance of uncertainties as real. Implications for theology and the relation between science and religion emerge from all three. Helpful presentations of these basic positions in Barbour (1997, chapter 7; 2000, chapter 3) and Ferris (1997, chapter 11) have guided the following summary.

Intepretations of Quantum Mechanics

The first school of thought, a minority view, holds that uncertainties occur in quantum mechanics because the theory is not complete. This view stems from the "classical realist" view that the theories represent the real, objective physical world, as it exists separate from the observer. In that world, strict cause and effect and known mechanisms operate, and one should be able to make exact predictions. The uncertainties of quantum mechanics are thus not inherent to the world but a product of our temporary inability to know; they are just a sign that the theory is still incomplete. Rigid causal and deterministic laws will still be found for subatomic phenomena. Both Einstein and Max Planck held this view. Despite being one of its architects, Einstein remained convinced that quantum theory was incomplete, although he acknowledged its great success. To a fellow physicist he wrote: "Quantum mechanics is certainly imposing. But an inner voice tells me that it is not yet the real thing. The theory says a lot, but does not really bring us any closer to the secret of the 'Old One.' I, at any rate, am convinced that *He* is not playing at dice" (Ferris 1997, 346n34). Another physicist who sought a deterministic resolution to quantum weirdness was David Bohm, who proposed that "hidden variables" govern particles at lower levels. The mechanism was a "guiding wave" or "quantum potential"—a subtle field that does not change with distance but guides particles in an exact way. They seem to act randomly because these unknown forces act on them in a variable way.

A second major school of thought, called the "Copenhagen" interpretation after the Dane Bohr, held that quantum indeterminacy represents a permanent and inescapable limit to our knowledge of reality; we can never say what happens to a particle in between observations. Knowledge is limited either because our observing process disturbs what is being observed or because our experiment choice limits us to one view—wave *or* particle, for instance. This interpretation also involves the "instrumentalist" view that quantum theory is simply a useful calculating tool to give probabilities for a certain observation occurring; it can say nothing about the "reality" of an atom, for example. The atom could be fully determined *or* have intrinsic indeterminacy, as in being in many "superposed" possible states at once—we simply cannot know. Only a measurement "collapses" the wave function of possibilities, as quantum language puts it, and gives exact knowledge. The magic bullet dispelling uncertainty is thus the human act of observation, a notion that gives rise to much philosophical discussion about the role of the observer and human consciousness in constructing reality. Many found this whole interpretation very strange and sought to refute it with such thought experiments as Schrödinger's cat and the EPR paradox. Einstein jokingly critiqued it by claiming that the moon only existed when people looked at it.

Bohr's way of discussing the role of the observer and resolving the wave-particle paradox was his principle of complementarity. He overcame the dualities of observer versus observed and wave versus particle by viewing them as complementary parts of a whole. The observer and the observer are related in the single process of observation, since the observer's choice of experiment determines what is seen; there is no sharp division between them. It is not just the mind of the observer but its interaction with the object of observation that is important. In the same way, wave-like and particle-like properties are complementary aspects of an electron's "whole being." Bohr and others would extend this principle to other realms of life, including the relation between science and religion.

The third interpretation asserts that quantum indeterminacy really exists in nature as an objective feature of reality. A particle has no determined position. Maybe it's not the kind of thing that *is* in just one place, or maybe it has many possible positions all at once, each with a certain probability. The act of observation, then, does not simply reveal an exact position that was unknown, but makes actual one of many possible locations. Thus Heisenberg referred to quantum indeterminacy as "the restoration of the concept of potentiality." By this he meant that nature has tendencies or inclinations, a whole spectrum of possibilities, and what will happen next is simply unknown. The future is open.

One version of this idea, Hugh Everett's "many worlds" view, is that the many possibilities of each quantum event *are* all actualized. Each time a system could follow many potential routes, the world divides into many discrete and separate other worlds. Within each one, one of the possibilities is realized. The number of worlds generated is truly mind-boggling! This interpretation eliminates paradoxical dualities and observer roles and other things that vex quantum mechanics; a particle becomes a wave *and* a particle in different worlds. In this sense, his view is classical and deterministic. A form of Everett's idea has become the basis for many of the multiple universe ideas in quantum cosmology.

Religious Reflections of Scientists

Talk of philosophy and religion among scientists became much more prevalent in the era of the new science. As philosopher-historian R. G. Collingwood commented in 1934, "modern scientific leaders talk about God in a way that would have scandalized most scientists of fifty years ago" (McQuarrie 2002, 242). And, according to science historian John Hedley Brooke, Eddington remarked that "religion first became possible for a reasonable scientific man about the year 1927." This shift in attitude was due in no small measure to the new understanding of subatomic physics, which allowed an opening for a less "scientistic"—that is materialistic and reductionist—notion of science. Brooke describes an argument Bohr once made to Einstein—that "recent developments in quantum mechanics demanded a complete renunciation of the classical ideal of causality and a radical revision of attitudes toward the problem of physical reality." All of the pillars of classical physics—"picturability, determinism, and reductionism"—were toppled, and in their place there emerged a more open world of unknowability, uncertainty, chance, and holism (Brooke 1991, 327–28).

Many quantum physicists expressed views about religion. Planck believed strongly in the importance of the religious element in man's being and felt that science fosters it by its "love of truth and reverence," which led to sincere striving for knowledge and humility before the mysteries such a quest revealed. Such a striving for comprehension of the real world and belief in its order and intelligibility was a significant aspect of Einstein's "cosmic religion." His own faith of birth, Judaism, he felt, was a harbinger of this type of spirituality in its rejection of superstition and anthropomorphic "graven" images. Heisenberg believed that the new physics ameliorated the old conflict between science and religion, which existed because the rigid materialistic, naturalistic nineteenth-century framework of "space," "time," "matter," and "causality" allowed little space for such deeply important notions as "mind," "soul," and "life." As he wrote in *Physics and Philosophy*, "modern physics has perhaps opened the door to a wider outlook on the relation between the human mind and reality" (McQuarrie 2002, 246). It has done so, however, by recognizing its own limitations—the incompleteness of its description of nature—a fact which Heisenberg feels has made science regain a lost "modesty" and bring more humility into the discourse with theologians (Brooke 1991, 330).

Eddington, perhaps the most philosophically minded of all, argues that the world of physics is above all a world of abstract symbols removed from everyday life—symbols like "electrons," "quanta," and "potentials"—which are the product of thought. In *The Nature of the Physical World*, he argued for the spiritual nature of reality:

Recognizing that the physical world is entirely abstract and without "actuality" apart from its linkage to consciousness, we restore consciousness to the fundamental position, instead of representing it as an inessential complication occasionally found in the midst of inorganic nature at a late stage of evolutionary history ... the idea of a universal Mind or Logos would be a fairly plausible inference from the present state of scientific theory. (Eddington 1928, 332)

Such a science-based universal Mind is a far cry, however, from the God of religion, who can be known only through mystical experience. However, science's abstract, limited approach creates much space for other views. By renouncing the concepts of "inert material substance" and "strict universal determinism," the new science fosters a spiritual perspective on the world and "lends support to the mystical insight" (MacQuarrie 2002, 247–248).

Many quantum physicists displayed an interest in Eastern religion, and a number of examples are given by historian J. J. Clarke in his book *Oriental Enlightenment: The Encounter between Asian and Western Thought* (1997, chapter 10). One leading quantum researcher, Wolfgang Pauli, collaborated with psychologist C. G. Jung in examining the notable similarity between the underlying assumptions of quantum physics and the *I Ching*. In an article entitled "Synchronicity," published alongside an article by Pauli, Jung wrote: "the latest conclusions of science are coming nearer and nearer to a unitary idea of being." Bohr was also quick to see strong parallels with ancient Eastern thought, especially in his own "re-introduction of consciousness into the scientific understanding of nature." His interest in Taoism, following a trip to China, led to his use of the Yin–Yang symbol on his family coat of arms—a design representing the union of opposites and suggesting his most significant idea—complementarity.

Other quantum physicists were more drawn to India. Heisenberg was once a guest of philosopher-poet Tagore (see Figure 2.11) while lecturing in India in the 1930s. Their talks helped his thinking about quantum theory and encouraged his belief that the wild ideas might indeed be true, since a whole culture had similar notions (Wilber 1982, 218). Such philosophical attunement of Eastern spirituality and new ideas in science could aid understanding of the latter, an influence perhaps indicated by the great success of Japanese theorists. He wrote: "It may be easier to adapt oneself to the quantum-theoretical concept of reality when one has not gone through the naive materialistic way of thinking that still prevailed in Europe in the first decades of this century" (Heisenberg 2007, 176). Probably the most involved of all was Schrödinger, whose intense study of Vedantic philosophy occurred while he was developing wave mechanics. While it is hard to determine the exact influence of this study on his scientific work, his biographer wrote that Vedanta came to be a "foundation for his life and work" (Moore 1992, 173). In *My View of the World* (1964) Schrödinger bemoaned that Western culture no longer had a solid metaphysical framework and felt that one could be provided by the Vedantic conception of the "ultimate unity of the world—self, nature, and God," a unified view harmonious with the new science of "inseparable probability waves" (Clarke 1997, 169).

A rich resource for the view of physicists on religion and mysticism is Ken Wilber's *Quantum Questions: Mystical Writings of the World's Greatest Physicists* (2001). In it he presents the writings of most of the major players in the quantum era.

PHILOSOPHICAL AND RELIGIOUS IMPLICATIONS OF QUANTUM THEORY

Many concepts in quantum theory raise theological questions and possibilities, among them indeterminism, holism, notions of complementarity and nonlocality, the role of the observer, and limitations to human knowledge. How can chance and indeterminacy exist if God is all-powerful and all-knowing, or does it produce the openness he needs to act freely? How can particles be two things at once or communicate instantly from far-flung locations, and what do these phenomena imply about the nature of reality? What does it mean that only an act of observation produces an "exact" measurement, and that nothing can be known—perhaps does not even exist—until observed? Does this elevate the role of consciousness to a fundamental or even primary constituent of reality? A vast literature abounds with responses to these questions and attests to the fact that analyzing philosophical and theological implications of quantum theory is a complex task, indeed. Confusion often arises over which of the several scientific interpretations is being used, and, in a curious parallel with the subject itself, issues are very often intertwined. For simplicity, the review below organizes religious response according to Barbour's four modes of interaction—conflict, independence, dialogue, and integration—and will follow his summary in *When Science Meets Religion* (2000, chapter 3).

Conflict

This relationship arises from the role of chance and how it relates to determination of events by natural law and by God. Determinism had been a feature of science for several centuries, but views as to its cause had changed. In Newton's time everything was predestined to follow God's will in a perfectly designed, machine-like universe, which he created and continually sustained. A century later God became "deistic clockmaker" who simply created the universe and then let it run on its own; natural laws determined the future. Newton's laws were so successful, it was reasoned, that eventually science might explain everything and even predict all future events, if only one knew the present positions and velocities of all particles. In the past centuries, then, one way or the other, the future was determined, by God or by nature's laws. Not so in quantum mechanics if indeterminacy is a real feature of nature. Which particular event from a range of possibilities actually occurs is determined purely by chance. The randomness of chance speaks directly against purposive direction by God and erodes belief in a deity. British philosopher-mathematician Bertrand Russell voices this view: "Man is the product of causes which had no prevision of the end they were achieving; his origin, his growth, his hopes and fears, his loves and his beliefs, are but the outcome of accidental collocations of atoms." French biologist Jacques Monod also held that the prevalence of chance ruled out God and created a cosmos without purpose. In quantum cosmology some multiple universe theorists also argue that chance, and not design, made our universe perfectly suited for life. Ours is simply one among a colossal, if not infinite, number of universes, where the right conditions just happened to exist.

Independence

Some views regard quantum mechanics as lacking theological implications—theology and quantum physics are independent. One such independence view derives from the instrumentalist interpretation of knowledge. It holds that neither science nor religion can know reality itself; our theories and beliefs are useful mental constructs that help us calculate and organize what we experience. Our limited concepts are unable to describe the subatomic world as it is in itself, and a similar conceptual barrier applies to understanding of God. Comparison is impossible because of the very different languages and functions of quantum theory and theology.

Another example of independence is Bohr's central notion of complementarity, the coexistence of apparently contradictory elements within the same entity or system, a concept which he and others felt could be generalized to other fields, including science and religion. Within quantum theory, complementarity applies to those pairs of properties that cannot be measured simultaneously, such as a particle's wave and particle nature. Bohr felt that the concept could fruitfully be applied to other fields, such as organicism and mechanism in biological models, clashing viewpoints in politics (the authority of the U.N. vs. the power of individual countries), and philosophical/theological models, such as "free will versus determinism" and "God's justice versus his mercy." In *Science and Christian Belief* (1955) C. A. Coulson argued that science and religion could be thought of as "complementary accounts of one reality," a view both supported and criticized by other scholars. One scholarly article (Loder and Neidhardt, 1996) finds "unexpected compatibility" between Bohr's complementarity in atomic physics and the "dialectic" in the theology of Karl Barth, who wrote of the paradoxes of God's divine versus human nature as Christ and of the eternal versus the temporal. Barbour sounds a cautionary note in applying complementarity beyond physics to other domains. Such a comparison should only apply to similar "entities" and "logical types," which science and religion are not, and should be represented by analogy and not direct connection. Furthermore, it should not preclude critical examination of inconsistencies or the search for deeper unity. Polkinghorne (2007) also argues against a "facile kind of direct transfer . . . between physics and theology," calling it "quantum hype" that invokes quantum thinking as a "licence for lazy indulgence in playing with paradox in other disciplines." Both theologians believe in a strong, even unified, relationship between science and religion, but they object to easy, unexamined comparisons. Complementarity is clearly a tricky concept, steering a course between outright inconsistency and a stronger union of opposites contributing to a unified whole. In the latter sense it is more expressive of dialogue or integration.

Yet another example of independence emerges in Richard Jones, study of physics in comparison to the Eastern mysticism of Advaita Hinduism and Theravada Buddhism (Jones 1986). While both have cognitive worth, they are very different. Science provides knowledge about the realm of multiple, objective discrete objects and lawful patterns in the world of change, while mysticism involves the inward perception of an undifferentiated, nonobjective, unified realm behind superficial appearances. There can be no comparison of such different realms. Jones argues that demonstrating imprecise parallels and lifting language out of context, such as physicist Fritjof Capra does, is not a valid way to compare the two.

The same independence note is struck by philosopher Ken Wilber in his analysis of scientists' writings on the subject in *Quantum Questions: Mystical Writings of the World's Greatest Physicists*. Interestingly, he found that they all "rejected the notion that physics proves or even supports mysticism, and *yet every one of them was an avowed mystic!*" Wilber explains this paradox by arguing that quantum theory finally made scientists realize that science was not showing them a picture of the real world but a "shadow world of symbols"—abstract mathematical formulations—which was only a "partial aspect of something wider." Since science could say nothing about this wider reality, scientists were led to approach mysticism directly. Although proving or supporting mysticism through science is part and parcel of New Age thinking and Wilber's early thinking included it, he later claimed that it is erroneous and harmful. It confuses different levels of reality, it misrepresents the task of mystical practice as just reading a book to learn a "world view," and it is thoroughly reductionistic, suggesting as it does that mystical holism is based on the holism of subatomic particles (Wilber 2001, ix, chapter 1).

Dialogue and Integration

Many, on the other hand, see that quantum theory and religion have distinct parallels both in methodology and content or, further, that they could be merged in a systematic integration. Similarities can be seen when considering the role of the observer and consciousness, indeterminacy, stages in the respective searches for truth, holism and nonlocality.

Dialogue and Integration: The Role of the Observer and Consciousness

The participation of the observer in some interpretations of quantum theory provides new support for the ancient Platonic philosophy of idealism, in which the highest reality is mental in character. In the 1930s physicist James Jeans expressed this idea in *The Mysterious Universe* (1948): "The universe begins to look more like a great thought than like a great machine. Mind no longer appears as an accidental intruder in the realm of matter" (Barbour 2000, 78). Eddington echoed these thoughts: "To put the conclusion crudely—the stuff of the world is mind-stuff" (Wilber 2001, 199). Such ideas are consonant with the greatly expanded role of the observer in quantum theory. A crucial problem is how to understand the act of observation, which reduces the range of superposed possibilities of a system to one actual state. Exactly when and how does this occur, and does it require a human mind? Physicist Eugene Wigner argues that it happens when the measured values enter an observer's consciousness, whose "introspection and self-reference" help to fix the state. For physicist John Wheeler, "intersubjective agreement [and] communication" are the main factors in his "participatory universe," in which, "as observers of the Big Bang and early universe, we have helped to create those events." Atoms existed in a nebulous "partially individuated" state before they were observed, able to undergo chemical interactions but not be "fully real."

Barbour questions such a role for consciousness, citing results where "decoherence" of a wave function occurred through interaction with an environment (a beam of atoms probed by laser pulses) rather than a human observer. He draws a different theological message from observer-participation: the importance of relationship. The observer's relation to the observed always affects the outcome, whether in quantum systems or in time and space measurements in relativity. So in religion we come to know God by relating to him.

The relationship of the mind to the physical domain and the role of the observer in creating the world have been rich topics of theological discussion in Buddhism for centuries. In *The Universe in a Single Atom* (2005) the Dalai Lama describes several views. On one side are the realists, who believe that the constituents of matter exist objectively independent of the mind. The idealists, on the other hand, believe that the material world is but an extension of the mind, which is the only reality. In the middle position, favored by Tibetan tradition, the material world has reality that is objective but also relative and dependent on the observer. Just as in quantum theory, matter cannot be defined separately from the perceiver—"matter and mind are co-dependent" (The Dalai Lama 2005, 63). Buddhist scholar C. P. Ranasinghe seems to incorporate the latter two beliefs in affirming that "mind builds and maintains the cosmos." He writes, "The Buddha declared that the primary factor of the universe is the mind and that the force of the mind is supreme . . . [but] mind, as we normally know it, is so interrelated and interlocked with matter that many people mistake mind also to be matter." It is through the action of mind that units of matter, which normally tend to disperse, are drawn together to form larger structures (Ranasinghe 1957, 35, 254). The Dalai Lama also points out that in some Buddhist cosmologies the karmic tendencies of "sentient" beings are deeply connected with the development of the "universe system" they will inhabit. Except for the part about karma, this sounds rather like Wheeler's view.

Modern Hindu scholars of the Advaita Vedantic school also affirm the primacy of consciousness and draw parallels with principles of quantum physics. In *Science and Spirituality: A Quantum Integration* (1997) Hindu scholar Amit Goswami holds that quantum theory reveals the physical world to be a "projection of human consciousness . . . " Manifestation of the universe is needed for consciousness to "see" itself. The cosmos grows and develops by a continual play of "quantum processes," in which a multitude of possibilities pour forth, and consciousness acts to realize some of them. For example, the interaction of consciousness and quantum possibilities for gene mutations help to drive biological evolution. The primacy of consciousness and its role in creating and evolving the universe and life are also integral features of the cosmology of Meher Baba (see Chapter 7).

Dialogue and Integration—Quantum Indeterminacy
and God's Freedom

One integrative idea is to say that God decides what the future state of a quantum system will be, as physicist-priest William Pollard suggested in the 1950s. No physical intervention is necessary, since the system has only potentialities, so no violation of physical law occurs. By working with many atomic systems together, God would govern all events. He would guide evolutionary change by actualizing certain mutations at the quantum level. This view re-introduces determinism by God, thereby diminishing human freedom and exacerbating the problem of evil, which God would now create. It also limits God to acting at the lowest level of matter and is therefore somewhat reductionist. To overcome such problems, Robert Russell and others have proposed that God might act upon only certain select quantum systems but would also influence events from the top-down as well.

Two theologians who have written thoughtful analyses of the theological meaning of quantum indeterminacy are Christopher Mooney and Mark Worthing. Mooney views God's presence in the quantum world as analogous to the way he works with free humans—in both cases, a "sign of the freedom given to creation to be itself . . . God's action meshes dynamically with the potentialities inherent in the behavior . . . at both levels, without disturbing the natural probability patterns at either level." This perspective aids theology to develop new models of divine action and deepen understanding of God's "active presence in the lives of free persons" (Mooney 1996, 108–9). Worthing draws his own analogy to illustrate how both physicists and theologians struggle to reconcile phenomena at different levels. Just as physicists grapple with the interface between quantum and classical levels, so theologians attempt to fit together the "micro" level of human freedom, miracle, and prayer with the "macro" level of God's predestined plan for the future. While science and theology cannot answer each other's questions, each might develop ways of reconciling paradoxes that can be illuminating for the other.

The analogies above illustrate one way that creative mutual interaction between science and theology works effectively, not by suggesting direct causal connections between the two realms but by stimulating thinking through analogies and metaphors that help each field resolve their own problems. In a recent book, *Quantum Physics and Theology*, John Polkinghorne provides many rich analogies in the ways science and theology each search for truth. He first clarifies differences and then presents five "cousinly relationships" in the stages of knowledge-seeking. He follows this with many examples. One is the need to bear "unresolved perplexities"—quantum theory in science and the problem of evil in theology. Another is the notion of Grand Unified theories—the quest for unified forces in physics and the quest to understand the three aspects of God in Trinitarian theology. Another such analogy was suggested by theologian Wolfhart Pannenberg, who likened field theories in physics— "space-filling immaterial forces"—with the "effective presence... [and] cosmic activity of the divine Spirit" (Olson 2000, 308).

Physicist Freeman Dyson went further, however, to suggest direct connections between quantum events, human mentality and God and make a new argument for divine design. The human mind can be viewed as a choice-making entity that magnifies the "quantum choices made by molecules inside our heads." Because features of the universe are so perfectly attuned to developing the mind, we can infer a higher mind that chose such conditions, and perhaps this is God. Just as choice-making molecules are part of the human mind, so our minds might be "parts of God's mental apparatus" (Olson 2000, 311).

Dialogue and Integration—Quantum Holism and Eastern Mysticism

One of the most notable features of quantum mechanics is its challenge to reductionism. No longer could one think of an atom's behavior in terms of its parts; parts lose their identity in the whole, and the atom acts as a whole system. What were once "hard little spheres" became undulating wave patterns temporarily emerging, joining, dissipating, rejoining elsewhere—"a local outcropping of a continuous substratum of vibratory energy" (Barbour 2000, 81). In a final blow to classical physics, particles once "entangled" in the same system are forever connected and able to communicate instantly across vast reaches of space. Perhaps all particles are deeply entangled since they were connected at the Big Bang. "Nonlocality" suggests a deep interconnection in the cosmos that transcends time and space but still affects the physical level. There are strong theological implications here. If even the physical world has dimensions beyond our own material level, could not a divine being have them as well?

Many have claimed that the new science and Eastern mysticism present parallel views of the universe. Bohr himself was intrigued with such parallels, although he was careful to note that they were simply illustrative and did not imply that science proper was mystical. A much stronger stance was taken by physicist Fritjof Capra in his book *The Tao of Physics* (1975, 1991), probably the best known treatment of these parallels. In it he summarizes his major argument:

In modern physics, the universe is thus experienced as a dynamic, inseparable whole which always includes the observer in an essential way. In this experience, the traditional concepts of time and space, of isolated objects, and of cause and effect, lose their meaning. Such an experience, however, is very familiar to that of the Eastern mystics. (Capra 1991, 81)

Parallel concepts are the basic unity of the universe, the integrated role of consciousness, the limitations of human thought and the dynamic, ever-changing nature of things. One especially strong point of convergence is the notion of complementary opposites. Eastern traditions recognize basic dualities within an overall unity—good/bad, light/dark, knowledge/ignorance, male/female, etc.—and science encounters matter being both wave and particle, destructible and indestructible, continuous and discrete. Capra's work has received criticism for several reasons: comparing incommensurable things, relating elements that were abstracted misleadingly from contexts, and allowing similarities to overshadow significant differences between methods, goals, practical applications, and cultural contexts. The unified whole is unstructured and undifferentiated for Asian traditions, but not so for science, and their concepts of time and timelessness differ also. In his defense, Clarke notes that Capra's goal is not to prove or bolster either via the other but to foster a paradigm shift in understanding physics and its philosophical implications and reconciling science and religion (Barbour 2000, 84–85; Clarke 1997, 170–71). Not long after, Gary Zukav wrote *The Dancing Wu Li Masters* (1980), whose last chapter included Eastern spiritual references.

Another integration of science and quantum philosophy came from physicist David Bohm, who proposed, in *Wholeness and the Implicate Order* (1983), that both matter and consciousness are merged in and project from an underlying unified realm, a notion Eastern religions affirm. In this realm, which he called "the implicate order," each element enfolds within itself the entirety of the universe in the same way that a fragment of a hologram can reproduce the whole, although it would look fuzzier. Because this realm transcends space and time, nonlocal interaction can occur within it. Ferris speculates that science might hint at such ideas. Could interconnection come from hidden passages between black holes, called "wormholes?" Since time slows down for objects nearing light speed, does it disappear for a photon itself? He further conjectures that the deep connection of all particles at the universe's origin means they *are* in fact one thing, just as Wheeler once speculated that all electrons might be "the same electron."

An energetic dialogue occurred in the period when the works by Capra, Zukav, and Bohm were first published. Many of the interviews conducted and articles written appear in Wilber's *The Holographic Paradigm and Other Paradoxes* (1982). Another fine collection of views can be found in Renee Weber's *Dialogues with Scientists and Sages: The Search for Unity* (1986), which presents interviews with leading scientific and spiritual figures, including Bohm, Hawking, and the Dalai Lama.

Dialogue and Integration—Quantum Theory and Buddhism

Quantum physics and Eastern philosophy and mysticism, then, each at their own level, seem to deliver parallel messages. There is a fundamental unity to all being: individual parts are impermanent—interchanging, merging, and losing their identity in the whole. Everything achieves meaning by relating to the whole, and the universe is a dynamic, ever-changing process more than a set of "things." Whereas Eastern philosophers discuss these as intellectual ideas, mystical practitioners attempt to experience their reality by meditation and spiritual practice.

The religion which in many ways seems most consonant with science, both in its practice and beliefs, is Buddhism. One of its distinguishing features is encouraging practitioners to learn about the physical world and their own inner mind through empirical investigation. In a "scientific" spirit, Buddhists regard their beliefs as revisable. Because of this commitment, Buddhists are especially open to joint research with scientists into the nature of physical reality and the workings of the mind, and such efforts occur regularly through the work of the Mind and Life Institute and other programs. A number of recent publications review the status of the joint enterprise. *Buddhism and Science* (Wallace 2003) contains scholarly articles, while *The New Physics and Cosmology: Dialogues with the Dalai Lama* (Zajonc 2004) and *The Quantum and the Lotus* (Ricard and Thuan 2001) present dialogues. The Dalai Lama gives his own thoughts and reviews the status of joint inquiry in *The Universe in a Single Atom: The Convergence of Science and Spirituality*.

A few examples of Buddhist thinking from these sources will illustrate the areas of consonance. The Buddhist belief in the interdependence of all things is likened to the quantum notion of inseparability (Ricard and Thuan 2001, chapter 4). Objects do not exist as things in themselves; they are insubstantial and neither possess nor are defined by their properties. They exist only in relation to other things—they have "dependent origination." Furthermore, all of reality exists in each of its parts. The Buddha once likened reality to a "display of pearls—each pearl reflects all of the others, as well as the palace whose facade they decorate, and the entirety of the universe" (Ricard and Thuan 2001, 61–62). Interdependence is at the root of Buddhist notions of the impermanence and emptiness of things and also at the root of key moral precepts of overcoming false separateness and fostering compassion and love—points which David Bohm also emphasizes. The Dalai Lama points out how Buddhist "emptiness" resonates with the lack of solidity and definitiveness of matter in the new physics (Ricard and Thuan, 50).

THE FUNDAMENTAL CONSTITUENTS OF
REALITY—RELIGIOUS PERSPECTIVES

Interesting parallels exist between current scientific notions of the most fundamental particle or constituent of reality and conceptions in some religious systems. A number of Buddhist writings discuss this subject, among them K. N. Jayatilleke's *Facets of Buddhist Thought* (1971, 5th section), Akira Sadakata's *Buddhist Cosmology* (1997, chapter 1) and, very briefly, the above-mentioned book by the Dalai Lama. Probably the most thorough is Ranasinghe's *The Buddha's Explanation of the Universe* (1957, chapters 1–3), a Pali text of the Abhidamma section of the Buddha's teachings, which was republished not long ago by Lawrence Reiter in a derivative work entitled *Lord Buddha's Explanation of the Universe*. The text gives a detailed description of the constituents and workings of minuscule units of inanimate matter, animate matter, and the mind. He attributes these writings to reports of lectures given by the Buddha that have been handed down over the centuries. Purportedly, the Buddha arrived at such knowledge through enlightened inner vision, which Ranasinghe asserts could "dissect" matter down to its tiniest elements.

In Ranasinghe's explanation, inanimate matter is composed of varying proportions of four "abstract" elements—earth, air, fire, and water—each of which endows a substance with a certain property: hardness, flow, expansion, and heat. All of these elements mix in different proportions to form different atoms, the "last possible particle[s] of any substance." The air element acts in two ways: as a constituent of atoms but also as space. The atom itself is described as a

feather-like fabric...knit together closely with the fine threads that each unit of abstract element turns into, the threads of one kind alternating with threads of the other three kinds...In other words, the atom is a structure formed by the intercoiling of the fibres and filaments which stretch from each unit of the abstract elements. (Ranasinghe 1957, 51–52)

Atoms lose their individuality when they unite to form a substance, as the threads of their constituent elements interweave with the fibers of other atoms.

All elemental units "exist by repetition"—they experience a beat, a rhythmical pulsing that has three separate phases of birth, existence, and death. "On this three-spoked wheel of repetition, the universe of matter turns on and on." The repetition of these cycles is extremely rapid—176,470,000,000 repetitions per the duration of a flash of lightning. It would appear, then, that each unit of matter goes through a minute "creation, growth, and destruction" cycle, beating on and off, with extreme rapidity. Of even greater power and vibratory frequency are units of mind, which beat or pulsate seventeen times faster—at a rate of 3 trillion per the duration of a flash of lightning. Every beat of the mind also passes through the same threefold birth-existence-death cycle, which is further subdivided into seventeen smaller waves. This tiniest unit of time appears to be of the order of a thousand trillionth of a second (assuming a flash of lightning to be one hundredth of a second). One physicist who examined this material speculated after quick calculations that such time frames might be comparable to oscillation rates of the time-dependent Schrödinger wave between real and complex or "imaginary" space. Although this may fall into the category of "facile parallel," it is nevertheless intriguing. The material in this book is quite fascinating, although difficult to comprehend, and goes into much detail about exactly how the mind works in these tiny time intervals with tiny units of living and non-living matter, as it penetrates and interweaves throughout all of material existence, directing its incessant flow.

The Dalai Lama and Meher Baba also describe the creation and nature of "particles." In the Buddhist Kalachakra system, before a universe manifests, all material elements exist as potentialities in a "state of emptiness... [which] is not a total nothingness, but... a medium of 'empty particles' or 'space particles... which are thought of as extremely subtle 'material' particles." When the consciousness of future sentient inhabitants is ready and a universe forms, particles of each abstract element manifest from subtle to gross form—first air, then fire, then water and finally earth. Still "potentialities," these elements become "real" properties of hardness, fluidity, etc. when aggregate matter develops. Eventually, they "dissolve from the gross level to the subtle and back into the empty particles of space." Thus space is the foundation of the whole process. He goes on to describe the four stages in the cycle of the universe: formation, endurance, destruction, and void or emptiness before the next formation. Only space particles persist into the void, and they become the source of all of material being in the next cycle. The Dalai Lama recognizes that the word "particle" is inadequate for the phenomena being described. He also acknowledges that this ancient Buddhist way of regarding matter may need revision in the light of modern science (The Dalai Lama 2005, 85–90).

These Buddhist descriptions have interesting parallels with Meher Baba's cosmology, which describes wave motion producing "drops" of "matter," "particles" descending from subtle to physical realms, and sanskaric "threads" constituting matter. In his system, God's original whim to know himself produced a clash of energy with space, and the two interwove together in wave movement that produced drops or bubbles. As his original fire descended down through mental and subtle levels of existence, it acquired bubbles or "bodies" of mental and subtle matter before manifesting as physical "gross" particles. Between the subtle and first recognizable gross forms of protons and electrons, there were several gas-like and semigaseous forms. No matter how tiny and insubstantial, however, each drop is of the whole and always united with God's being in its essence. A further similarity lies in Baba's description of sanskaric threads as the constituents of matter and mind. Every object or being carries with it a set of sanskaras, or impressions, which are like threads forming the tissue of matter or consciousness. These threads actually constitute the bodies at the different levels. Impressions of experience are recorded on them, and the soul carries the subtle and mental sanskaras forward from lifetime to lifetime. (A fuller description of Baba's cosmology is given in Chapter 7.)

There are many concepts in the above descriptions that resonate with science: the role of mind, the interplay of space and energy, the wave nature of matter, matter in "potential" form, the emptiness of the atom, descent of matter through stages, and strings constituting matter. In science, the observer, or possibly consciousness, "actualizes" an observable event; for Baba, God's thought or desire brought the universe into existence, while for Ranasinghe, mind elements have a gravitation-like force that brings inanimate elements together to build structure. In science space and energy constitute everything, and string theory tells us that the shape of space makes energy and matter particles what they are—all is space! The atom is mostly empty space filled by insubstantial waves, in "potential" until actualized. In the universe's first second, matter and energy went through several stages and phase transitions and finally settled down to the more "stable, permanent" particles of the atom. So with Baba, the opposites of energy and space appeared, clashed, and synchronized into wave motion. The waves produced bubbles, near empty mixtures of energy and space, which gradually descended through rarefied subtle levels, where they were "in potential," and finally became physical atoms. In the Kalachakra, space gives rise to potentialities that descend through subtle to gross levels. "Emptiness" and insubstantiality characterize material reality. For Ranasinghe, a pulse or wave beats in every unit of matter and mind, cycling it from birth to death and back—back and forth between potential and actual existence—in each tiny instant of time.

Strings are a very significant element in both current science and in spiritual accounts. In science, strings may comprise the entire material world and are coded to vibrate in certain ways by the shape of space— vibrating one way to be an electron, another for a quark, etc. Just as in Buddhism, a particle's properties are not inherent to it but derive only from its string's vibration which changes through interaction. As string theorist Greene says, "a string is like a chameleon" (Ricard and Thuan 2001, 109). Strings show that everything is a unity but interchangeable. In Buddhism, fibers of different elements interweave and interlock to form the atom or bind atoms to make substances. For Baba, sanskaric threads are the information-carrying entities that record and hold all experience and form the tissue of different bodies—gross, subtle, and mental. Just as the strings of DNA build physical tissue, so sanskaric threads of experience build the tissue and fabric of consciousness. Renowned Romanian-born religious scholar Mircea Eliade also writes of threads and chords in religion and mythology, which connect heaven and earth or bind the universe as one living unity. A Cosmic Weaver—which can be the Sun, Brahman, or the personal gods—weaves the universe like a spider weaves a web. The threads give life and unity to all but also bind, and thus symbolize that from which we seek release.

QUANTUM COSMOLOGY AND CREATION FROM NOTHING

Quantum cosmologies seem to suggest that the universe could have arisen from the "nothingness" of the vacuum. This appears to accord better with with the Christian doctrine of *creation ex nihilo* than older Judeo-Christian interpretations, but closer inspection reveals that quantum cosmological ideas do not exclude more traditional Judeo-Christian theism. Quantum cosmologies do accord well with origin accounts in Kabbalah, Buddhism and Meher Baba's cosmology.

Scientific Theories

Before quantum cosmology, Big Bang theories assumed emergence of the universe from a singularity, or black hole. Since laws break down or are unknowable inside black holes, no information about a real beginning in time can be gleaned. Even if it could, and the universe emerged at time zero, what did it come from? Some kind of energy, ultradense matter, or physical laws must be assumed to exist. Traditional cosmology, therefore, presented no challenge to the Christian doctrine of creation from nothing.

Quantum cosmologies, on the other hand, seem to challenge the Christian doctrine in offering a "creation from nothing" using just the laws of physics. Edward Tryon's 1973 idea held that the universe "appeared from nowhere . . . and was as a 'fluctuation of the vacuum' of quantum field theory" (Worthing 1996, 98). Such a universe had to have zero energy—with motion energy offset by negative gravitational energy—but could last an infinite time. Although highly unlikely, such a thing could actually happen because quantum field theory allows for every possibility to happen at some time, statistically. However, the idea is really not a creation from nothing because the fluctuation happens within some larger space, and the fluctuation itself is not really nothing. This objection applies also to Vilenkin's quantum cosmology, in which the universe is "spontaneously created from literally nothing." He uses the phenomenon of quantum tunneling, wherein particles can penetrate through energy barriers, to propose that the universe tunneled from some state that had "no classical space–time" and tunneled into a state whose vacuum energy could make a universe. Again, the quantum tunneling is itself not really "nothing," nor is the wave function he uses, nor is the previous state from which it came. Stephen Hawking and James Hartle came up with a quantum gravity origin theory that devised a "wave function for the universe," whose fluctuation brought about the cosmos. They show the possibility for a universe without a boundary in space or time. Hawking talked about how this eliminated God, since there was no creation moment, but others have shown it only seems to eliminate a deist creator God. Again, the wave fluctuation and the laws governing it constitute something, so there is no true creation from absolutely nothing, in the philosophical sense of a true void. Absolute creation from nothing is totally avoided in the eternal cosmic inflation scenarios of Linde and Vilenkin. The process has been going on forever, creating inflating bubble universes from some preexisting megauniverse, and the question of ultimate beginning is meaningless (Worthing 1996, chapter 3).

Theological Responses from Christianity and Judaism

With quantum cosmologies, science failed again to produce a real *creatio ex nihilo*, and theologian Worthing concludes that only God can do that. He further asserts that the real meaning of *creatio ex nihilo* is not a message about our physical origins but about our total dependence on God. It also implies a transcendent creator (Worthing 1996, 98–106). Another theological implication Worthing discusses is whether God has freedom to act if science discovers a perfect theory of everything that specifies exactly how everything had to be. Theologians respond that God's creativity might have discovered that theory (Worthing 1996, 98–109).

Quantum cosmologies have interesting parallels with Kabbalah, as discussed in Chapter 2. In this philosophy, God did not create the universe from a total void, but rather from a kind of mystical "nothingness," which cannot be named or labeled but is the oneness of all things from which all things emerge. Jewish scholar Daniel Matt (1996) makes a strong connection here with the quantum vacuum. Astrophysicist Howard Smith (1996, chapter 5) makes detailed comparisons of the development of particles and forces in the very early universe, before it was a second old, with aspects and stages of the divine creative process symbolized in some of the ten Sefirot, or states of God's being in the Kabbalah. In "In a Beginning . . . Quantum Cosmology and Kabbalah" Joel Primack and Nancy Abrams explain how the first three Sefirot deal with creation and parallel closely ideas of both single and eternal inflation in quantum cosmology: *Hokhmah,* for instance, means "the bursting through of our universe;" *Binah,* means the "female womb in which creation expands." Together, they seem to describe the Big Bang, although, as the authors mention, these terms have a much different and deeper spiritual meaning for the followers of Kabbalism.

Comparison with Buddhism and the Cosmology of Meher Baba

In both Buddhist thought and Baba's system there is something akin to the quantum vacuum state from which everything emerged. The notion of "creation from nothing" is difficult for Buddhists, and while Meher Baba speaks of creation emerging from the unconscious, "nothingness" aspect of God, it cannot be truly "nothing" if it came from the divine being.

There is no creation from nothing in Buddhist philosophy. The Buddhist concept of emptiness does not equate with nothingness. It means the absence of anything permanent that has intrinsic existence. Things can only arise by "dependent origination," in relation to something else. There is no such thing as something passing from nonexistence to existence, and nothing can be born in that way. Thus, the idea of the radical Christian "creation from nothing" is an impossible notion for Buddhists. As the Dalai Lama expressed it, "Buddhism does not posit an absolute beginning. A causeless absolute beginning flies in the face of Buddhist logic" (Zajonc 2004, 183). Nevertheless, there could be an "unfolding of the world." Buddhist monk Matthieu Ricard suggests that, "perhaps the Big Bang can be interpreted as the process of the world of phenomena springing forth from an infinite but nonmanifest potentiality, which is metaphorically called 'particles of space'" (Ricard and Thuan 2001, 30). The Dalai Lama describes these "particles" as the "ultimate cause of material objects." They existed before the Big Bang as the "residue of the preceding universe that has disintegrated." For the Dalai Lama, the theory of these particles resonates with the notion of the quantum vacuum, which likewise has the potential to become all material things (Ricard and Thuan 2001, 86–87). As described above, Ricard paints a similar picture of how these particles created a void that became "'full' of five 'winds' or energies . . . [which] appeared as lights of five different colors that gradually materialized into five elements . . . When combined they formed a 'soup,' an ocean of elements that was whipped up by the primordial energy, thus producing the heavenly bodies [galaxies, stars, planets], continents, mountains, and finally living beings." This describes simply how one universe developed among the infinitude that exists. Thus, there is no initial creation (Ricard and Thuan 2001, 33).

Meher Baba writes often about "nothing." Creation is the emergence of countless forms out of nothing, he says. The nothing is contained in the "everything" of God. Like quantum cosmologists and Kabbalists, Meher Baba has a very rich idea of nothing—for him, it emerged from God himself. In the beginning, God was in a state of pure existence, like deep sleep, where there was no duality, and no knowledge. He suddenly had a "whim" to know himself fully—a causeless "chance fluctuation" in God's being—and this thought disturbed the great ocean of his being. As he awoke, all the opposites within him manifested—energy and emptiness, everything and nothing, knowing and ignorance. The aspect of his being that became "everything" was fully conscious of itself as God, while the other aspect that became the "nothing," was totally ignorant of what it was, but yearning to know. It sought answers by becoming all the forms in creation, slowly evolving its consciousness with each one, until, individual by individual, it finally gained full knowledge of itself as God. This picture has a rather complex view of nothingness. As in quantum cosmology, it is a realm with zero total energy that nonetheless has forward drive, and it is the source of all things in creation. It also has produced an infinite number of universes. In sharp contrast to both science and Christian theology, however, the "nothingness" that became the universe is an aspect of the being of God himself.

Chapter 4

⁓

Cosmic Evolution: Reflections of Scientists

In the latter half of the twentieth century a new creation epic sounded from the halls of science, made possible by a host of significant new advances. The discovery of the cosmic microwave background radiation in the 1960s firmly supported the idea that the cosmos had evolved over billions of years from a homogeneous state very different from today's universe. In the 1990s very slight temperature variations were found in this radiation, revealing that seeds of later galaxy formation had existed near the very beginning. Both of these developments put the Big Bang theory on a sound observational footing. The chemical evolution of matter in stars and even events in the first second of the universe came to be understood, or at least speculated about, through the theories of stellar evolution and elementary particle physics. When these discoveries were linked to the astronomical theories of the formation of the Earth and the whole of biological evolution, the conceptual picture of science's creation story was nearly complete: human beings have emerged as the flowering of the natural, 14-billion-year growth and development of the universe itself. Furthermore, since the ingredients and processes for developing life exist everywhere, it may be possible, if not highly likely, that a multitude of living forms and intelligent beings inhabit the universe.

The cosmic evolution story has brought science face to face with religion and produced a great variety of reactions and much philosophical and religious reflection in the scientific domain. Implications scientists draw from it seem almost contradictory. At almost all the stages along the way, the evolution appears to be explainable in terms of natural laws and chance events, with no recourse to any supernatural agent. On the other hand, many physical features of the early universe seem perfectly fine-tuned to

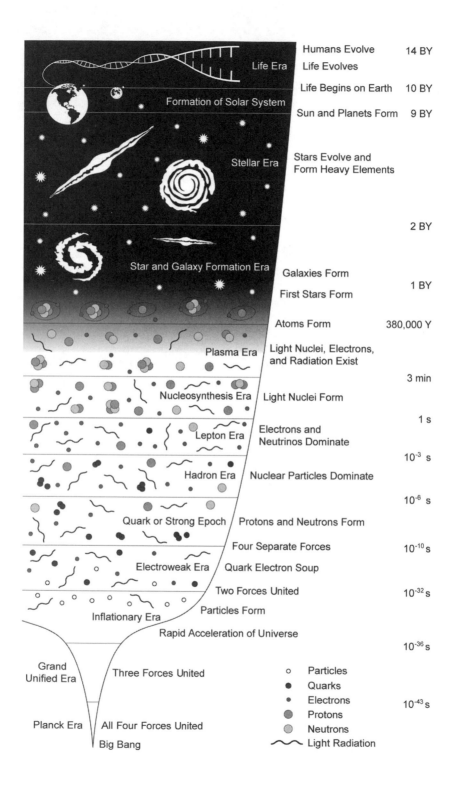

Humans Evolve	14 BY
Life Era — Life Evolves	
Life Begins on Earth	10 BY
Formation of Solar System — Sun and Planets Form	9 BY
Stellar Era — Stars Evolve and Form Heavy Elements	
	2 BY
Star and Galaxy Formation Era — Galaxies Form	
First Stars Form	1 BY
Atoms Form	380,000 Y
Plasma Era — Light Nuclei, Electrons, and Radiation Exist	
	3 min
Nucleosynthesis Era — Light Nuclei Form	
	1 s
Lepton Era — Electrons and Neutrinos Dominate	
	10^{-3} s
Hadron Era — Nuclear Particles Dominate	
	10^{-6} s
Quark or Strong Epoch — Protons and Neutrons Form	
Four Separate Forces	10^{-10} s
Electroweak Era — Quark Electron Soup	
Two Forces United	10^{-32} s
Inflationary Era — Particles Form	
Rapid Acceleration of Universe	
	10^{-36} s
Grand Unified Era — Three Forces United	
	10^{-43} s
Planck Era — All Four Forces United	
Big Bang	

Key:
o Particles
● Quarks
· Electrons
● Protons
○ Neutrons
〜 Light Radiation

produce life, almost as if the universe were designed with living beings in mind. A debate has arisen in science over one answer to the fine-tuning dilemma: the "anthropic principle," which suggests that the universe had to be the way it is to produce us, or, more strongly, that it had to produce us. This idea has evoked a wide spectrum of scientific reaction from outright hostility to thoughtful exploration of its ramifications and a call for a wider cosmology that incorporates the meaning of human existence.

COSMIC EVOLUTION AT THE MILLENNIUM

At the dawn of the new millennium, all the pieces were in place for a fairly complete picture of the grand sweep of cosmic evolution, from the tiniest fraction of a second after the Big Bang to the present time (see Figure 4.1). While the boundary conditions of the very first instant were still a matter of intense speculation, the stages in between were considered by most scientists to be fairly well understood, at least in theory. The understanding of evolution starting from near the very beginning and leading up to the origin of life arose from work in the physical sciences, and these stages are emphasized here.

The First Nanosecond (10^{-43} to 10^{-10} seconds)

The Planck Era (10^{-43} seconds)   The story begins at an infinitesimally tiny moment after the absolute beginning, the Planck time—one ten million trillion trillion trillionth of a second after time zero. The universe at that moment had a temperature of 10^{32} degrees, beyond which the known laws of physics cannot yet theorize the state of matter and energy. Before this instant, all four forces governing the interactions of matter particles are unified in one grand "superforce" or "superparticle." The physical universe thus began in a state of undifferentiated primordial unity, but an unstable one.

The Grand Unified Era (10^{-43} to 10^{-36} seconds)   At the moment our understanding begins, three dimensions of space began to grow large,

←——

Figure 4.1 Diagram of Cosmic Evolution. Almost all fields in natural science have contributed to the understanding of the epic of cosmic evolution, shown in pictorial form to the left. It sketches the development of matter from its coalescence into particles during the first second to the formation of the sun and Earth 9 billion years later and the final emergence of humans at the 14 billion year mark. (Illustration by Jeff Dixon)

while, if superstring theories are correct, the other six or seven remained curled up at the Planck scale of 10^{-33} cm. The superforce divided into two streams, as the force of gravity diverged from the three subatomic forces, which still acted as one "grand unified" force. The perfect symmetry of the original state was broken; duality entered the picture. The entire universe was in a state of vacuum, where quantum fluctuations ruled the day. But the bubble was expanding and cooling; at the end of the grand unified era, the temperature had dropped to 10^{27} degrees.

The Inflationary Era (10^{-36} to 10^{-32} seconds)   An infinitesimally short time after the Planck time—approximately 10^{-36}, or one trillion trillion trillionth of a second after time zero—the hot, dense bubble universe is theorized to have undergone an explosive, incredibly short burst of accelerated expansion, dubbed "inflation." According to the theory, the energy locked up in the vacuum transformed into a maelstrom of particles and energy. In a period of time lasting barely a hundred million trillion trillionth of a second (10^{-32} seconds), the space of the universe enlarged from the size of an atomic nucleus to something very much larger, perhaps even the size of our solar system. Quantum fluctuations existing at the moment inflation began were preserved during inflation and became large-scale fluctuations in density and temperature—a fact of great importance in explaining the later development of structure in the universe. Had they been slightly smaller or larger in amplitude, large structures could never have formed.

In this early era another important milestone occurred. The forces of nature further divided as the strong nuclear force froze out from the electroweak force. It was perhaps this process that fueled the inflationary expansion.

Electroweak Era (10^{-32} to 10^{-10} seconds)   At the end of inflation the universe had a temperature of 10^{27} degrees, and the expansion rate settled down to a slower, steady, noninflationary rate, which would allow for the growth of the universe to very large proportions over an enormously long span of time. In this short epoch, the electromagnetic and weak nuclear forces still operated as one, mediated by the same family of particles. By one ten billionth of a second, however, the energy of the universe dropped below the threshold for electroweak unification, and the symmetry between them was broken. The final separation of the superforce into four distinct forces became complete and the ratio of their strengths set for all time.

Thus, before the universe was one billionth of a second old, all the factors which make our universe conducive to life were in place. Three dimensions were growing larger slowly, density ripples existed that would lead to galaxies and stars, and force strengths and particle masses became fixed at values that would allow long-lived stars to manufacture the heavy elements necessary for life and provide steady radiation and warmth to nearby planets. The temperature at the billionth of a second mark—a million billion (10^{15}) degrees—is a milestone in another way for current research, for such energies can now be achieved in particle accelerators. Hence, it is the earliest moment which has experimental verification.

Particle Era (10^{-10} seconds to 380,000 years)

The Quark or Strong Epoch (10^{-10} to 10^{-6} seconds)   During and following the electroweak epoch, the entities of the universe engaged in a wild and frenzied dance, as matter-antimatter particles—quarks and antiquarks, electrons and neutrinos (leptons), and their antiparticles—and powerful radiation continually transformed into each other and back again. But as the universe cooled, the quarks and antiquarks lost energy and could no longer resist the strong nuclear force, which drew them together to form the strongest stable particles in existence—the proton and the neutron—each composed of three quarks—and their antiparticles. The frenzied dance continued, however, as protons, neutrons, electrons, neutrinos, and their antiparticles annihilated and reconstructed themselves in the subnuclear ocean. The era of free quarks had ended.

The Hadron Era and the Victory of Matter (10^{-6} to 10^{-3} seconds)   The creation-annihilation dance continued but a short while for protons and neutrons—particles called "hadrons" that would later compose atomic nuclei. With further cooling photons of light became too weak to transform back into these heavy nuclear particles after annihilation. If protons had exactly equaled antiprotons in number, they would have undergone a final annihilation into radiation and disappeared from existence. Fortunately, elementary particle processes show a preference for interactions that decay into protons rather than antiprotons. The preference is slight, indeed. Once the annihilation was complete, one proton was left for every billion photons, but that was enough to make a whole universe! Thus, matter triumphed over antimatter, even as most nuclear particles (hadrons) disappeared forever. Their era ended at a temperature of a trillion (10^{12}) degrees.

The Lepton Era (10^{-3} to 1 second)   The cosmic brew at the first millisecond contained numerous electrons and neutrinos and their antiparticles (leptons and antileptons), photons, dark matter particles, and the more sparse protons and neutrons left over after the last annihilation. Lepton pair creation and annihilation dominated the scene, until additional trans-formations came to pass. In successive stages and with further expansion, densities fell below the threshold needed for neutrino and, most proba-bly, dark matter particle interactions. They "froze out" from the cosmic ocean, separating forever. Neutrinos have ever since roamed free, rarely interacting with other matter. Dark matter, greatly outweighing ordinary matter yet unbound to the ever expanding radiation field, was free to heed gravity's call much sooner and lead the way in forming large-scale structure.

Sometime around the first second, the last major annihilation occurred. As the universe grew too cool for electron–positron creation, the remain-ing pairs disappeared in a last burst of radiant energy, leaving a solitary electron for every billion pairs that annihilated. The loss of electrons also reduced neutron numbers: they were no longer available to bond with protons to form new neutrons. As neutrons underwent their normal de-cay into protons, they could not be replaced, resulting in a permanent imbalance in neutron–proton numbers.

By the end of the first second, the fundamental particles that comprise today's universe were in place, in just the right proportions for further development of the universe and life—the protons, neutrons, and electrons that form atoms, and dark matter particles that form large-scale structure. Processes that occurred in this first brief moment set the stage for and made possible almost all later development in the evolution of the universe and life—a stunning accomplishment for one second of work! As expressed by Trinh Xuan Thuan: "This 1 second, by virtue of all the various events that occurred within it, is of greater significance than all the other 10^{17} seconds in the 15 billion years that followed" (Thuan 1995, 130).

The Era of Nucleosynthesis (First 3 Minutes)   During the first few min-utes of the infant universe, temperatures cooled down to the level where energetic photons no longer prevented the nuclear forces from fusing neu-trons and protons into helium nuclei. Over a period of about three minutes all the neutrons merged with protons to form one helium nucleus for every twelve protons—by mass one quarter of the total matter. As astronomers survey the universe today, approximately 25 percent is observed to be helium, far more than can be accounted for by later nucleosynthesis in stars—but exactly what is predicted by the Big Bang.

The Plasma Era (3 Minutes to 380,000 Years)   For the next 380,000 years or so, the universe remained a plasma—a separated gas of pro-tons, helium nuclei, electrons, and photons. Should electrons happen to combine momentarily with protons, a powerful photon was sure to destroy their bond soon after. Radiation was bound up in constant interactions with electrons, making the universe opaque. Slowly but surely, however, photons were losing energy as the universe grew bigger and cooler.

The Era of Atoms (380,000 A.B.B.): The Great Decoupling and the Birth of Atoms   A crucial milestone was reached when the temperature dropped to around three thousand degrees. Photons no longer had sufficient energy to dissociate atoms, and electrons could easily pair with protons and helium nuclei. Atoms were born. Photons that formerly appeared and disappeared quickly in constant interactions were now set free, and the universe became transparent—filled with visible light. Thus decoupled from matter, this radiation has pervaded the universe ever since, slowly losing energy to become finally microwaves—the famous "cosmic background radiation" so important in cosmology today.

This milestone event marks the first moment in cosmic history that humans can directly observe, and its study is reaping rich rewards. The slight temperature variations now measured may actually reveal information about a truly primordial moment of the universe—10^{-36} seconds!—the preinflationary epoch when they formed as tiny quantum fluctuations. Preserved and enormously enlarged by inflationary expansion, they became large-scale density fluctuations, seeds of the largest structures in the universe—clusters of galaxies.

The Era of Galaxy Formation (the First 2 Billion Years)

The release of matter from its bondage to radiation allowed the density fluctuations imprinted at the dawn of time to grow. Ordinary matter now joined forces with dark matter in gravity's cosmic battle against expansion. According to the currently favored bottom-up, cold dark matter model of structure formation, ordinary matter fell toward the agglomerations of dark matter, which had already begun forming massive galactic halos. Losing energy through radiation and collisions, the ordinary matter in most cases collapsed into spinning disks to form today's most plentiful spiral galaxies. Sometime during this process, star formation began, initiating a new era of nucleosynthesis, and the universe's journey toward creating more complex forms of matter began anew.

The Stellar Era (the First Billion Years to the Present)

An understanding of the formation and distribution of chemical elements during the life cycle of stars is one of the great achievements of

The Life Cycle of Stars

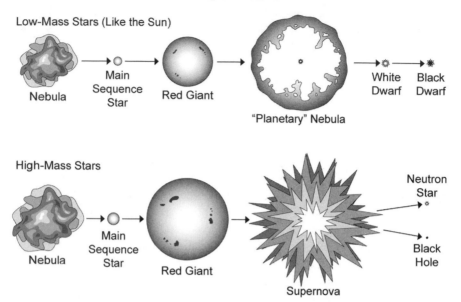

Figure 4.2 The Life Cycle of Stars. In cosmic evolution stars perform the essential role of creating all the elements needed for life through thermonuclear fusion in their cores. Main sequence stars synthesize hydrogen nuclei to form helium, which red giants later fuse to form carbon and oxygen. From there, light stars shed outer layers in gaseous shells, while heavy stars fuse elements up to iron and distribute them back to space in gigantic supernova explosions, which create elements heavier than iron. (Ilustration by Jeff Dixon)

astrophysics during the twentieth century (see Figure 4.2 and Figure 4.3). It is now well understood that in successive stages of nucleosynthesis, stars have manufactured first helium, then carbon and oxygen, and in heavy stars and stellar explosions, nuclei of all the remaining chemical elements. They distributed them back to the interstellar medium through quiescent stellar winds or in violent supernova explosions (see Figure 4.4 and Figure 4.5). In a grand cosmic recycling scheme, new generations of stars collected this enriched matter and built more heavy elements. Gradually, the chemistry of the universe has changed and continues to do so. At the present time, in our region of the universe, the mixture contains 1–2 percent elements heavier than hydrogen and helium.

The Formation of the Solar System and the Evolution of Life
(9–14 Billion Years)

Some 4.5 billion years ago, one interstellar nebula of gas and dust is thought to have formed our sun and planets. The theory, first proposed

Figure 4.3 Where New Stars Are Born: Pillars of Creation in the Eagle Nebula. This most famous of all Hubble photographs portrays a region of active star formation—columns of gas which will eventually become hundreds of stars. The small fingers of gas protruding from the top of the left column show pockets of gas contracting under gravity to form stars. (NASA, ESA, STScI, J. Hester and P. Scowen, Arizona State University)

by Immanuel Kant in the eighteenth century, suggests that the process is but another chapter in the cosmic battle of gravity and heat. A spinning nebula slowly contracted, the central part growing ever smaller, denser and hotter to form a star, while the outlying material fell into a disk and formed smaller bodies of rock and gas. Near the proto-sun, temperatures prevented all but the heavier metallic and rocky substances from sticking together and accreting into the seeds of planets. There, the rocky, Earth-like planets formed—almost totally lacking in the plentiful hydrogen and helium that comprise 98 percent of the rest of the universe. Such an unlikely composition, however, is just what the development of life required.

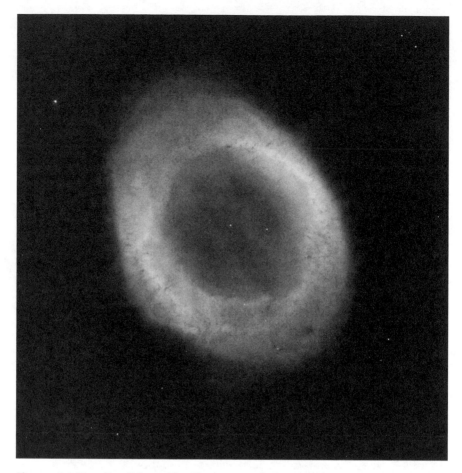

Figure 4.4 How Small Stars Die: Planetary Nebula. This best known of all "planetary nebulae"—the Ring Nebula in the summer constellation Lyra—shows an expanding bubble of gas expelled by a dying star. The core visible at the center will gradually cool and shrink to become a highly condensed, Earth-sized white dwarf. (The misnomer "planetary nebula" came about because early observers mistook them for planets.) (NASA The Hubble Heritage Team, STScI, AURA OD The Ring Nebula)

For a 100 million years, the infant planets suffered heavy bombardment by rocky and icy "planetesimals"—leftover seeds of planets—which caused reheating, melted their interiors, and set the stage for the life-enhancing geological activity of epochs to come. This activity caused outgassing of the oceans and atmosphere, which Earth's moderate temperatures allowed to survive. Sometime near the first billion year mark, processes occurred that led to the development of the first forms of life. Numerous experiments have repeated the now famous discovery of Urey and

Figure 4.5 How Heavy Stars Die: Crab Supernova Remnant. The Crab Nebula in Taurus is the remains of a gigantic stellar explosion first seen on Earth in 1054 A.D. At the exact spot in the sky recorded by Chinese and Native American observers, astronomers centuries later observed this expanding cloud of gas with telescopes. Such explosions spew into space a rich brew of newly formed heavy elements, to be incorporated into the next generation of stars and planets. (NASA, ESA, J. Hester and A. Loll, Arizona State University)

Miller in 1953 that the most abundant chemicals of the early Earth, when energized and allowed sufficient time, readily develop into the chemicals of life—amino acids, the building blocks of proteins. Other experiments showed that the same process also yields nucleotides, the building blocks of DNA and RNA.

Exactly how these chemicals transformed into living entities is not completely understood by science, but the fossil record shows that sometime 3–4 billion years ago, the first microscopic life form emerged from the primordial broth. Once established on Earth, plant life, in league with vast water oceans, transformed the atmosphere of Earth into a unique oxygen-rich mixture and thus paved the way for the next crucial step—animal life. A billion years ago macroscopic life forms developed and experienced a rapid proliferation 500 million years later, in the Cambrian explosion. Large land animals came to dominate the landscape by 200 million years ago, and the age of reptiles continued until their sudden extinction—perhaps caused by the impact of an asteroid. Their demise opened a niche for small land creatures, and the first mammals flourished. Two million years ago marks the appearance of the first humans.

The understanding of cosmic evolution sketched above is one of the great achievements of science in the last century. It represents the synthesis of results from many scientific disciplines pertaining to widely disparate phenomena in vastly different epochs of time and realms of the cosmos. Elementary particle physicists, astrophysicists, chemists, and geophysicists together have shown how matter could have evolved from elementary particles in the early universe, collected into vast galaxies of stars that forged all the chemical elements, and finally congealed near stars into planets. Molecular and evolutionary biologists, synthesizing Darwin's original theory with twentieth-century understanding of DNA reproduction, have suggested a process by which life might have emerged and developed from simple molecular beginnings to the rich complexity existing today. All of these findings have been integrated into a single, unbroken chain of events stretching from near the very beginning over billions of years right up to the present time.

This stunning result elicits from scientists themselves a host of varied reactions and reflections. Foremost, perhaps, is the desire to tell the story to general audiences, and the literature abounds with many fine narratives, all written by firsthand practitioners or science-trained journalists. Some of the first were accounts by Robert Jastrow (1967, 1978, 1992), Steven Weinberg (1977, 1988), Carl Sagan (1980), Joseph Silk (1980, 1989, 1997, 2001), and James Trefil (1983)—followed soon after by accounts from Heinz Pagels (1985), John Gribbin (1986), and Stephen Hawking (1988, 1993, 1998). During the same decade, stunning video creations by Sagan (*Cosmos* 2000) and Timothy Ferris (*The Creation of the Universe* 1985) sought to inspire the general public with the new vision. More recently, George Ellis (1993a), Trinh Xuan Thuan (1995), Martin Rees (1997), Ferris (1997), and Craig Hogan (1998) have contributed volumes. Among the newest descriptions are Neil deGrasse Tyson's *Origins* (2004), which accompanies a four-part NOVA series and accounts by Fred Adams (2004) and Silk (2005). This list is by no means exhaustive but is representative of both the quantity and quality of excellent presentations on this subject readily available to the public. The account described above was drawn from a number of these sources, in particular Silk (2001), Thuan (1995), and Hogan (1998).

PHILOSOPHICAL AND RELIGIOUS IMPLICATIONS SEEN BY SCIENTISTS

In presenting the new cosmology to general audiences, virtually all of the above writers feel compelled to discuss, at least briefly, its wider philosophical and religious implications. Their reflections focus on three areas. The first is a reaction to the magnitude and comprehensiveness of the vision and its clear metaphysical overlap with religion and philosophy. The second is an epistemological reflection about the power of the scientific method. The third is a focus on the relation of human beings to the universe and questions of design (teleology) and ethics.

Metaphysical and Religious Reflections

All of these authors express a certain awe at the "sweeping vision of the entirety of creation from the beginning" (Hogan 1998, vii) revealed by science. Its grandeur has religious proportions for many authors. Even those who are cautious about metaphysical extrapolation pepper their narratives with religious language. Fred Adams refers to "the miraculous chain of events" and getting to "glimpse the face of creation" while subtitling his text "A Book of Genesis in 0+7 Chapters" (Adams 2004, 5). Joseph Silk speaks about the "Big Bang Gospel" (2005, 170), while Neil deGrasse Tyson names a chapter "The Greatest Story Ever Told" (2004). Carl Sagan comments that "science has found . . . that the universe has a reeling and ecstatic grandeur" (Sagan 1980, xii), a comment less religious in content, but equally so in feeling. Although such resounding language is no doubt intended to dramatize the accounts for general consumption, and probably attract readers, it nonetheless suggests a clear overlap with religion.

Beyond suggestive language, virtually all of the above writers acknowledge that cosmology now approaches ultimate questions that were formerly the exclusive province of metaphysics and religion. As Thuan expresses it,

Modern cosmology . . . has tackled subjects that for many years were the exclusive property of religion, and has thrown a completely new light on them. By attacking the wall surrounding physical reality with the powerful tools of physical and mathematical laws, cosmologists and astronomers have found themselves face to face with theologians. (Thuan 1995, 248)

Scientists deal with this meeting in a variety of ways, from avoidance and unease to direct grappling with questions of the cause of the universe, the nature of reality, and the existence and role of God. According to Weinberg, some cosmologists try to "avoid the problem of Genesis" by subscribing to the cyclic version of the Big Bang (1988, 154). In *God and the*

Astronomers, Jastrow acknowledges that "the astronomical evidence leads to a biblical view of the origin of the world" (1978, 116). He describes physicists' frustration at encountering a state of existence whose cause cannot be determined by science and that necessarily invokes the question of God (Jastrow 1978, 1992).

As the synthesis of quantum physics and cosmology developed in the 1970s and 1980s, a new optimism arose that science could finally "settle" certain metaphysical questions and either address directly or avoid altogether the moment of creation and the notion of a creator God. John Gribbin claims that science now answers all three of the great metaphysical puzzles—the nature of reality, the origin and nature of life, and the origin and cause of the universe—thus "blurring" the distinction between science and metaphysics. For him, this signals the death of metaphysics (Gribbin 1986, 392).

The enormous success in combining particle physics and cosmology gave momentum to the hope that science will understand why the universe exists. James Trefil describes ways some theorists are logically approaching this question (Trefil 1983, 203–8). For him, the newest advances herald the "ultimate triumph of reductionism" and reconfirm the separation of object and observer, traditional scientific principles recently challenged both from within and from outside of science. Particle theory and unification of forces reduce the natural world to a set of "complex relationships between simple objects" with an "underlying reality (that is) simple and beautiful" (Trefil 1983, 220). The separation of object and observer finds new confirmation in gauge symmetry theories, which delve deeper than quantum mechanics to demonstrate that many properties do not vary under transformation but are independent of observers' states of mind (Trefil 1983, 221). "These advances," Trefil argues, "represent the final working-out of an age-old scientific and philosophical goal" (219).

Stephen Hawking sounds a similar triumphal note in suggesting that quantum cosmology might eliminate the notion of a creator God with free choice in how to make the universe. He proposes that our four-dimensional universe may have no boundaries of space or time, even if space is finite (as with the surface of a sphere). Furthermore, a single theory of unification may explain all of physical reality. A universe without a beginning does not allow for a "creation by God" or for God's determination of the "initial conditions," and a single set of laws that can make the universe function does not allow for God's choice in how the universe operates (Hawking 1998, 190).

In further elaborations, both Trefil and Hawking modify their triumphal stances, conceding that confirmation of unique unified theory will still leave crucial questions unanswered. What caused the universe's coming into being? Who created the laws or the laws of logic that make them necessary? For Hawking, "What is it that breathes fire into the equations and makes a universe for them to describe?" (1998, 190). When asked whether he had dispensed with God, he replied, "All that my work has shown is that you don't have to say that the way the universe began was the personal whim of God. But you still have the question: Why does the universe bother to exist? If you like, you can define God to be the answer to that question" (Hawking 1993, 173). Seeking the answer to this most basic "why" question is a quest for all human beings together, not just scientists (Hawking 1998, 191). Trefil believes that scientists will overcome their unease at "encroaching" on religion's territory, just as they did with Darwin's theory of human evolution, and that there is ample room for both scientific and religious perspectives (Trefil 1983, 222–23).

Some cosmologists and science writers, notably Rees, Ellis, Ferris, and Thuan, counter triumphal thinking by contending that science cannot deliver metaphysical truths—it cannot answer basic cosmological questions underlying its own framework. According to Ellis, scientists who try are "knowingly or unknowingly transgressing the bounds of science . . . through wishful thinking, or perhaps through trying to use the prestige of science to claim certainty in buttressing some form of personal prejudice or strongly held philosophical opinion." An example is the assumption that physical laws were valid before the universe of space and time existed and brought the universe into being; this is a theological statement, as it implicitly recognizes "a transcendent reality." He feels that this "transgression" merits more serious attention than it has gotten and people need to recognize that neither science nor religion can offer "intellectual certainty about ultimate issues" (Matthews et al. 2002, 167–68). Rees finds scientists' "incursions into theology or philosophy . . . embarrassingly naive or dogmatic." They can contribute little to the science and religion dialogue, especially regarding the more "sophisticated [religious] world-views." Cosmology may have deep implications for religious thought, but it is not the job of scientists to elucidate them (Rees 1997, 6; Matthews et al. 2002, 22).

Ferris and Thuan both devote whole chapters of their accounts to discussions of God and cosmology and heartily agree that science has nothing to say about God. Ferris writes,

Cosmology presents us neither the face of God, nor the handwriting of God, nor such thoughts as may occupy the mind of God. This does not mean that God does not exist, or that he did not create the universe, or universes. It means that cosmology offers no resolution to such questions. (Ferris 1997, 304)

Both authors review traditional arguments for the existence of God and note how, at the hands of science, each fails. But this is not conclusive; Ferris

argues that atheism is likewise unsupported by cosmological science, and
for Thuan there is still doubt. For both, the choice of what to believe is
simply a matter of faith.

Science may, however, rule out certain conceptions of God. The "naive
creationism," which asserts a 6,000–10,000 year age for the Earth, is clearly
falsified by modern cosmology. Worldviews that integrate physical cosmology with spirituality, such as Ellis has proposed, are also clearly inconsistent with "religious traditions that are militant, monarchical, or tyran-
nical" (Matthews et al. 2002, 173).

Several authors speculate about the nature of a God who would be
compatible with modern cosmology. Trefil prefers "a God who is clever
enough to devise the laws of physics that make the existence of our marvelous universe inevitable (to) . . . the old-fashioned God who had to make
it all, laboriously, piece by piece" (Trefil 1983, 223). Ellis proposes a conception of God as "creator and transcendent and also as active in the universe
in an immanent kenotic (or letting go) way" (Matthews et al. 2002, 173).
Ferris envisions a God who fashioned an unpredictable universe with
"spontaneous creativity" and beings who are themselves creative. Such
a God works in silence, not revealing himself but allowing his creative
beings independence to pursue the truth. All seekers of truth, mystics and
scientists, must bear the ambiguity of not knowing final answers about
God or the universe, but they are united in having reverent faith, which
"respect(s) the eloquence of silence" (Ferris 1997, 312). Rees sets a similar
reverential tone in endorsing this comment of Silk: "Humility in the face of
the persistent great unknowns is the true philosophy that modern physics
has to offer" (Rees 1997, 6).

Epistemological Reflections

Closely woven with the metaphysical issues are the epistemological
claims about what science can know, how it generates knowledge, and
what kind of knowledge it obtains. Many authors implicitly assume and
some boldly assert that science, and only science, will come to know everything there is to know about the universe, and, furthermore, science
determines what is ultimately real. The logic is as follows. The cosmos
is all that exists, and cosmology, now an empirical science, will eventually answer all questions about it. Furthermore, what science learns is the
"real" picture of how things are, and science is the only avenue to such
knowledge.

This "naïve realism" and "ontological determinism," as philosophers
and theologians label them, are expressed in a variety of ways by several authors. Sagan opens his volume *Cosmos* with the ringing sentence,

"The Cosmos is all that is or ever was or ever shall be" (Sagan 1980, 4). For him, the reality of the cosmos is describable only by scientific inquiry. Pagels reflects, "Some day ... the physical origin and the dynamics of the entire universe will be as well understood as we now understand the stars. The existence of the universe will hold no more mystery for those who choose to understand it than the existence of the sun" (Pagels 1982, 376). If science doesn't find something, then it doesn't exist. Steven Weinberg applies this reasoning to the question of purpose in his famous statement: "The more the universe seems comprehensible, the more it also seems pointless" (Weinberg 1988, 154). In a more optimistic but equally triumphant vein, Hawking hopes that human reason will uncover the purpose of our existence and, thus, know the mind of God.

In the minds of some authors, science's ability to know is humanity's supreme gift and hope—the highest form of seeking truth. As Sagan expresses it, "I believe our future depends on how well we know this Cosmos" (Sagan 1980, 4). For Weinberg, "The effort to understand the universe is one of the very few things that lifts human life a little above the level of farce, and gives it some of the grace of tragedy" (1988, 155).

Others make no claims for science's almighty power and either bemoan or accept that ultimate knowledge of creation may be forever beyond the reach of science. Robert Jastrow describes frustration with this limitation in his famous mountain climbing analogy:

At this moment it seems as though science will never be able to raise the curtain on the mystery of creation. For the scientist who has lived by his faith in the power of reason, the story ends like a bad dream. He has scaled the mountains of ignorance; he is about to conquer the highest peak; as he pulls himself over the final rock, he is greeted by a band of theologians who have been sitting there for centuries. (Jastrow 1978, 116)

Others, as noted above, accept science's limitations and sometimes react strongly against triumphal epistemological claims. Ellis is very clear in his view that basic underlying cosmological questions are beyond science and that as time progresses, it becomes clearer what science can and cannot do. Gleiser echoes this view: "It is the question of the origin of the laws of physics that truly deals with 'the Beginning.' And the answer to this question is beyond the scope of physical theories, at least as they are formulated at present" (Gleiser 1997, 282). Others willingly admit that such issues may forever remain a mystery and a matter of personal belief, and that humility is the appropriate stance before the grandeur of the cosmos and in the face of our innate limitations.

Teleological Implications and the Anthropic Principle

Perhaps the most powerful aspect of cosmic evolution is its suggestion of a deep connection between the appearance of intelligent life and the development of the physical universe billions of years before such life appeared. In exploring this connection, various twentieth-century scientists became intrigued with how conditions in the very early universe had to be "just right" for intelligent human life to develop. For complex organic molecules to develop from simple ones, a planet needed to receive constant warmth from a stable star. The universe had to be very expansive in both time and space to allow for the evolution of stars and the creation of sufficient carbon and other heavy elements in several generations of stars. For such conditions to exist, the ratio of nature's force strengths and other physical laws had to be very finely tuned. As described earlier, all these factors were set within the first nanosecond of the universe. Had they been different even by minuscule amounts, life, much less intelligent life, would never have developed. The large number of factors that must be in perfect alignment supports the stance taken by some that the odds of intelligent life developing in the universe are virtually nil.

In the latter twentieth century, these "cosmic coincidences" gave rise to a grand new principle of the universe: the anthropic cosmological principle. Discussed in a large body of literature, the anthropic principle continues to be an intriguing and puzzling idea and a source of ongoing debate in both scientific and theological circles. Whether scientists like it or not, even those deeply opposed to its fundamental suggestion have to admit that it has brought teleological considerations of design and purpose front and center into the middle of scientific debate.

Early recognition of fine tuning came from Harvard biochemist Lawrence Henderson in his 1913 book *Fitness of the Environment*. There he showed that the special properties of hydrogen, carbon, and oxygen greatly facilitated the evolution of living organisms. In the 1930s a new cosmological perspective entered the picture with the work of Eddington and Dirac, who both explored certain large number coincidences, which were later understood anthropically. One example was that the size of the universe exceeds the size of the proton by the same factor that the electromagnetic force exceeds gravity. Believing that these ratios are more than coincidentally related and remain equal over time, Dirac argued that, as the universe grew, gravity would weaken. The amount of weakening, however, was not confirmed by observations. The dilemma was later resolved when Princeton physicist Robert Dicke noted that humans observe the universe when it is approximately the age of a typical star—an age when there has been sufficient time for stars to evolve but not all die out. Dicke calculated that both stellar masses and stellar lifetimes depended on the ratio of electromagnetism to gravity, and it is no surprise to find Dirac's ratios equal at our epoch.

Also in the 1950s, astrophysicist Fred Hoyle realized that the nucleosynthesis of life's key element, carbon, in aging red giant stars was a very unlikely process. That it could happen at all depended on several fortuitous coincidences. Fusion of the carbon nucleus is accomplished when three alpha particles, or helium nuclei, fuse, but it is highly unlikely for three to fuse at once. An intermediate stage of beryllium synthesis from two helium nuclei has to occur and last long enough for a third helium nucleus to join and form carbon. For this to work, the total energy of the colliding beryllium and helium nuclei has to match closely an energy (resonance) level of the carbon nucleus, which it does. For the carbon to persist, reactions fusing it and helium into oxygen have to be less likely, which they are. For Hoyle, it was as if someone had "fixed" these resonance levels to be just what was needed to produce the roughly equal amounts of carbon and oxygen needed for life.

Numerous other examples relating to element production exist. One crucial coincidence is the balance in atomic nuclei between electrical and nuclear forces. A weaker nuclear force would have prevented production of stable elements beyond hydrogen; stars would lack an energy source to sustain their long lives; and the carbon needed for Earth life would never have been produced. A stronger nuclear force would have fused hydrogen into helium much more quickly, shortening stellar lives and making the slow evolution of life impossible. Likewise, a stronger gravity would have produced smaller, faster burning suns and a smaller universe. Only a weak gravity allows for the development of a large, long-lasting, life-bearing universe. Another crucial coincidence involves the interaction rate of neutrinos. A slightly different rate would have prevented optimum helium production in the early universe and the ability of massive stars to explode and thus distribute the heavy elements they produced back to the universe. Further examples involve the number of dimensions, the amplitude of the density fluctuations in the early universe, the small size of the cosmological constant, the amount of dark matter that seeded the clustering of ordinary matter to form galaxies and stars, and the total number of particles able to form stars. Many scientists offer detailed enumeration of the many anthropic coincidences, among them Davies (1982), Greenstein (1988), Rees (1997), and Barr (2003).

In 1970, Australian physicist Brandon Carter stimulated widespread interest in the subject by detailing many examples of this fine-tuning and formally delineating the "anthropic principle." Challenging the traditional Copernican principle that humans do not occupy a privileged position in the universe, the anthropic principle asserts: "Our location in the Universe is necessarily privileged to the extent of being compatible with our existence as observers (Barrow and Tipler 1988, 1). Carter distinguished two forms: the "weak" and the "strong" anthropic principles, which were later explicitly defined by physicists Barrow and Tipler as follows:

Weak Anthropic Principle (WAP): The observed values of all physical and cosmological qualities are not equally probable but they take on values restricted by the requirement that there exist sites where carbon-based life can evolve and by the requirement that the universe be old enough for it to have already done so.

    Strong Anthropic Principle (SAP): The Universe must have those properties which allow life to develop within it at some stage in its history. (Barrow and Tipler 1988, 20)

Barrow and Tipler identify three distinct interpretations and one major extension of the SAP:

1. There exists one possible Universe 'designed' with the goal of generating and sustaining observers. (Barrow and Tipler 1988, 21)

This interpretation follows in the tradition of natural theology of past centuries, suggesting as it does that all the "coincidences" were, in fact, purposefully adjusted by a Designer. Barrow and Tipler point out that this "teleological version . . . does not appear to be open either to proof or disproof and is religious in nature. Indeed it is a view either implicit or explicit in most theologies" (1988, 22).

    Two additional interpretations of the SAP were derived from the consideration of quantum mechanics. Physicist John Wheeler offered the following version:

2. *Participatory Anthropic Principle (PAP):* Observers are necessary to bring the Universe into being. (Barrow and Tipler 1988, 22)

The "many-worlds" interpretation of quantum mechanics suggests yet another possibility:

3. An ensemble of other different universes is necessary for the existence of our Universe. (Barrow and Tipler 1988, 22)

Assuming the SAP is true, it appears to make no sense for life to evolve and die out quickly. Such considerations led to the following generalization of the SAP:

Final Anthropic Principle (FAP): Intelligent information-processing must come into existence in the Universe, and, once it comes into existence, it will never die out. (Barrow and Tipler 1988, 22–23)

The Anthropic Principle is one of the most controversial ideas in science in the last fifty years. Reactions from scientists range from distaste, hostility, and ridicule to strong interest and support. Opposition revolves around several themes: that the anthropic principle is nonscientific and trivial, that it avoids "real" explanations, and that it is too religious, controversial, and dangerous. Some of its staunchest former detractors, however, are now supporting the idea for its usefulness as an adjunct to currently favored multiple-domain and multiple-universe theories. Other supportive arguments focus on its potential to stimulate new scientific thinking, raise deeper questions about a fundamental mystery, and create a new post-Copernican paradigm that incorporates the possibility of a significant connection between life and the universe. A few tackle the implied teleological issue head on and explore the implications of a Grand Design and the nature of the Designer from a modern perspective.

Critics point out that the weak anthropic principle is an empty, trivial statement—a mere tautology—unworthy of further consideration. It simply draws attention to an obvious selection effect. Humans could not observe any other kind of universe than the one suited to produce us. The stronger form of the principle—that the universe must produce us—is simply not scientific. It radically departs from the time-honored process of science, which, four centuries ago, liberated itself from just this kind of teleological explanation of final causes or a grand design. It defies the normal logic of cause and effect by making a posteriori statements—suggesting that something that succeeded another event in some way caused it. The death knell for these critics is that the principle defies the scientific process by not being in any way predictive. It is not, therefore, falsifiable and cannot be determined to be right or wrong (Tyson and Goldsmith 2004; Thuan 1995; Silk 2005). Because it is not predictive, critics contend, the anthropic principle is sterile—"one of the greatest swindles in physics" (Silk 2005, 83). It has neither engendered scientific findings nor revealed any significant truths. Although it appears to be "saved" by theories invoking other universes, this resolution is unsatisfying and suggests a wasteful process. Paul Davies argues the point: " . . . it flies in the face of Occam's razor, by introducing vast (indeed infinite) complexity to explain the regularities of just one universe. I find this 'blunderbuss' approach to explaining the specialness of our universe scientifically questionable" (Davies 1992, 218–19). Others brand it purposeless and futile, since there is no way to interact with or validate such universes.

An additional argument against anthropic reasoning is that conventional physical explanations exist and will be found for all the coincidences. In the illustrative example described earlier Dicke eventually explained one of Dirac's large number coincidences by exact calculations of force ratios and stellar timescales. Likewise, the resonances in the carbon and oxygen nuclei found so remarkable by Hoyle in the 1950s have been explained in a natural way by theory. On a more fundamental level, if the search for a unified theory of everything succeeds, all the laws and values of constants would derive from physical principles with no consideration of our own existence.

Invoking anthropic reasons might also prevent or postpone investigation of the true causes of observed coincidences. Ferris cites an example where Stephen Hawking and Barry Collins offered an anthropic reason for the temperature invariance, or isotropy, of the cosmic background radiation: it was necessary for stars, planet, and life to develop. A few years later, however, the theory of inflation explained the isotropy in a "much more natural and elegant fashion." This illustrates how "scientists confuse *constraining* a phenomenon with *explaining* it. If they think they have explained it by showing it to be necessary to life, they may be discouraged from seeking a deeper and more productive explanation" (Ferris 1997, 300). Susskind labels this "surrender, a giving up of the noble quest for rational answers" (2006, 21).

There might be many "nonanthropic" reasons for the coincidences. Some might emerge when the origin of life and its requirements are better understood. At the present time, we may have too narrow a conception of what produces life. Some critics prefer to emphasize life's adaptation to the universe rather than the reverse. Perhaps when we more fully understand life's truly remarkable adaptive capacities, the picture will be clearer. Sagan voices all of the above objections in his charge that the anthropic principle represents a "failure of the imagination" (2006, 57). One segment holds firm to the belief that the coincidences are just what the word implies—chance occurrences that need no explanation.

A major objection to the anthropic principle is that it is simply too religious and metaphysical. Some would argue that even the weak form smacks of destiny and design, when it is used to explain how our local region in the multiverse is one of the few with conditions propitious for life (Silk 2005, 83). Others argue that the weak form is too conservative and bypasses the significant issue. The strong form, on the other hand, deals with the main issue but is unverifiable and highly controversial as science. Its only support in science is from one interpretation of quantum mechanics, but the ultimate foundation of this interpretation and of quantum mechanics itself is unverifiable and subject to debate (Ellis 1993a, 93.).

Some scientists react intensely to the anthropocentrism and religious implications of anthropic ideas and regard them as dangerous. Greenstein recounts his feelings as he became convinced that the coincidences could not be explained as chance phenomena:

It was an intense revulsion, and at times it was almost physical in nature. I would positively squirm with discomfort. The very thought that the fitness of the cosmos for life might be a mystery requiring solution struck me as ludicrous, absurd. I found it difficult to entertain the notion without grimacing in disgust, and well-nigh impossible to mention it to friends without apology. To admit to fellow scientists that I was interested in the problem felt like admitting to some shameful personal inadequacy. (Greenstein 1988, 25)

Scientists' apparent indifference to anthropic ideas only "masks an intense antagonism." The implied anthropocentrism strikes a deeply felt and negative chord, for the modern scientific method was born in an attempt to overthrow this perspective and free intellectual inquiry from the clutches of religion. The issue raises anew the unwelcome and fearful specter of Galileo and Bruno.

Leonard Susskind concurs. The anthropic principle has been, until recently, a concept "hated by most physicists... a goofy misguided idea... and a creation of inebriated cosmologists, drunk on their own mystical ideas... To some it smells of supernatural creation myths, religion, or intelligent design" (Susskind 2006, 14, 21). Herein lies a major problem for scientists. It brings the question of God once again front and center into the scientific debate, where such questions do not belong, and fuels the ongoing clash between scientific and religious factions. The anthropic principle and coincidences can much too easily be bent and co-opted by religious extremists to suit their own ends. Susskind cites several examples of how scientists' writings have been so misused (2006, 7–8).

Those reacting in a more lighthearted vein regard the more highly speculative forms of the anthropic principle with amusement and brand them ludicrous or absurd. Science writer Martin Gardner epitomizes this perspective in his presentation of the ladder of ever more speculative anthropic principles. He tops off the WAP, the SAP, the PAP, and the FAP with the CRAP—the Completely Ridiculous Anthropic Principle (Ferris 1997, 350n).

Despite these many objections, a number of scientists nevertheless support the anthropic principle. One large and growing contingent recognizes that something like the anthropic principle is necessary to explain our existence if, in fact, there are many separate universes within a larger "multiverse," as many current hypotheses suggest. An enormous variety of different laws of physics and values for constants might then exist, and we live in one of the very few regions or universes where conditions for life just happened to develop. For many scientists this approach has the happy advantage of challenging or at least avoiding distasteful design implications. It incorporates anthropic thinking into the traditional scientific framework of applying probabilistic reasoning and thus avoids finalism and the need to discuss a grand plan.

Many multiple universe scenarios now fill the landscape of cosmology. They are discussed more fully in Chapter 3 and reviewed briefly here. In chaotic inflation theory, an eternal and infinite "multiverse" produces inflating "pockets" or bubbles that grow into separate universes. Successive cycles might occur in the closed universe scenario, with each new cycle being reborn, phoenix-like, from the destructive collapse of the last. In more speculative hypotheses, universes may spawn new ones through black holes (Smolin 1997), or many parallel universes may float as "three-dimensional branes" in a higher dimensional realm (Steinhardt and Turok, 2007). Yet another version emerged from Everett's "many worlds" interpretation of quantum theory. Whenever quantum indeterminacy allows for different possible outcomes in a system, in fact they are all realized in separate, branching universes. Most recently, developments in string theory are suggesting that a staggeringly enormous "landscape" of possible environments or vacuums is possible. In all these cases, the basic vacuum parameters—the laws of physics, the constants of nature, the types of particles, etc.—would vary from one universe or possibility to the next. It is in no way remarkable that at least one of the inordinate configurations had the right characteristics to produce life and intelligence. The weak form of the anthropic principle becomes necessary to explain our existence. We just happen to live in one of the few that might have the correct parameters to evolve life and intelligence. Even the strong form is supported; with an inordinate number of possibilities, life must evolve somewhere, just by the laws of chance. While many criticize the uselessness of multiverse speculation, some assert that certain versions could, in fact, be falsified (Rees 2001, 177–78).

The strong anthropic principle also finds support in a tenet of positivist philosophy—that only what is observed is real. The same proposition emerges from one interpretation of quantum mechanics—that "the observation itself . . . brings the physical world into existence" (Greenstein 1988, 222). It seems to be a mind-boggling mystery how conscious beings arriving on the scene late in the universe's history could ever have produced, in any causal way, coincidences in its early epochs, but according to proponents, the idea may not be ruled out by modern theory. It forms the basis of the Participatory Anthropic Principle and Greenstein's proposal of a "symbiotic universe." Both redefine the "observed" and "observer" of traditional quantum mechanics. The "observed" need not be an atomic level phenomenon but could be larger structures or the whole universe, and the "observer" must be a conscious mind and not an instrument. As Greenstein describes it, "The very cosmos itself depends for its being on the uttermost mystery of consciousness. And thus the symbiosis, the union between the physical world and mind, the great metaphysical dance by which each brings into being the other" (Greenstein 1988, 238). While Greenstein is at pains to distance himself from anthropocentrism, he nevertheless embraces "anthropism," the suggestion of a deep relationship—an "immense, symbiotic unity" melding the universe, life, and mind.

This seemingly mystical explanation of anthropic coincidences is a clear step into metaphysics, but not the old style teleology of direct design by God. It is a metaphysics derived directly from quantum theory and centering on philosophical issues raised and addressed by it—questions of the ultimate nature of being and the relation of the mind to reality. The proposal is tentative, for sure, because a full understanding of the philosophical underpinnings of quantum mechanics continues to elude both scientists and philosophers.

The anthropic principle, supporters argue, can be valuable as a stimulus to new ideas. While it cannot replace rigorous testing of physical laws with hard data, it can guide intuition and steer scientists toward correct understanding. An example is the way Dicke used anthropic considerations to discover the underlying physical reason for a seemingly remarkable coincidence. Barrow and Tipler present many such examples in their book and argue that the principle is a fertile concept that "leads to synthesizing insights that deepen our appreciation of the unity of Nature" (1988, 17). Further, in examining the implications of the anthropic principle, scientists must ask what could be different in other universes and what the requirements for life really are and may thereby come to understand better the laws that govern our universe and the connection of those laws to the living world. Researching the apparent drive toward complexification in the cosmos may, according to Davies, yield new conceptions that "supplement the laws of physics" (Davies 1988, 203).

Building on this idea, exploration of anthropic coincidences may raise deeper and more fundamental questions and rightfully move the discussion closer to metaphysics. Even if physical law explains all the fine-tuning examples, the question remains as to why physical law itself is so perfectly adjusted for life. Likewise, if a single unified theory of everything, such as superstring theory, successfully determines all the laws and values of constants, one is forced to ask why the superstring framework is the only one that works to make both a universe and life. And if it's not, what does determine the kind of other frameworks that might produce universes and living beings? By raising such "why" questions, which are unanswerable within the domain of science, the anthropic principle stimulates dialogue and joint inquiry among scientists, philosophers, and theologians.

A stronger position favoring the anthropic principle holds that it in fact has real significance. The coincidences simply cannot be brushed aside. Fred Hoyle expressed this view after discovering the coincidences in carbon nucleosynthesis:

Is that another put-up, artificial job? . . . I am inclined to think so. A common sense interpretation of the facts suggests that a superintellect has monkeyed with physics, as well as with chemistry and biology, and that there are no blind forces worth speaking about in nature. The numbers one calculates from the facts seem to me so overwhelming as to put this conclusion almost beyond question. (Hoyle 1981, 12)

Many others agree that the coincidences reveal a significant connection between life and the universe that offers a new meaning to our existence. For Greenstein, the habitability of the universe points to a "great and profound mystery . . . of immense significance" (Greenstein, 21), and in Dyson's famous phrase, "The universe in some sense must have known we were coming" (Davies 1995, 128). Paul Davies expresses a similar idea at the conclusion of his book *Cosmic Blueprint*:

The very fact that the universe is creative and that the laws have permitted complex structure to emerge and develop to the point of consciousness . . . is for me powerful evidence that there is 'something going on' behind it all. The impression of design is overwhelming. Science may explain all the processes whereby the universe evolves its own destiny, but that still leaves room for there to be a meaning behind existence. (Davies 1988, 203)

Knowing that human existence is not an accident provides a "deep and satisfying basis for human dignity" (Davies 1988, 203).

This position faces squarely the teleological implications of the anthropic principle and admits there might be a plan. After all, it is no more scientific to deny the existence of a plan than to allow for its possibility. While the notion of a Grand Design necessarily evokes the question of God, which is inappropriate to discuss in science proper, science itself has now embarked on a quest to probe mysterious and invisible realms formerly contemplated only by theologians and philosophers. Once again, as with metaphysical questions of the very beginning and the "why" questions discussed above, the scientific quest itself creates an unavoidable overlap and a strong need for investigation by both sides and dialogue between them.

A number of scientists make the point that whatever its merits, the anthropic principle represents a welcome shift away from the extremes of strict materialism and Copernicanism. With anthropic coincidences still unsolved, materialists can no longer claim that it is totally out of the question for the universe to have been designed with life and intelligence in mind. The idea that humans may in fact be the focal point of the universe's existence is an abrupt shift away from a five-century-long trend that rendered humanity's physical position in the cosmos ever more insignificant. Recent multiverse concepts represent an even more extraordinary leap in the Copernican direction, but now they themselves are finding it necessary to incorporate the anthropic principle in order to account for our existence. Thus, the pendulum may finally be swinging back and humans regaining a central status.

Exactly what might replace the Copernican perspective is not yet clear, but a few scientists are forging new paths that accept the significance of fine tuning and explore ways to explain it beyond purely material terms. They neither reject design outright nor embrace the old "watchmaker" version, which has been out of vogue in scientific and theological circles ever since the time of Darwin. Intimations of a new paradigm appear in many scientists' reflections, including Davies, Young, Greenstein, and Ellis.

First and foremost, the old notion of a Designer who directly creates the specific structures has given way to a Designer of laws or processes. The design is not a rigid plan that forces structures to develop in a predetermined way but is a general "pattern of development" that "predisposes" constants and laws that are amenable to life. It is a creative process in which the universe organizes its own self-awareness, as if responding to some drive toward complexity and consciousness (Davies 1988, 203). Louise Young expresses it well:

As we view the groping, exploratory nature of the process . . . it is apparent that we are not witnessing the detailed accomplishment of a preconceived plan. "Nature is more and better than a plan in course of realization," Henri Bergson observed. "A plan is a term assigned to a labor: it closes the future whose form it indicates. Before the evolution of life, on the contrary, the portals of the future remain wide open". . . . The process of transformation has all the earmarks of a great creative work in progress, bringing into being something that has never existed before. (Young 1986, 201–2).

The creative processes suggested by Davies and Young may also relate to the symbiotic interplay of cosmos and mind envisioned by Greenstein and the Participatory Anthropic Principle. Young's analogy of artistic creation highlights similar elements. Creation of art involves both an urge or drive and a guiding idea, but not at first the choice of details. In a similar way, quantum processes involve sets of probabilities, like the ideas swirling in an artist's head, but no predictability of the exact outcome. In all these conceptions the developmental process involves randomness and uncertainty but also the gradual unfolding of a whole—in an overarching framework that partakes of both chance and subtle design.

For some, the achievement of satisfactory answers to anthropic questions requires a new framework of inquiry beyond the normal scientific method. A possible broader approach—an "expanded scientific process"—is suggested by Ellis. It would have its own logical principles but not the usual scientific requirement of strict verification with hard physical data. Science alone, he argues, can never deliver answers to anthropic issues, because its process of inquiry is limited to studying the observable properties of physical objects. Anthropic issues, however, involve metaphysical questions that lie beyond this domain—why anything exists at all and what underlies physical laws and determines their nature or efficacy. These deeper questions are at the root of science but beyond its competency to answer.

The wider cosmology Ellis proposes bravely tackles the broader question of why the cosmos is so perfectly tuned to produce not just our physical existence but the full depth of our being, including the basis for ethics and meaning (Ellis 1993a, 125). Science should attempt to "come to terms" with such phenomena, Ellis argues, "in a greater synthesis, stepping out of its rigid frame of logic and proof, and the unattainable desire for a world view that is strictly verifiable" (Ellis 1993a, 102). Using an expanded inquiry method, the wider cosmology must explore how the universal human quest for truth, meaning, beauty, and morality is incorporated into the very design of the evolutionary framework. It must accept as evidence the findings of genuine religious figures through the ages—that a transcendent level of reality exists and that voluntary self-sacrifice and true love for others is of supreme importance.

Using this perspective to analyze the anthropic principle and explain the anthropic coincidences, Ellis first outlines physical properties the universe must have to permit moral and spiritual development: intelligible order from physical laws, conditions for developing beings with free will, and the apparent absence of verifiable divine action so that free response is possible. He then suggests that of all the explanations for fine tuning, only the Design concept, conceived in broad religious terms, can give a satisfactory answer to why the universe is conducive not only to our physical existence but also to the full range of our spiritual qualities. He frames the controversial answer as follows: "The Universe exists in order that humankind (or at least ethically aware self-conscious beings) can exist . . . this is done so that unselfishness and love may make [themselves] manifest, an obviously and patently worthwhile purpose" (1993a, 127).

The Epic of Evolution: Mythological Dimensions

In the latter half of the century a variety of scientists sought to elevate cosmic evolution into a grand mythology for our time—this epoch's creation myth. While some merely suggested it or called for it, others developed a full-blown "story" replete with poetic renaming of events in cosmic history and guidance for how to live in a way that honors our deep relationship with the whole.

Suggestions that such a new cosmology is needed have come from a great variety of figures. Notable among them is Joel Primack, a physicist deeply interested in the relationship of cosmology and culture. Noting that the cosmological world views of the past were inextricably tied to the religious beliefs of their era, he believes this will happen again. The new cosmology of today will likewise inspire a new kind of spirituality. As expressed by Primack,

We may see in the first decades of the 21st century the emergence of a new universe picture that can be globally acceptable, and with this and the contributions of image-making writers, artists, and spiritual visionaries, it is possible that the painful centuries-long hiatus in human connection with the universe will end. (Primack 1997,12)

Expanding on this theme is the work of Brian Swimme and Thomas Berry—a mathematical cosmologist and cultural historian. With ideas first developed by Swimme in his *Hidden Heart of the Cosmos* and later by him and Berry in *The Universe Story*, a call is put forth to engage in a personal way with the magnitude and splendor of our new story of creation. They focus on the deep significance of the "story" and proceed to recount the rich and sweeping narrative of our origins from the "primordial flaring forth" through the evolution of galaxies and stars to the coming of humans. They trace the history of civilizations up to the present "Ecozoic Era," as they call it, where we face an ecological and cultural imperative. We must recognize our interdependence with all other creatures and with the universe and Earth that have grown us, celebrate reverently the knowledge that has brought this new awareness, and deepen our commitment to our new story of origin and our role within it. While there is almost no mention of God, a deep reverence for the natural world and processes of origin pervade the work.

A similar spirit infuses a major work by cell biologist Ursula Goodenough, *The Sacred Depths of Nature*. Goodenough focuses more than anything on the mystery of existence and the epic of evolution, which she recounts from the origin of the universe through the origin of Earth and then of life. Describing everything in very materialistic terms, she nevertheless professes a deep faith, not in a deity but in the very sacredness of the living world itself. For her, ultimate meaning and value derive from deep appreciation of the beauty and complexity of nature and her ability to comprehend it. Her nature spirituality incorporates a commitment to life's continuation above all else.

Chapter 5

———

Life and Intelligence in the Universe

No aspect of cosmic evolution has raised a broader range of scientific, philosophical, and religious issues than the possible existence of life and intelligence on other worlds. And none has posed a greater challenge to scientific theory and observation. We have only one sample of life and intelligence to study, no accepted theory of life's origin, and enormous challenges to observation and communication with other possible inhabited worlds. For all these reasons, scientists have found it impossible so far to assess empirically or agree theoretically about the probability or characteristics of life elsewhere. For the last century a lively debate persisted about whether such life exists, what the philosophical and religious implications would be, and whether searching is worthwhile. Belief and support for search projects waxed and waned but grew stronger in the second half of the century when a handful of passionate proponents established an ongoing search program, arguing that the benefit to humanity could be momentous, both materially and spiritually. The search itself, some claim, has religious overtones, and success would have immense implications for humanity's self-image and for religious doctrines.

TWENTIETH-CENTURY DEVELOPMENTS

As the century opened, habitable planets were thought by astronomers to be plentiful, and public interest was keen in response to the alleged observation of Martian "canals" by astronomers Giovanni Schiaparelli and Percival Lowell, who saw them as evidence of a superior, peaceful, intelligent race (Basalla 2006, 66, 86–87). A decade later, more accurate observations dispelled the notion, but fictional depictions of Martian life continued to fuel public fascination and deeply impress a generation of future space scientists (Basalla 2006, 9). During the next two decades, however, skepticism grew among scientists, not only as a backlash against the Martian canal fervor but also because of a shift in the dominant view about the solar system's origin and the possible abundance of planets. In this period, a new "tidal" or close encounter theory gained favor over the earlier nebular hypothesis. As formulated by English physicist and astronomer Jeans, this theory suggested that a passing star could have produced tidal distortions in the sun, causing the ejection of a long stream of hot gas that condensed into planets. Calculating that perhaps only one encounter would occur in 30 billion years, Jeans argued that solar systems and life must, indeed, be very rare (Dick 1998, 74, 76). Jeans' colleague Eddington echoed this assessment, asserting that nature may have a single purpose—to create million of stars so that just one might "provide a home for her greatest experiment, Man" (Dick 1998, 78–79; Basalla 2006, 127).

Mid-century, the tide dramatically shifted again with new developments in theory, observation, and experimentation in both physical and biological sciences. Criticisms of the tidal theory on dynamical grounds caused its demise in the late 1930s, and in the early 1940s a modified nebular hypothesis gained new acceptance, reviving the idea that planets form as a natural accompaniment to star birth. Around the same time, two independent research projects claimed to have detected planet-sized bodies around two nearby stars, further suggesting that planets might be abundant. Additional support came from the new cosmology, which presented an expanding universe of tens of billions of large galaxies, suggesting that even if planetary systems were relatively rare in our galaxy, their numbers in the universe were still staggering. Finally, studies of the biochemical origin of life also fostered support for its extraterrestrial existence. The experiment by Urey and Miller showed that in a hydrogen-rich and oxygen free "reducing" atmosphere amino acids can be readily synthesized from simple inorganic compounds believed abundant on the primitive Earth and elsewhere in the universe. This lent support to the Oparin-Haldane hypothesis of life's very natural origin from a primordial broth of prebiotic chemicals and strongly implied that life would easily develop elsewhere as well. Even though such processes were assumed to occur by chance interactions, the enormous time available, then estimated at 2 billion years, made life's development inevitable. Biologist George Wald summarized the standard conclusion: "Wherever life is possible, given time, it should arise" (Basalla 2006, 130). In his estimate 100,000 Earth-like planets exist in our Galaxy, each one a potential host for living things.

These developments gave rise to what has become the dominant modern conception of the place of life and humanity in the universe, held to this day by the great majority of physical scientists. It strongly reconfirmed the Copernican paradigm, now enlarging its scope to incorporate a much vaster universe and the realm of living beings. It also presented a clear challenge to any religious doctrine that regarded the Earth or its life as unique or special. Astronomer Harlow Shapley, one of the architects and heralds of the new vision, gave a poetic summation:

[Earth and its life are] on the outer fringe of one galaxy in a universe of millions of galaxies. Man becomes peripheral among the billions of stars in his own Milky Way; and according to the revelations of paleontology and geochemistry he is also exposed as a recent, and perhaps an ephemeral manifestation in the unrolling of cosmic time. (Dick 1998, 86)

He further elaborated its implications for extraterrestrial intelligence: "We are a little vain or anthropocentric if we consider ourselves the center of life and the highest being in the universe." With such a high probability that other intelligent beings inhabit the cosmos and share our aspirations, he remarked, "to deny them this privilege of having philosophers talking about the universe is not fair" (Witham 2003, 11).

Shapley saw deep implications for religion. Once critical of the traditional religious biases of some scientists, he later found "modernist religious people" to be valuable partners in world political causes he cherished, and he also sought to share with them his deep convictions about humanity's non-privileged place in the modern universe. In 1954 he helped establish a forum dedicated to a "rational, almost scientific, religion"—the Institute on Religion in an Age of Science (IRAS)—where science could "strengthen religion and not upset it." The dialogue was a valuable "confrontation of religion and science," although he felt that scientists fared much better and that "theology was in a bad way" (Witham 2003, 10).

Another scientist who explored theological implications of extraterrestrial life at mid-century was the British astrophysicist and cosmologist E. A. Milne. In a lecture on modern cosmology and the Christian idea of God, he asked, "Is it irreverent to suggest that an infinite God could scarcely find the opportunities to enjoy Himself, to exercise His godhead, if a single planet were the sole seat of His activities?" Furthermore, he argued, "there is nothing to prevent each [one] being the seat of a unique process of biological evolution." For Milne, God did not act all at once to "wind up the world and leave it to itself" but continually guided the evolutionary process in an infinite number of instances in an infinite number of locations. "That is of the essence of Christianity, that God actually intervenes in History" (1952, 152–54).

The early 1960s to the mid-1970s was an expansionary period in the search for extraterrestrial life and intelligence. Microbe testing of Martian soil was planned, radio searches and transmissions initiated, and meetings on the topic held, now attended by origin of life specialists as well as physicists and astronomers. The interdisciplinary field of "exobiology"—now termed "astrobiology"—was born. Several scientists emerged as chief proponents and spokesmen—physicist Philip Morrison and biologically inclined astronomer Carl Sagan, who both coauthored seminal publications, and astronomer Frank Drake, who was the first to conduct a radio search and developed the famous "Drake Equation." This formula estimated the number of radio-communicating civilizations in the Galaxy by multiplying together several astronomical, biological, and social factors, such as the number of habitable planets, the number that actually support life, the fraction that develop radio communicating intelligence, etc. Though often criticized for being too speculative and imprecise, the equation has remained useful as a concise summary of the important issues in the field and as a focal point for discussion, often elaborated by additional factors. The results, as estimated by the first researchers, ranged between one thousand and one billion, with Sagan later settling on and typically using a figure of one million. Such large estimates greatly encouraged the advocates for contact, and at the close of this period, Drake sent his famous interstellar message, a binary code pictorial representation of the solar system, the human form, and the four nucleotides of DNA (Basalla 2006, chapter 8).

As at all periods, however, the enthusiasm was not universal, and during this time increasing doubts were expressed by some scientists about the probability of the first living forms emerging by chance and of anything like human intelligence evolving elsewhere. A few mathematicians argued that chance interactions could never have led to complex life forms, and the chemist Michael Polanyi claimed that the DNA molecule with all of its information content could not have been synthesized by random processes (Witham 2003, 100). Additional challenges came from the discovery of 3.8-billion-year-old fossilized microorganisms and the emerging possibility of a non-hydrogen rich early atmosphere, suggesting that there had been very much less time—perhaps only 100 million years or less—and possibly much less hospitable conditions for ready synthesis. The "chance plus time" argument for life's inevitability was beginning to be challenged. Evolutionists such as George Gaylord Simpson and Theodore Dobzhansky argued that human-like species would never evolve twice, given the unpredictable, inventive nature of evolution. Even if other intelligent species had evolved, they would be so different, chances of communication were nil (Dick 1998, 194–95; Basalla 2006, 178–81).

Another development of this period was the revival of theories of an extraterrestrial origin of life, fueled by the problems with earthbound origin theories discussed above and also by new space discoveries. Improved techniques enabled detection of amino acids in meteorites and simpler organics in comets and molecular clouds. Some even claimed to find the remnants of algae-like organisms in meteorites, although the finding was widely disputed. These developments sparked new interest in "panspermia," an idea first proposed in the early twentieth century by Swedish chemist and physicist Svante Arrhenius, who hypothesized that the pressure of stellar radiation had caused life's "seeds" to drift from one world to another. The new theories mostly assumed that only complex organics, and not life itself, had an extraterrestrial origin, although there were the following notable exceptions. Studies by Hoyle and Chandra Wickramasinghe led them to believe that amino acids and nucleotides not only formed in interstellar dust clouds but also then developed into living forms on asteroids and comets and came to Earth during the period of early bombardment by those bodies. Past epidemics and periods of rapid proliferation of new life forms, they later argued, might have been caused by bacteria or viruses or new genes from space. Nobelist Crick and biologist Leslie Orgel advanced as a "logical possibility" the even more radical idea of "directed panspermia"—that extraterrestrials had seeded Earth with life via spaceships. While such wildly speculative ideas never widely caught on, the idea that prebiotic chemicals might have an astronomical origin is still seriously considered today (Basalla 2006, 131–32; Dick 1998, 170, 179–86).

During the last thirty years, the search for extraterrestrial life and intelligence has experienced major ups and downs in support from both the scientific community and from the government. The first significant crisis occurred in the late 1970s with the failure of the Viking mission to detect microbes on Mars, continued problems with origin of life theories, and a mounting disbelief in the existence of extraterrestrials. Challenge came from astronomers who raised anew Enrico Fermi's famous question: "Where are the extraterrestrials?" Since our solar system is so young, the argument went, older civilizations must surely exist, and since interstellar travel is by no means impossible, where are they? Believing that empirical results were the surest way to silence critics, proponents persisted in developing and securing NASA funding for a radio search program called SETI (Search for Extraterrestrial Intelligence) and sustained it through private funds when they lost NASA support for a decade (Basalla 2006, chapter 9; Dick 1998, chapter 7).

From their different perspective, evolutionists have continued to exert an influence on the program, agreeing among themselves about the non-prevalence of human-like aliens but disagreeing as to whether some kind of non-humanlike intelligence might exist beyond the Earth. Paleontologist Stephen Jay Gould argued for the possible evolution of intelligence on other worlds as a phenomenon of "convergence," whereby numerous pathways can lead to the same function, as with eyes for vision on Earth.

Accepting that there was some possibility of intelligence elsewhere, he joined four other evolutionists in 1982 signing a pro-SETI petition. In contrast, biologist Ernst Mayr argued strongly against the possibility, noting the long time it took for intelligence to develop here and the inapplicability of convergence, since intelligence only developed once on Earth (Basalla 2006, 178–85; Dick 1998, 196–97).

Most recently, two opposite trends have emerged. Tantalizing new discoveries have increased the hope of finding life beyond Earth at the same time that new hypotheses from physical science have suggested that Earth may be unique as a bearer of complex life and intelligence. One discovery was the detection of numerous planet-sized bodies circling other stars, and another was the possible finding of fossilized bacteria in a Martian meteorite found in Antarctica. While the latter is now questioned, the success of the former has continued unabated with over 200 extrasolar planets detected to date. Although none have Earth's mass or distance from its sun, present techniques could not find such planets. Hope for finding life in the solar system also increased with the discovery of liquid environments on the Jovian satellites Europa and Titan. Continued study of "extremophiles"—microorganisms that can live in such inhospitable conditions as extreme heat, cold, salinity, or acidity—was further evidence that life beyond Earth might be more plentiful than previously thought. Encouraged by all these developments and in response to continued public interest, NASA created the Astrobiology Institute in the late 1990s and in 2001 reinstated SETI scientists as one of its teams (Basalla 2006, 172).

In contrast to these hopeful signs were new hypotheses from a few physical scientists that Earth's complex life and intelligence might be very rare. In their book *Rare Earth*, geologist Peter Ward and astronomer Donald Brownlee argued that while simple life developed quickly on Earth and is probably common in the universe, complex life and intelligence are not. This is due to the many rare astronomical and geological conditions required to maintain Earth's stability over the multi-billion year span of time it took complex life and intelligence to develop. Such conditions, they assumed, were the result of chance. In *The Privileged Planet* astronomer Guillermo Gonzalez and philosopher Jay Richards also analyzed these rare conditions leading to intelligence and further argued that they have also made the Earth a supremely fit location for observing the universe. The simultaneous occurrence of multiple conditions required for both habitability and measurability, they claim, is so improbable that they could never have occurred as a chance fluke of cosmic evolution. While the thrust of their argument is to challenge Copernicanism and make design arguments, and they do not rule out the possibility of alien intelligence, the implication is that it is very unlikely.

At the present time, then, not a single confirmed discovery of life beyond Earth has been made, multiple ideas about life's origin are being pursued, and views of life's prevalence range from plentiful to singularly rare. Nevertheless, the last century saw notable progress in the field as physical and biological scientists established interdisciplinary programs to search for both life and intelligence elsewhere and conduct origin of life studies. The notion has also become deeply entrenched in the public imagination with many polls showing that half or more of the general public believes in the existence of some form of extraterrestrial life (Dick 2000, 195). Among origins of life researchers, astronomers, astrobiologists, and no doubt many other scientists, there is widespread acceptance, although certainly not consensus, that life beyond Earth is a natural outgrowth of cosmic evolution, with biological evolution inevitably following chemical evolution in many places in the universe. This acceptance is based, more than anything, on philosophical assumptions.

PHILOSOPHICAL AND RELIGIOUS IMPLICATIONS SEEN BY SCIENTISTS

As with the anthropic principle and all of cosmic evolution, the possibility of life and intelligence throughout the cosmos raises philosophical and religious issues that underlie all questions of origin and the relation of humanity to the universe and to God. These issues are the role of chance, necessity, and design in originating life and evolving it to consciousness and the cosmic significance and spiritual status of the human being. Related to the latter are issues of a more religious nature—the effect that an abundance or absence of other intelligence-bearing worlds would have on religious doctrines and concepts of God and the notion that searching for extraterrestrial intelligence may itself be a form of religious quest.

CHANCE, NECESSITY, AND DESIGN IN LIFE'S ORIGIN

Scientists today, for the most part, no longer debate whether life had a natural versus miraculous origin, at least in public discussions. Mainstream discourse now revolves around the more philosophical notions of chance and necessity. Although all scientists working in the field acknowledge the role of both chance and necessity in life's origin and evolution, many emphasize the dominance of one over the other. Others argue for a weak determinism, in which supplementary laws of self-organization and rules of information theory now being developed would play a major role. In almost all of these philosophical positions, life and intelligence might be common, the exception being chance causation by a rare contingent event. A few, even when speaking as scientists, argue for design—either by a traditional God or some unspecified "superintellect." The implication of this position for life's existence beyond Earth is indeterminate.

Life by Chance

The position that life originated as a rare chance occurrence has been held by many physical scientists and biologists over the last century but was most famously argued by French Nobel biologist Jacques Monod in his book *Chance and Necessity* (1972). He wrote:

[Life was] the product of an enormous lottery presided over by natural selection, blindly picking the rare winners from among numbers drawn at utter random... The universe was not pregnant with life, nor the biosphere with man. Our number came up in the Monte Carlo game. (Dick 1996, 382)

He estimated that life's chance of originating "was virtually zero"—a view supported by various researchers' calculations of the time required for random molecular shuffling to produce simple proteins, RNA or a bacterium. These estimates exceeded the time available in Earth's history by factors of several hundred to several hundreds of millions—staggeringly improbable numbers that bolstered Monod's idea that life's cause was a freakish accident. Not all researchers doing these calculations drew the same conclusion. French scientist Lecomte du Nouy and astronomer Hoyle invoked design by God or a superintellect, and physicist Harold Morowitz and biochemist Robert Shapiro were led to explore different mechanisms for molecular biogenesis (Dick 1998, 187–89). Nevertheless, a large number of life scientists came to agree that life's origin was driven by chance and was dauntingly improbable—at least if caused by random molecular assembly. Mayr summarizes well this view and its clear implications for extraterrestrial life:

A full realization of the near impossibility of an origin of life brings home the point how improbable this event was. This is why so many biologists believe that the origin of life was a unique event. The chance that this improbable phenomenon could have occurred several times is exceedingly small, no matter how many millions of planets in the universe. (Dick 1998, 169)

The "life by chance" hypothesis posits further that even if simple life could readily form, the likelihood of its evolution into human-like intelligence, or perhaps any intelligence at all, is virtually nil. Humans are the product of a staggering number of distinct variations brought about by chance and selected by whatever environment an organism happens to inhabit. As expressed at the dawn of the last century by evolution's co-founder, Alfred Russell Wallace, "The total chances against the evolution of man, or an equivalent moral and intellectual being, in any other planet, through the known laws of evolution, will be represented by a hundred million of millions to one" (Dick 1998, 193). Two decades later paleontologist W. D. Matthew agreed with Wallace's assessment and his view that such estimates almost surely excluded extraterrestrial intelligence. Even if it did arise elsewhere by a fluke, such intelligence would most assuredly neither resemble ours nor be accessible to us. Mid-century anthropologist Loren Eiseley chimed in, "Of men elsewhere, and beyond, there will be none forever... Every creature alive is the product of a unique history. The statistical probability of its precise reduplication on another planet is so small as to be meaningless" (Dick 1998, 194). In the 1960s and 1970s Simpson and Dobzhansky both emphasized the extreme unlikelihood and nonrepeatability of humanity's emergence, while Mayr argued that any kind of intelligence was unlikely. Gould later joined the chorus, calling evolution

... a staggeringly improbable series of events, sensible enough in retrospect and subject to rigorous explanation, but utterly unpredictable and quite unrepeatable. Wind back the tape of life to the early days of the Burgess Shale; let it play again from an identical starting point, and the chance becomes vanishingly small that anything like human intelligence would grace the reply. (Dick 1996, 394)

A corollary of the chance hypothesis is that the process of evolution can in no way be interpreted as progressive or advancing. Because each individual change happens purely by "chance caught on the wing," as Monod put it, the fact that more complex and even conscious beings developed from simpler ones is itself pure accident. Gould was especially forceful on this point: "Progress is a noxious, culturally embedded, untestable, nonoperational, intractable idea that must be replaced if we wish to understand the patterns of history" (Davies 1995, 76).

Life by Necessity

In sharp contrast to the chance hypothesis is the position that life has developed as a necessary outcome of the working of known physical laws, wherever conditions were suitable. Life and mind were not a fluke; they were built into the fabric of the universe. Chance still plays a part, but the odds are overwhelmingly in favor of life's formation. An early twentieth century argument for necessity came from biochemist Leonard Troland, who proposed an enzyme theory for life's genesis:

The striking fact that the enzyme theory... necessitates the production of only a single molecule of the original catalyst, renders the objection of improbability almost absurd... and when one of these enzymes first appeared, bare of all body, in the aboriginal seas it followed as a consequence of its characteristic regulative nature that the phenomenon of life came too. (Dick 1996, 380)

Around mid-century British Astronomer Royal Sir Harold Spencer Jones and organic chemist Melvin Calvin, an early exobiologist, both argued for life's inevitable appearance, given the proper conditions, Calvin going so far as to say that estimating the probability of cell life in the universe required only a knowledge of the number of planets with Earth-like conditions. The codeveloper of the "primordial broth" hypothesis for life's origin, Russian biochemist A. I. Oparin, concluded in 1975:

There is every reason now to see in the origin of life not a "happy accident" but a completely regular phenomenon, an inherent component of the total evolutionary development of our planet. The search for life beyond Earth is thus only a part of the more general question which confronts science, of the origin of life in the universe. (Dick 1998, 269)

In recent years Belgian biochemist Christian de Duve has been a strong voice for biological determinism. To say that the universe was not pregnant with life, he argues, is to invoke miracles, but modern biology demonstrates that divine intervention was not required. Here he is noting the philosophical similarity of origin by chance and origin by divine act—they are both contingent events—and rejecting them both. For him,

Life and mind most likely developed through purely natural events rendered possible by the prevailing physical-chemical conditions or perhaps even imposed by these conditions... The "pregnancy" that was erroneously negated by Monod is in fact the outcome of very special features built into the natural structure of the universe. (de Duve 2000, 11–12)

Rejecting the "gospel of contingency" by which humanity is a "meaningless outcome of chance events in a pointless universe," de Duve sees our emergence as a "watershed" rich in significance. A species has now developed that is beginning to discern the "reality behind the appearances"—not only the nature and history of the physical universe, but also such abstractions as "truth, beauty, goodness and love... the closest we can get with our feeble means to the ultimate reality to which many give the name of God." Superior beings of the future will no doubt have clearer vision of these realities, but for him "the glimpses we are afforded already are immensely rewarding" (de Duve 2000, 12–13). In his view the universe, through us, is growing toward greater consciousness of an ultimate spiritual reality. Although superior aliens may exist today and the possibility should certainly be included in our cosmology, according to de Duve they in no way detract from the meaningfulness of what has happened here.

Another argument for the significance of emerging consciousness comes from Davies. In his book *Are We Alone: Philosophical Implications of the Discovery of Extraterrestrial Life*, he points to the remarkable fact that the organ of greatest complexity in the universe, the human brain, is capable of grasping the laws that govern the simplest level of matter. This "deep and still mysterious . . . cosmic connection . . . between minds that can do mathematics and the underlying laws of nature that produce them" surely suggests how fundamental and prevalent consciousness must be. To find extraterrestrial intelligence would be a confirmation of this essential role of mind in the cosmos and of the progressive nature of the evolutionary process (Davies 1995, 127–29).

The idea of evolutionary advancement is a common feature of deterministic thinking. In contrast to the firm denial of directionality by most evolutionists, a few biologists argue that the development from simple to complex to conscious life is clearly a "ladder of progress." Christian de Duve holds that while individual variations are random, there is nevertheless a general, inherently predetermined direction toward greater complexity and eventual consciousness, an idea supported by the occurrence of evolutionary convergence, according to British paleobiologist Simon Conway Morris. In convergence, independent genetic pathways invent similar functions in widely different creatures, such as wings for flight or eyes for vision, or animals filling similar niches in different locales, such as placental and marsupial mammals. In the same way, Conway Morris argues, wherever life arises, a humanlike niche will eventually evolve, and on other planets intelligent beings would even resemble us in appearance (Davies 2003).

Philosophically, exobiologists, most of them biochemists or physical scientists, subscribe to a version of necessity. Life is inevitable not so much because it is intrinsic to physical laws but rather because of the enormous temporal and spatial range where pure chance can operate. Astronomer and science historian Dick summarizes their position: "Given enough time or space, or a simple enough entity, or the need for only a single first molecule, exobiologists could argue that an event governed by chance was transformed into necessity when the laboratory was the immense, and immensely old, universe" (1998, 187). Over the past half century they have persisted in this view, extending it to include the inevitable evolution of intelligence. Nearly all exobiologists would probably agree with Sagan's assessment:

Once life has started in a relatively benign environment and billions of years of evolutionary time are available, the expectation of many of us is that intelligent beings would develop. The evolutionary path would, of course, be different from that taken on Earth. The precise sequence of events that have taken place here . . . have probably not occurred in precisely the same way anywhere else in the entire universe. But there should be many functionally equivalent pathways to a similar end result. (Dick 1996, 397)

Some physicists and exobiologists, however, went further than Sagan, arguing that even human-like intelligence would necessarily arise. Such prominent physicists as Nobel laureates Glashow and Weinberg, as well as astronomer Drake, all asserted that "humanoids" on other worlds would have "alien mathematics and science . . . essentially like ours," and Drake speculated that even in their physical form they "won't be too much different from us" (Basalla 2006, 176–177, 198).

Commentators have noted diverse philosophical implications in the positions of exobiologists, among them anthropomorphism, adherence to the universality of physical law, and Copernicanism applied in contradictory ways. Historian Basalla (2006, 199) sees belief in the necessary development of human intelligence as clear anthropomorphic provincialism—even labeling Sagan an "electronic chauvinist" for assuming radio communicating ability in aliens. Dick sees it as stemming from a normal assumption physical scientists hold—that all laws, including biological ones, are universal and bound to produce the same result everywhere. In fact, Dick identifies the quest for a universal biology as a major part of all extraterrestrial life research—even though it is not a goal shared by most of the biological community (1998, 192). For Gonzalez and Richards, exobiologists' mixture of necessity and chance amounts to a philosophical contradiction, an "uneasy amalgam," well illustrated by a statement of astrobiologist and geologist Bruce Jakosky:

Finding non-terrestrial life would be the final act in the change in our view of how life on earth fits into the larger perspective of the universe. We would have to realize that life on earth was not a special occurrence, that the universe and all of the events within it were natural consequences of physical and chemical laws, and that humans are the result of a long series of random events. (Gonzalez and Richards 2004, 411n)

Thus, two seemingly opposite ideas—that life is natural and common but is still a chance occurrence—are both drawn from the Copernican assumption that Earth life is in no way special. For Gonzalez and Richards, both design advocates, this contradiction simply argues against Copernicanism. In their defense, exobiologists would probably acknowledge the difficulty of philosophical resolution and urge action on the only path promising sure solution—empiricism.

Life by Self-Organization—Weak Determinism

A middle position between chance and determinism, called by some "weak determinism," has arisen in recent years from theories of self-organization and information science. It posits that life arises with a high probability, not because it is "written into the laws of physics," but because matter and energy have a natural inclination to "self-organize and self-complexify," according to principles that are supplemental to but consistent with ordinary physical law (Davies 2000, 18). Study of chaotic systems has demonstrated that self-organization can occur when new energy input forces a system to the "edge of chaos" and away from equilibrium, and it then reacts by unexpectedly and quickly developing greater complexity and order. At the chemical level, certain kinds of self-organizing, interconnecting cycles, called "hypercycles" were found by chemist Manfred Eigen to produce greater complexity much more quickly than simple random molecular interaction. He also found that in "autocatalytic cycles" a system of organic molecules can reach a minimum level of complexity to trigger their own creation in a "self-reinforcing loop" (Davies 1995, 34, 79). Such processes seem to increase greatly the probability that living molecules could form in the short time available in Earth's history. Biophysicist Stuart Kauffman has applied these ideas to both the origin and evolution of life, offering an alternative or supplemental theory to neo-Darwinism. He proposed that Eigen's integrated cycles attained some degree of advancement before nucleic acids appeared; when DNA and RNA developed, they simply took command of a preexisting order. With regard to evolution, he argued that the inherent propensity of complex systems to freely create new levels of order is an additional and perhaps more powerful force for change beyond Darwinian natural selection. Nature selects from among systems where spontaneous increase in complexity has already occurred and thus "moulds an already existing biological order . . . As these forces tangle and vie in coevolving populations, so selection tends to drive the system towards the edge of chaos, where change and adaptation are most efficient." In this way, Kauffman suggests, something like a "law of increasing complexity" may supplement the known laws of nature (Davies 1995, 78–79).

Concepts of information theory appear to be fruitful and perhaps crucial in forging this middle path between the extremes of pure chance and strong determinism, since living matter, with its structure and function encoded in genetic data, is essentially "an information processing and propagating system" (Davies 2000, 18). Biogenesis and evolution in this context are seen not as questions of how some "exotic chemistry" works but of how biological information originates and develops greater and greater functional complexity. In information terms the structure and function of a macromolecule such as DNA correspond to the "syntax" (the defined sequence of base pairs) and the "semantics" (the meaning or function assigned to the sequence pattern). To pose the question of how life originates and evolves is, then, to ask how each of these aspects arises—first, how macromolecules with biologically useful base sequences are selected from an enormous number of structurally equivalent ones and second, how particular functions or "meaning" are actually assigned to sequence patterns.

The hypothesis of "strong" determinism is ruled out when concepts of information theory are applied to the problem, according to physicist Davies and biophysicist-philosopher Bernd-Olaf Kuppers. If strong determinism held, then there would have to be a physical explanation of how biological macromolecules develop their complex, random, and highly specified sequences that have meaning and function in an organism. Physical law can produce randomness, and self-organization can even select biologically meaningful sequences from among many possible random ones, but no physical law can predict or determine what the sequence will be—that is, what the actual information content is. As algorithmic information theory expresses it, simple physical law, which has low information content, cannot create richer, more complex information than it possesses itself, so the laws of physics and chemistry could not themselves "inject the complexity necessary for a structure" (Davies 2000, 21). The information content of a macromolecule—the "message" in the medium—can only come from unique historical interactions of a biological entity and its environment, which are not expressible in a general law. Darwinian processes can provide both the randomness and functional complexity but are only known to operate on something already living.

Although information science certainly exposes the limitations of strong determinism, there is promise that further developments will support a weaker determinism for biogenesis. New information laws may be discovered that can generate information-rich molecules—actually creating information rather than just moving it around. As Eigen expressed it, "Our task is to find an algorithm, a natural law that leads to the origin of information" (Davies 2000, 24). Some scientists call such ideas mysticism, but others regard such an accusation as unjustified since information theory is still being developed. Better understanding of information "dynamics"— how information flows, relates to the movement of physical matter and energy and connects the parts of a system to the whole—may yet provide clues. Davies speculates that some new principle related to quantum computation may be found that will supplement and expand known laws of local dynamics. Kuppers focuses on the possibility that the origin and development of informational blueprints may be explained by a "general theory of historicity" but concedes that the many small details of historical interactions over time can never be explained by law, since they contain so many "bifurcation points governed by chance" (2000, 42),

Yet another version of weak determinism is a simple extension of Darwinism. It suggests that Darwinian process may have operated even at the prebiotic level on a simple random replicator, which formed by chance. In this "Darwinism all the way down" scenario, variation and natural selection increased both structural and functional complexity of the replicating molecule until nucleic acids and proteins emerged and achieved dominance (Davies 2000, 22). The dilemma is that a replicating molecule complex and large enough to be such a prototype has to emerge by chance, and at the present time, that chance is estimated to be very small. Thus, the genesis of the original information-carrying molecule is still problematic. Kuppers makes a similar point in emphasizing that even in the self-organization scenario explanation of information blueprints has to start from earlier ones, which must simply be posited like initial conditions, so that pursuing their ultimate origin is to fall into an endless "regression cycle" and reach a limit to knowledge. For him, the whole issue is then "open to metaphysical speculation" in the same way that questions of the origin of the universe are (2000, 43).

Life by Design

One last but important philosophical position is causation by design. It is noteworthy that design arguments have traditionally appeared far less in scientific discussions of biogenesis than in speculations about the origin of the universe itself. This is perhaps understandable since the genesis of the cosmos involves more all-encompassing questions of origin— the very beginning of matter and energy themselves and the laws which govern them whereas biogenesis presumably occurred in an already existing material universe. Furthermore, explanations invoking design by God are usually scrupulously avoided, vulnerable as they are to being overturned when natural causes emerge. Nevertheless, design arguments for life's origin have been offered over the years by a number of scientists, with increasingly varied interpretation of the concept, and more recently they have become the core focus of the Intelligent Design (ID) movement.

In the first half of the twentieth century two influential scientists were among those presenting arguments for life's purposive or divine origin. Wallace, a key figure in evolutionary thought, invoked the notion of purpose in his 1903 book *Man's Place in the Universe* to explain the unlikely appearance of man in the cosmos.

Our position in the material universe is special and probably unique, and . . . lend[s] support to the view . . . that the supreme end and purpose of this vast universe was the production and development of the living soul in the perishable body of man. (Dick 1998, 22)

At mid-century du Nouy went further in interpreting life's tiny chance of appearing. His much acclaimed 1947 book *Human Destiny* presented calculations of the staggeringly improbable time proteins would take to form by chance but then concluded that probability arguments were not applicable in biology. For him, the findings of science "lead inevitably to God" (Dick 1998, 188).

In the latter part of the twentieth century many scientists continued to propose or believe in design by a purposive creator God, as envisioned above, while others began to interpret design in broader terms, as related to concepts of emergence or the product of an unspecified "intelligence" or "superintelligence." In his book *By Design* journalist Larry Witham illustrates the variety of design conceptions that emerged among scientists as theories of biogenesis involving random molecular shuffling over immense time periods began to fail. Biochemist Edward Pelzer, a research scientist interested in evangelical faith and traditional design by God, saw no conflict between science and religion, commenting, "As long as God is the first cause, the mechanism wasn't important" (Witham 2003, 104). With so many approaches and so little evidence, Pelzer feels that one must simply decide between life's deliberate creation or natural emergence and for him "the idea that it was purposely designed is becoming ever more apparent" (Witham 2003, 112). Pelzer is exemplary of a large number of scientists who profess firm Christian belief but also adhere firmly to the integrity of the scientific process as practiced by mainstream science.

A different journey—toward softer design views—was followed by the biophysicist Morowitz, who moved from "hard-nosed reductionist" to "mystic scientist" through his pioneering work in applying information theory to simple biological systems. He later labeled as "hubris" his early attempts to calculate all the information and assembly time for the parts of a cell. When he found the assembly time so improbably long as to move the question "outside of science," creationists were pleased, but he was led to a different realization—that life's origin needed a significant input of energy that provided new information, something like the "negentropy" conceived by quantum physicist Schrödinger in his famous book *What Is Life?* These explorations led him into the developing fields of chaos, complexity, and emergence theories, which he feels have "changed science at a deeply epistemological level." The old mechanistic adding up of parts was inadequate in the face of the "combinatorily explosive" character of living systems, whose game-like processes enhance what works and eliminate what is not useful, giving rise to new and unpredictable emergent properties. Morowitz sees this as a form of design that has no Designer. When queried about his beliefs, he said, "I'm a pantheist in the tradition of Spinoza . . . I do think of the universe and the divine as being somehow the same or overlapping." Morowitz is clear, however, as is Pelzer, in distinguishing when he is speculating about religious implications of his work and when he is conducting research as a deterministic scientist (Witham 2003, 108–10).

Perhaps the best known design advocates working today are scientists and philosophers associated with the ID movement, whose core argument is that the complexity of biological forms cannot be accounted for by natural causes and must be explained as the product of intelligence. While most scientists clearly label their design views as personal belief or philosophical speculation, ID advocates present them within the framework of science itself, thus challenging its methodological assumptions.

Two prominent ID proponents, biochemist Michael Behe and philosopher-mathematician William Dembski, both focus on complexity as evidence for design. For Behe, evolution has occurred, but the "irreducible complexity" of numerous cellular structures and processes with finely tuned, interdependent features could never have developed by the gradual, one-at-a-time changes of Darwinian process but had to have been designed by an interventionist intelligence. In his book *No Free Lunch: Why Specified Complexity Cannot Be Purchased without Intelligence*, Dembski posits three conditions that test for design—contingency, complexity, and specificity. An event or process is contingent if produced by both law and chance, complex if produced by more steps than cosmic time and available particles allow, and specified if possessing a "detachable" pattern—one that can be recognized independently from the event or process. Dembski argues that such evidence for design explains natural phenomena more adequately than law or chance and also better than self-organization processes, which fail because they produce repetitive patterns instead of the nonrepetitive, information-rich ones living organisms possess. In applying Dembski's design test to the appearance of intelligence, Gonzales and Richards claim that the correlation of Earth's fitness for developing intelligence with its supreme suitability for scientific discovery *is* such a specified pattern that reveals intelligent design. However, they follow other ID proponents in making no claim about who or what the "designer" is.

British astrophysicist Hoyle, on the other hand, is quite willing to offer speculation about the nature of the designer. His springboard for design arguments are the many anthropic coincidences in physical laws that allow life to evolve and which seemed to him too contrived to have arisen by chance-driven physical processes. Some "superintellect" must have been at work. In *The Intelligent Universe* he argues that the extraterrestrial microorganisms that seeded life on Earth were the handiwork of advanced intelligent beings—craftsmen similar in concept to Plato's Demiurge. Just as the Demiurge was directed by a higher power—The Good—so Hoyle's advanced beings operate under the direction of a "superintelligence" abiding in an eternal realm (Davies 1995, 136–37).

As these various positions have shown, scientists' philosophical debates about life's origin and cosmic prevalence present a more complex picture than previous debates centering on the simpler polarity of nature versus miracles. Debates today revolve primarily around the concepts of contingency and necessity, but these terms are not uniquely defined. Contingency suggests two very different possibilities—origin by chance and origin by design, and both positions appear to have strengthened in recent decades. Similarly, necessity relates to both strong and weak forms of determinism and even to exobiologists' "pure chance" operating over long enough times. Design positions themselves have proliferated into a number of varieties from the pure contingency of interventionist divine action to the potentially more law-like "design" implied by complexity and emergence theories.

The relation of these various positions to belief in God has also grown somewhat more complicated. Paul Davies argues that while there is a general correspondence between atheism and seeing life's origin as a "freak accident," and also between determinism and belief in "meaning, purpose, and design in nature" (Davies 2000, 26–27), these categories are not rigid. He writes, "It's perfectly possible to be an atheist and believe that life is built ingeniously into the nature of the universe. It's also possible to be a theist and suppose that God engineered just one planet with life, with or without the help of miracles" (Davies 2003). All of these positions, however, have clear implications for the cosmic status of humanity.

EXTRATERRESTRIAL INTELLIGENCE, HUMAN SIGNIFICANCE, AND GOD

Scientists voice a very wide range of views as to how alone, unique, and advanced humans are in the cosmos and what religious significance these possibilities hold. Their views no doubt come from many influences—from personal religious belief or lack thereof to modes of thought inherent in particular scientific disciplines.

Are We Alone?

The idea that humans are solitary inhabitants of the cosmos is a view shared by scientists at opposite ends of the religious spectrum. Both agree that the evolution of intelligence is unlikely in the extreme, but their interpretations could not diverge more. For Monod, the improbability makes intelligent beings a complete fluke and denotes nothing at all about purpose or action by God. His following well-known statement clearly reveals his atheism: "The ancient covenant is in pieces: man at last knows that he is alone in the unfeeling immensity of the universe, out of which he has emerged only by chance. Neither his destiny nor his duty have been written down" (1972, 167). In his footsteps follow many evolutionary biologists and current spokesmen for atheistic evolution. For Wallace, on the other hand, humans are probably alone because God created us so as the universe's "supreme end and purpose," and he may have had no need of other intelligent beings to accomplish his aims (Dick 1998, 193). Biologist Conway Morris' view is clear from the subtitle of his 2003 book: "*Inevitable Humans in a Lonely Universe.*" He holds that evolution's convergence on human intelligence only happened here but is part of a universal plan directed by a "lord of all creation," as Basalla expresses it (2006, 185). In a perspective similar to Wallace's, but only hinting at God, Gonzalez and Richards suggest that searching for extraterrestrial signals may be missing a grander truth:

Perhaps we have also been staring past a cosmic signal far more significant than any mere sequence of numbers, a signal revealing a universe so skillfully crafted for life and discovery that it seems to whisper of an extra-terrestrial intelligence immeasurably more vast, more ancient, and more magnificent than anything we've been willing to expect or imagine. (Gonzalez and Richards 2004, 335)

As discussed earlier, most physical scientists and exobiologists, both theists and atheists, took the opposite view that extraterrestrial life and intelligence are abundant. For theist Milne an "infinite God" can express his "godhead" far better on an infinitude of planets (1952, 152–54). From a very different religious perspective, Shapley viewed the highly probable abundance of life as an obvious extension of the Copernican paradigm and evidence for our unimportance. Earth's relegation to the outskirts of one galaxy among millions (now billions) made man "peripheral" too. Today historian Dick asserts that most astronomers and origin of life researchers accept life's cosmic abundance as the most likely scenario and "the working hypothesis of those in the growing hybrid fields of bioastronomy and astrobiology" (2000, 191).

Are We Unique?

But are we unique among our cosmic cohabitants? A number of highly respected physicists and astronomers and a small minority of biologists hold that we are not—extraterrestrials must resemble us—but the majority of biologists disagree. Physicists such as Weinberg and Glashow emphasize similarity in terms of mental capacity and ability to perceive the same universal physical laws, while others claim physical similarity. Drake writes, "They won't be too much different from us ... [A] large fraction will have such an anatomy that if you saw them from a distance of a hundred yards in the twilight you might think they were human" (Basalla 2006, 198). Biologist Robert Bieri agrees that "they will look an awful lot like us" (Basalla 2006, 184). Primack reports the general thinking that intelligent aliens are probably our size, which is optimal for complexity and fast thinking, and may possibly share our fractal circulatory system, rates of energy use, and even lifespans (Primack and Abrams 2006, 224–28). Many biologists, on the other hand, argue for uniqueness on the grounds that the many unpredictable historical steps leading to intelligence could never be duplicated. Eisley gives poetic expression to the special aloneness that uniqueness carries:

Nowhere in all space or on a thousand worlds, will there be men to share our loneliness. There may be wisdom; there may be power, somewhere across space great instruments, handled by strange, manipulative organs, may stare vainly at our floating cloud wrack, their owners yearning as we yearn. Nevertheless, in the nature of life and in the principles of evolution we have had our answer. (Dick 1998, 169)

Exobiologists and SETI astronomers generally believe that most aliens are physically different from humans but intellectually and technologically superior. This follows simply from the fact that humans are a young species in the universe; others must be more advanced. Furthermore, to be plentiful enough to afford us a chance of detecting them, they must be very long-lived. According to SETI astronomer Jill Tarter, "For one of the nearest 1,000 solar-type stars in our galaxy to host another technology, the average longevity L must be measured in tens of millions of years." To live so long, such societies must have greater wisdom, knowledge, and social stability than ours. They "either never had, or have outgrown, organized religion," which Tarter sees as a source of so much conflict on Earth (Dick 2000, 145). Drake speculated on another advanced trait—immortality, which aliens achieved by curing all disease. For physicist Jastrow alien scientists achieved immortality by figuring out "the secrets of the brain" and "uniting mind with machine" (Basalla 2006, 13, 160–61). Sagan conjectured that intelligent extraterrestrials have experienced and solved Earth-type social and environmental problems and have established a communication network throughout the Galaxy to spread their knowledge. Sagan's fictional alien in his book *Contact* spoke volumes about his hope for what they are like: masters of galactic travel, loving and concerned mentors, and immortals with spiritual, or at least magical, powers.

Are We Special?

Clearly, in such a universe teeming with advanced extraterrestrials, humans have no special status; in fact they have an inferior one. Shapley expressed it well in his comment about our vanity and anthropocentrism in considering ourselves the center of life and the highest being in the universe (Witham 2003, 11). SETI researchers agree that humanity lacks privilege, not only in spatial position but also in technology, knowledge, and probably wisdom. For those believing in life's cosmic abundance, only religious faith seems to temper this view, as Milne's example demonstrated. For him, an infinite number of inhabited worlds did not make our world less special in any way; in fact, Christ's incarnation here may imply just the opposite. In this sense, theists supporting life's abundance share a strong belief with theists or design advocates who argue that we may be alone: Earth's location and humanity itself may indeed have a very special status.

RELIGIOUS IMPLICATIONS OF SEARCH AND DISCOVERY

Religious Dimensions of the Search

When SETI researchers wax eloquent about the messianic promise of wise, immortal beings from beyond Earth, an obvious question arises: Is their search at heart a "quasi-religious quest"? Does the same religious strand that inspired medieval belief in supernatural angels now inspire some scientists to postulate natural immortal guardians in the heavens? Many commentators have said yes. Basalla argues that old religious and philosophical ideas about superior celestial beings continue to lie beneath current scientific investigation, unacknowledged for the most part by scientists themselves, who mostly profess atheism. These notions became part of modern science, according to one historian, when Western humanity first faced the lifeless emptiness of the Copernican-Newtonian universe and created rational superior beings to fill the void. The ideas persist, according to one psychologist, because yearnings for heavenly beings have deep emotional roots in the human being (Basalla 2006, 12–14). At the present time, speculates historian Dick, SETI may be a current example of what he calls a "universal religion... the never-ending search of each civilization for others more superior than itself." As such, SETI is "science in search of religion" (Dick 2000, 205).

Davies agrees that the current scientific search for extraterrestrials is "part of a long-standing religious quest," providing for some a larger framework for meaning in our lives and filling the void left by the decline of traditional religions (1995, 138). He identifies quasi-religious themes in the speculative writings of both Sagan and Hoyle. In Sagan's *Contact*, aliens share with a chosen human special knowledge about the universe, hinting at a great intelligent design hidden in its structure. In so doing, they act as angelic-type intermediaries between humans and some overarching, all-knowing presence. Aliens are a kind of "halfway house" to God in Hoyle's *The Intelligent Universe*, where two levels of advanced beings are described: cosmic engineers who arranged conditions for life's development on Earth and a much greater superintelligence who directs them from some timeless realm. Davies describes the obvious religious appeal of this idea:

This powerful theme of alien beings acting as a conduit to the Ultimate . . . touches a deep chord in the human psyche. The attraction seems to be that by contacting superior beings in the sky, humans will be given access to privileged knowledge, and that the resulting broadening of our horizons will in some sense bring us a step closer to God. (Davies 1995, 137–38)

For Davies, the religious streak is natural, given that theology helped give birth to science and elements of a theological world view are still accepted by working scientists, no matter what their beliefs are about God or extraterrestrial life. Although in current discourse a clear division exists between religious and scientific perspectives, for Davies "this separation is only skin deep" (1995, 138).

A few scientists have been willing to accept this analysis, but many have not. Frank Drake freely acknowledged that his childhood faith in fundamentalist Christianity had inspired him at the outset to join the search and that many of his coworkers "were either exposed or bombarded with fundamentalist religion." Although he abandoned the faith in early adulthood, its influence can perhaps be felt in his belief in alien immortality and their desire to teach us how to live forever. Sagan admitted no such inspiration, but his biographer, Keay Davidson, labeled his conviction about advanced benevolent creatures a "quasi-religious belief in alien super-beings . . . secular versions of the gods and angels he had long since abandoned." When an earlier interviewer proclaimed, "What you postulate is Angels. Faith, the same old faith," Sagan disagreed. "Not faith. Calculation. Extrapolation." His own son and science writer Dorion agreed with the interviewer, calling the scientific quest for alien intelligence a substitute for religion in secular times (Basalla 2006, 13, 198).

Effect of Discovery on Earth's Religions: Scientists' Views

Numerous scientists have ventured their opinions about the effect that discovering alien intelligence would have on Earth-based religions, on concepts of God, and on human spirituality in general. In 1951 British inventor and science-fiction author Arthur C. Clarke wrote that some people think that "contact with intelligent but nonhuman races, may destroy the foundations of their religious faith" (Dick 2000, 198), and Davies put it even more strongly in 1983, declaring that discovery "would have a profound impact on religion, shattering completely the traditional perspective of God's special relationship with man" (Davies 1983, 71). Tarter offers a more detailed analysis of what this effect might be. A contacting civilization would be millions of years old and very stable, she assumes, and the effect of their message on us depends on whether God exists and what kind of message they send. If God exists, then they must have developed a universal religion compatible with science; if they shared secrets of the universe and God, and we could verify it, we might become converted to their belief system. If God doesn't exist, presuming they could convince us of that, it would undermine our religions. A brief "Hello, we're here" message, on the other hand, might induce a slow change on Earth religions, as they adapt to the reality and perhaps develop a more universal common belief system here (Tarter 2000).

A discovery that alien intelligent beings might be more advanced spiritually as well as intellectually presents a special challenge to some Earth religions. Davies claims, "The difficulties are particularly acute for Christianity, which postulates that Jesus Christ was God incarnate whose mission was to provide salvation for man on Earth" (1983, 71). If intelligent and spiritually aware aliens exist, then a Christian must face the question of whether Christ died to save only humans on Earth or to save all beings everywhere. And if he died for all, was this accomplished by his one sacrifice on Earth, or does it need to be repeated on an endless number of worlds? Astronomer Milne addressed this question in the mid-twentieth century, reasoning that the Christian would, he imagined, find multiple Incarnations and crucifixions an intolerable thought and reason enough to regard Earth life as unique. His own resolution had an unusual and distinctly modern twist: he speculated that between the stars and even galaxies a vast radio communication network might develop and carry the message of Christ's salvation to all beings everywhere, thus making multiple atonements unnecessary (Milne 1952, 152–54). Many regard the idea of multiple Incarnations and crucifixions absurd or even comic, while others see it as close to heretical in making automatic and commonplace a special and holy one-time event. Nonetheless, the question is considered seriously by a number of theologians, as described in the next section, and, surprisingly, most display far more flexibility of thought than scientists predict, fulfilling Clarke's pronouncement that a robust faith has nothing to fear from "collision with the truth" (Dick 2000, 198).

Davies considers that of the other major world religions, Judaism and Islam have fewer but still not negligible difficulties with the existence of extraterrestrials, while Buddhism and Hinduism find the idea the least problematic. He quotes a verse from the Koran that seems to accept the idea: "And among his Signs is the creation of the heavens and the earth, and the living creatures that He has scattered through them" (Davies 2003). Since both Islam and Judaism emphasize the special status of human beings and even particular groups who are the faithful, there is a question whether they could easily accommodate extraterrestrials theologically. For Hinduism and Buddhism, however, the vast conceptions of the universe and greater universality of their religious concepts make accommodation to life beyond Earth seem much more natural (Davies 2003). A similar assessment is made by Dick (2000, 202).

Some reflect on how discovery would affect humanity's self-image and search for self-understanding. Davies sees the search as a test for the world-view that posits progress and the cosmic preeminence of mind. To find another example of intelligence would confirm our self-worth and "restore to human beings something of the dignity of which science has robbed them ... [and] give us cause to believe that we, in our humble way, are part of a larger, majestic process of cosmic self-knowledge" (Davies 1995, 129). For Primack and Abrams, this might happen whether we find aliens or not. If we do find them, perhaps their wisdom will "encompass" ours, much as the theory of general relativity encompassed but did not entirely overthrow Newtonian gravitational theory. Living long enough to experience an encounter, however, we may develop such wisdom ourselves. Even if we never find alien intelligence, or if we are truly alone, to identify those qualities aliens should have to make us not feel alone is a way of studying ourselves. Those qualities are "the essence of humanity ... [and] *what a long-lived civilization on Earth should aim to cultivate in ourselves*" (Primack and Abrams 2006, 234–35).

Historian Dick believes that the widespread acceptance of cosmic evolution and abundant life and intelligence is ushering in a new era of cosmic consciousness in which "cosmotheology" must transform older theologies. Its general principles are the noncentrality of humanity in the universe—either physically or biologically—the low-ranking of human intelligence in relation to others, the need for new conceptions of God "grounded in cosmic evolution," and the need for a moral dimension of reverence for all life. This perspective will become ever more natural as humanity moves out into the cosmos and reorients itself to a "biological universe." Actual contact would, of course, accelerate the transformation. He speculates that "in learning of alien religions, of alien ways of relating to superior beings, the scope of terrestrial religion will be greatly expanded in ways that we cannot foresee" (2000, 202). What might require the greatest transformation is our present conception of God. In place of the transcendent creator God of traditional monotheism, a new "natural" God is envisioned—a superior intelligence with many of the same attributes as the Judeo-Christian and Islamic God but a God abiding in nature and not separate from it. Such a "natural God of cosmic evolution and the biological universe," he claims, holds the promise of harmonizing religion and science and becoming the "God of the next millennium" (Dick 2000, 202–4, 208).

Christian and Jewish Theologians Respond

Christian theologians who address the issue of life beyond Earth generally reject scientists' assessment that Christianity would be threatened by the discovery of intelligent alien beings. In view of how fully the subject has permeated public consciousness, Christian scholars have considered it too little, according to Lutheran theologian Ted Peters. When they do, however, most openly accept the idea of life beyond Earth and "have routinely found ways to address the issue of Jesus Christ as God incarnate and to conceive of God's creative power and saving power exerted in other worlds" (Peters 2003, 126, 131). This conclusion applies across the board to Christian scholars, except for those of fundamentalist persuasion.

In the post–World War II era a number of Christian theologians and educators spoke positively about the possibility of life on other planets. German Catholic theologian Hans Kung wrote that "we must allow for . . . quite different living beings . . . on other stars of the immense universe," and the eminent Catholic scholar Rahner considers "the many histories of freedom which do not only take place on earth." A prominent educator proclaimed that Catholics should know that their beliefs are "entirely compatible with the most startling possibilities concerning life on other planets." Evangelical minister Billy Graham agreed, announcing his belief that "there are intelligent beings like us far away in space who worship God . . . [and] are God's creation." Certain Protestant theologians chime in enthusiastically on the subject. Krister Stendahl enthusiastically welcomes the idea of communication with extraterrestrials for the way it enlarges God's universe, and A. Durwood Foster points out that beliefs that accept God's mystery should find such ideas far from surprising (Peters 2003, 126–27).

In contrast, many conservative and evangelical clergy rejected such beliefs, influenced by fundamentalist literature in the 1970s warning followers against belief in extraterrestrials. According to Peters' research, three arguments underlie their position. Aliens are not mentioned in the Bible, belief in them assumes affirmation of evolution, which they reject, and the whole business is the work of Satan, tempting the faithful to believe in a source of salvation other than Christ (Peters 2003, 129–31).

A number of theologians also address the thornier issue of the meaning of Christ's incarnation in the light of other possible civilizations in the universe. Interestingly, views on this subject do not fall along denominational lines. Renowned Protestant theologian Paul Tillich believes that "man cannot claim to occupy the only possible place for incarnation." In his theology, divinity's redemptive process is merged with creation and is already active in whatever civilization we might encounter, dispelling the need for us to spread the word to them. From a process perspective theologian Lewis Ford also believes in multiple incarnations of divinity. Because God is present as a persuasive force in every event within the whole evolutionary unfolding, his redeeming action is part of creation and applies to all intelligent beings, wherever and whenever they may exist and be open to it (Peters 2003, 127–29).

Two Roman Catholic and one Anglican theologian also support, or at least do not exclude, multiple incarnations. Rahner agrees that multiple incarnations are conceivable, although knowledge of them is unattainable, limited as we are to revelation pertinent to our own salvation (Peters 2003, 127–29). In *Many Worlds, One God*, Catholic priest Kenneth Delano expounds on God's caring love for all intelligent beings in the cosmos and affirms incarnation on other worlds for "any one or all three Divine Persons of the Holy Trinity." This is a much greater likelihood than a one-time redemption of a "cosmic Adam," although the latter could provide opportunity for human missionary activity to spread the word. Along these lines, Anglican theologian E. L. Mascall rejects astronomer Milne's suggestion of spreading the Gospel via radio communication as being insufficient to bring about the special act of union that redemption involves. There is no reason to assume it could not have happened where those needing salvation lived. Mascall delights in "theological flexibility," declaring "how wide is the liberty that Christian orthodoxy leaves to intellectual speculation" (Dick 1998, 248–50). Such wide-open speculation can lead in many directions, however, and one Catholic philosopher, McMullin, warns against simplistic arguments that fail to recognize the divergence of theological interpretations of the meaning of the Incarnation itself. For him, the cautious answer is "maybe" there could be multiple Incarnations. (2000, 171–72)

The opposite position is also held by both Roman Catholic and liberal Protestant theologians. Jesuit journalist L. C. McHugh and scholar Edgar Bruns both affirm that Christ's one Incarnation on Earth has universal applicability. For Bruns, "[Christ] is the foundation stone and apex of the universe and not merely the Savior of Adam's progeny," a belief echoed by Protestant scholar Wolfhart Pannenberg, who sees Christ as the nexus of being aiming to bring the entire cosmos to a consummate fulfillment (Peters 2003, 127–29).

Regarding the views of Jewish theologians on the subject, one Jewish scholar, Rabbi Hayim Perelmuter, gave a definite answer: "Contemporary Jewish theology would have no difficulty whatsoever in accepting new knowledge regarding the existence of extraterrestrial life." Such knowledge would but widen the horizons of their conception of what God had brought into being (Peters 2003, 129).

Chapter 6

Cosmic Evolution: Christian Religious Perspectives

The cosmic evolutionary story has evoked a rich array of responses in the Christian religious domain. Its cosmic narrative of growth and change, its apparent fine-tuning for living beings, and possibilities for abundant life raise long-standing religious questions of divine action, design and purpose, and God's relation to the universe and to humanity. On the one hand, cosmic evolution has clearly presented a deep challenge to religion, especially for some conservative Christians. The apparent absence of a divine creator, the immense time scales involved, and the suggestion of a random, chance-driven evolutionary process all fly in the face of literal readings of Genesis I. This perceived conflict galvanized some creationists into political action in the arena of public education, a cause more recently and avidly pursued by the Intelligent Design (ID) movement. On the other hand, many in the religious community view cosmic evolution as compatible with belief in a God who created and still acts in the universe. They hold positions which range from the more conservative "progressive creationism" to the more liberal "theistic evolution" (Scott 2004). Most mainstream and liberal theologians and laypersons, as well as a significant number of evangelical Christians, fall in the latter category. Typically, they endeavor to explore common ground with science by examining traditional religious concepts of divine action, natural theology, design, cosmic purpose and God's relation to the world in the light of the new cosmology.

CREATIONISM

Young Earth Creationism

Cosmic evolution is most firmly opposed by groups which have, since the 1960s, been calling themselves "scientific creationists." The appeal to scientific legitimacy, and not Biblical authority, was necessitated by several historical factors, which are well-described in a definitive study by historian Ronald Numbers, *The Creationists* (1992). At mid-century antievolutionists had been successful in eliminating evolution from school textbooks for thirty years following the 1925 Scopes trial but nonetheless found their movement waning, due to internal disagreements about time scales and the paucity of scientifically trained members. Moreover, many young newcomers sought accommodation of their religious views with mainstream science. A creationist group that formed in 1941—the American Scientific Affiliation—took no position with regard to belief, and many members began to shift their thinking away from strict creationism. A second factor was the reinvigoration of science curricula in the post-Sputnik era and the return of evolution to textbooks. Antievolutionists were galvanized to respond but faced new constraints. Culturally, science had become more powerful, and a 1968 Supreme Court decision striking down Arkansas' antievolution law necessitated a shift to new tactics—focusing on the scientific rather than biblical aspects of creationism and promoting equal time for evolution and scientific creationism in the classroom.

This was the tack taken by the late Henry Morris, generally acknowledged as the founding father of the "Creation Science" movement and responsible more than anyone for the creationist revival of the 1960s. Trained originally as a hydraulic engineer, he brought strict creationism back on the scene with the 1961 publication of *Genesis Flood*, which he coauthored with theologian John Whitcomb. The book revived the early twentieth-century flood geology theory of Seventh Day Adventist geologist George McCready Price and laid down the fundamental tenets of Young-Earth creationism: Biblical literalism, recent creation for Earth, a fall that initiated the second law of thermodynamics and a global flood that in a year's time distributed all the geological layers. He founded institutes of trained scientists who adhered to these beliefs, including the Institute for Creation Research (ICR) near San Diego in 1972, and two years later published a textbook, *Scientific Creationism*, with Christian and public school versions. Although later court cases also banned the teaching of creation science, its beliefs have exerted an enormous influence among conservative Christians both in the U.S. and abroad (Numbers 1982, 543–44). Today a number of organizations promote strict creationist beliefs, among them Creation Research Society, the Seventh Day Adventist Geoscience Research Institute, and Ken Ham's worldwide ministry Answers in Genesis. Efforts of these organizations have no doubt contributed to the apparent rise in antievolution sentiment in the later twentieth century. When the creationist revival began, a survey of California churchgoers revealed that around 30 percent opposed evolution; by the early 1980s 44 percent of the U.S. population agreed with the statement that "God created human beings in their present form within the last 10,000 years"—a Gallup Poll statistic that has remained steady at 44–47 percent for the last 25 years (Numbers 1992, 300).

The basic tenets of Creation Science, as published by the ICR (Morris 1980), can be summarized as follows. *First, the physical universe and all life forms were "supernaturally created by transcendent personal Creator" and each life form, including human, was created in its mature "functionally complete" form. The soul of man was specially created separately from his physical form.* Thus creationists share with all orthodox Christians, Jews, and Muslims the basic *creatio ex nihilo* belief in "one self-existent eternal Creator, who called the universe itself into existence [from nothing] in the beginning, as well as all its laws and systems" (Morris and Parker 1987, 20). The belief about special creation of the human soul, apart from the physical body is, interestingly, shared by the late Roman Catholic Pope, John Paul II (Peters 1998). Creationists diverge from mainstream Christian theologians in believing that "creation was 'mature' from birth" and the "universe had an 'appearance' of age right from the start" (Morris 1980, 209). For some, this belief applies not just to living things but also to stars and galaxies, which are believed to be "unchanged" and "constant" (Morris 1985, 13).

Clearly, there is no room in this picture for any phase of cosmic evolution, even in stars where fixed physical laws operate, but there is some agreement with science's anthropic thinking, or at least its implications for design. Young Earth Creationists certainly agree that the Earth "is uniquely designed for life," for a benevolent God would naturally design the cosmos and Earth perfectly for its inhabitants. They mostly avoid the anthropic principle, however, because physicists formulate it in an "atheistic," old-universe framework and reject design. Morris and Parker bemoan that this "popularization of evolutionary pantheism . . . is not accepted as a testimony to divine design, but as a deterministic outcome of cosmic mind" (Morris 1985, 23). One creationist astronomer asserts that it is time for creationists to "retake this argument" (Faulkner 1998).

Second, there has been no evolution of one kind of creature into a different kind. Some microevolution has occurred within kinds, but such changes are "horizontal," allowing survival, or are "downward," impairing it. Strict creationists argue that macroevolution is not scientific because it has never been demonstrated, and it furthermore violates the second law of thermodynamics, which decrees that closed systems always tend toward greater disorder and not to greater complexity. When evolutionists reply that the earth is an 'open system,' where local order can increase with outside input of solar energy, creationists argue that this requires a blueprint and energy conversion processes, which evolution's mechanisms cannot provide. Mutations are negative or neutral, and natural selection can only sift out the harmful ones and preserve the present order, not create new complexity. In theory, evolution might possibly occur in open systems, but no one has ever observed such overcoming of the second law (Scott 2004, 143). As further support for the impossibility of evolution, they often cite Gould, who likewise criticizes slow, gradual change as insufficient to form new species, but suggests instead rapid periodic evolutionary bursts, or "punctuated equilibrium."

Third, the geological record in rocks and fossils can be accounted for by catastrophic events in the past that obey natural laws. Much scientific evidence points to a "relatively recent creation of the earth and the universe" Other evidence exists that most of the fossil-bearing rock layers" formed in an even more recent global hydraulic cataclysm." Strict creationists argue for a young Earth in a number of ways: disproving theories that require an old universe, identifying phenomena inconsistent with an ancient age, finding flaws in age-dating techniques, and presenting evidence consistent with alternative theories. Much evidence simply allows for a young Earth rather than requiring it, however, and the whole issue is secondary to the crucial question of whether creation or evolution occurred. In fact, the most important reason that a young Earth has become part of Creation Science orthodoxy is theological (Morris and Parker 1987, 254; Peters and Hewlett 2003).

A young universe is possible if the Big Bang never happened, and Morris attacks the Big Bang theory on thermodynamic grounds: "The very idea that a primeval cosmic explosion could somehow generate a highly ordered and complex universe seems preposterous on the very face of it. Explosions produce disorder, and this ultimate explosion would surely have generated the ultimate in disorder, as the primeval state of the universe" (Morris and Parker 1987, 260). Numerous other arguments proposed for a young universe and solar system were reviewed and evaluated not long ago by creationist astronomer Danny Faulkner—motions of galaxies in clusters, the paucity of supernova remnants, the population of bright comets, interplanetary dust, planetary magnetic field decay, and the moon's recession rate from Earth (Faulkner 1998). Morris and Parker list over sixty pieces of evidence supporting a young Earth, including several mentioned by an "old-Earth evolutionist," which mostly involve rates of buildup versus depletion of chemicals in Earth's atmosphere, oceans, or crust (Morris and Parker 1987, 288–91). A central argument questions methods of radiometric dating, whose flaws, they claim, rest in unproven assumptions and poor verification. The method assumes a closed system, no possible initial contamination by the end product of decay, and a constant rate of decay over time—none of which Morris agrees are valid, as they are unproven and not provable (1985, 138–39). Furthermore, creationists argue, the method fails to verify things of known age, and different techniques yield inconsistent results for the same sample (Scott 2004, 154–55).

To argue for past catastrophic events and a recent worldwide flood, creationists use fossil distribution in rock layers consistent with global flood activity, layers showing sudden change, and evidence consistent with the presence of an atmospheric "vapor canopy" before the flood, which produced a forty-day global rainfall. Higher polar temperatures, increased atmospheric pressure and higher oxygen content would have prevailed before the flood and a sudden, permanent decrease in polar temperatures afterwards. All of these are consistent with fossil evidence—tropical plant fossils in arctic and Antarctic regions, a very large number of tropical animal fossils which conceivably perished all at once, and very large earth-bound and flying reptiles whose existence could have been facilitated by the preflood atmosphere of greater density and oxygen content (Scott 2004, 147–48).

In sum, assuming a very great age for the Earth and universe is simply unnecessary, unless one is trying to "accommodate evolution and the uniformitarian interpretation of the geologic column." If evolution never happened and the geologic record is explainable by the action of a catastrophic flood, "then there is no need to think the Earth and universe are much older than humankind and the beginnings of human history" (Morris and Parker 1987, 273). And there are theological reasons to favor this idea. An old Earth in which animals suffer pain and death millions of years before man arrives suggests a cruel God who doesn't regard death as the "wages of sin" and punishment for Adam's disobedience. Christian theology holds that death and destruction, including thermodynamic decay, are all the result of Adam's fall into sin. If they occurred before man was here, it invalidates the meaning of Christ's dying on the cross to redeem us from sin—perhaps the most central of all Christian doctrines.

This type of theological influence on science is at the heart of the debate between creation scientists and mainstream science and theology, but some scholars question the real nature of the conflict. Peters and Hewlett argue that despite its popular image and some scholarly assessments, the debate is not a simple case of science versus religion—of Christian theists versus materialistic scientists; it is more a debate within science and within religion. In the science struggle two rival theories are competing to explain scientific facts. A debate also ensues over what constitutes proper science—what kinds of theories and explanations are acceptable and how data is selected and used. Creationists limit "scientific explanation" to what can be observed and known in the present and label any extrapolations or speculation about the past or origins nonscientific or metascientific. Mainstream scientists demand that hypotheses be testable and revisable. Many, both atheists and believers, soundly criticize creation scientists and reject them as scientists, for careless and selective use of data in service of an ideology, refusal to allow evidence to take them where it will, inconsistent use of extrapolations, and proposing untestable hypotheses. Creationists level exactly the same criticisms at evolutionary scientists.

The religious aspect centers on what constitutes good theology and which of two ideologies is correct. Creationists argue that Earth's ancient appearance is not real, but evangelical Christian physicist and astronomer Howard van Till criticizes this as "poor theology," suggesting a God who is "deluding us." They also argue that anything but a young Earth interpretation destroys essential Christian doctrines of redemption. Theologian Langdon Gilkey noted that creationists confuse "scientific language about facts of the present world" with "theological language about transcendent God" and therefore make a category error, conflating different levels of knowing (Peters and Hewlett 2003, 90–91).

Creationists see that they are fighting an ideological battle against all the forces supporting evolutionary thinking, which include liberal Christianity, Judaism and Islam, and virtually all other world religions as well as secular materialism and atheism. For creationists, the degradation of humanity underlying evolutionary thinking has produced all the immorality, evil, and loss of values in the world today, and they see themselves as waging a spiritual struggle to save the soul of modern man. Opponents counterattack that the black and white thinking and fervent exclusivity that characterizes fundamentalist beliefs is itself a source of intractable world, societal, and personal conflict.

Old Earth Creationism

Young Earth creationism lies at the extreme end of a wide spectrum of creationist views. Although there are not always sharp boundaries, along the spectrum one observes decreasing adherence to biblical literalism and increasing acceptance of science and its picture of cosmic evolution (Scott 2004, 57–58). What all versions of creationism share is the belief that God brought each "kind" of living being into existence by a special act of creation.

Two forms of old Earth creationism which preserve some degree of literal adherence to the biblical six-day creation but accommodate geological time scales are the Gap (or Ruin and Restoration) and the Day-Age theories, which were most popular in the nineteenth and early twentieth centuries. According to the Gap theory, the creation of matter and life in Genesis I:1 was followed by a long time gap, in which multiple catastrophes and creations took place that laid down the geological strata. About 6,000 years ago, a new creation occurred in six 24-hour days that culminated in Adam and Eve. Adherents of this view include leading Pentecostal ministers and the Christian Geology Ministry, an Internet-based fundamentalist Bible study group. A position held more widely in the last two centuries was Day-Age creationism, in which each day represents a long time period (thousands or millions of years). As with young Earth and Gap theories, however, there is no evolution within or between species. Adherents point to a rough correspondence between creation stages in Genesis and life's evolutionary path, but ignore certain contradictions. Day-Age adherents included the famous Scopes lawyer William Jennings Bryan (Numbers, 1992; Scott 2004, 61–62).

Today the stance taken by the largest number of creationists is "Progressive Creationism," a view first named and elucidated in 1954 in a seminal book by evangelical Baptist theologian Bernard Ramm, entitled *Christian View of Science and Scripture*. Originally drawn to the Day-Age theory, Ramm believed that Genesis I gave a general historical sketch but not reliable scientific information. He sought to synthesize aspects of evolutionary science and the Bible by merging the "pictorial-day" theory of Genesis I, wherein creation was "revealed pictorially" but not actually made in six days (or ages) with "progressive creation," wherein God occasionally acted directly to create new "root-species," which "radiated" into today's species. In this way God prepared, over millions of years, a fitting home for man, the pinnacle of creation. Ramm's ideas helped many evangelical Christian biologists accept evolution fully, although he himself never did (Numbers 1992, 185–87).

Progressive creationists accept the scientific findings about the Big Bang, Earth's age, and the long span of time for Earth's geological and biological development but reject major parts of biological science, especially macroevolution, or any naturalistic development of one "kind" from another. The appearance of each kind was due to unique, special creation by God from nothing, using means outside the realm of naturalistic science. To make this argument, British physicist Alan Hayward uses studies by non-Christian scientists to criticize Darwinian gradualism and natural selection. In the same book, *Creation and Evolution* (2005), he interprets Genesis to include both divine creation over long epochs of time and the figure of Adam as an historical figure.

A prominent spokesman for progressive creationism is Hugh Ross, a trained astronomer who now operates his own ministry, Reasons to Believe. A central theme in such books as *The Fingerprint of God* (2000) and *The Creator and the Cosmos* (1995) is that the physical universe reveals God's existence, character, and purpose, and that since both the cosmos and the Bible are revelations from God, they will never conflict, when properly understood. His writings are both scientific and devotional, celebrating advances in science as tools for expanding and deepening our understanding of God. He finds the anthropic coincidences compelling evidence for divine design. After listing almost six dozen examples of fine-tuned properties in the universe and in the Galaxy–Sun–Earth–Moon system and calculating a staggeringly low probability that they will all occur together—10^{-53}— he says there is no possible conclusion but a Creative Designing God at work throughout time. In a more recent work, *Creation as Science* (2006), Ross presents his own "scientific creation model" as a testable hypothesis alongside those of young Earth creationism, naturalistic evolution, and theistic evolution. While he finds many critics among both strict creationists and scientists, he himself implores all scientists, theologians, and philosophers to overcome their rivalries and engage in interdisciplinary dialogue to interpret scientific facts and understand cosmic purpose (Ross 1995, 15).

INTELLIGENT DESIGN

Design by an intelligent being is also the hallmark of the newest player in the creation/evolution arena—Intelligent Design (ID). ID is often considered a variant on progressive creationism, since both reject macroevolution, accept microevolution and support the idea that the origin of life and new species required design by an intelligence operating outside of natural laws. ID proponents, however, reject the creationist label, as they make no claims about the nature of the designer (Peters and Hewlett 2003, 103). Because they do not offer their own origin theory but aim primarily to critique Darwinism and naturalism in science, it is impossible to assess their views on cosmic evolution, and most probably there is a very wide range of views from acceptance to denial in the ID community. What they all agree on is that the totality of evolution could not have happened on its own without the agency of an intelligent designer.

The ID movement originally grew out of developments in biology in the later twentieth century: critiques of reductionism, natural selection, and molecular evolution in neo-Darwinian theory and hints that something more than naturalistic Darwinism was needed to explain life's origin and complexity. Many in the movement were influenced by the insights of chemist-philosopher Michael Polanyi, who argued in the 1950s against biological reductionism and came to believe that something "beyond physics and chemistry encoded DNA" (Witham 2003, 115). His ideas influenced an early book, *Mystery of Life's Origin* (Thaxton 1984), which examined problems with biogenesis research, including the origin of information in complex molecules, and introduced the possibility of interventionist intelligent design. A year later biochemist Michael Denton critiqued a central aspect of Darwinian theory in *Evolution: A Theory in Crisis*, which revealed problems in producing more complex life forms by random changes in genes. Although not a design enthusiast himself, Denton argued that inferring design was not based on religious assumptions but emerged inductively by strict use of the logic of analogy. By the late 1980s the term "intelligent design" was coined and being promoted in a supplementary high school textbook by Dean Kenyon, *Of Pandas and People: The Central Question of Biological Origins*. Later, in *Darwin's Black Box* (1996) biochemist and evolutionist Michael Behe made the most extensive case yet for why "irreducibly complex" biochemical processes must be the product of intelligent design and not gradualist natural evolution. Thus, as the movement was developing, its first major goal was to challenge neo-Darwinian theory.

A second focus of ID's current leading group—Fellows at the Center for Science and Culture (CSC) at the Discovery Institute in Seattle—is to establish a sound theoretical and observational basis for detection of design and to use it to find examples in nature. Developing a method for inferring intelligent design has been the main contribution of philosopher-mathematician William Dembski who proposes the "explanatory filter" method of detecting "specified complexity" (described in the last chapter). Because intelligence is difficult to define, detection of one of its products—information—has been used in his arguments. By analogy, where "complex specified information" exists in nature, an intelligent agent is inferred. Other key players who make the design argument and elucidate numerous examples in nature include Behe, who focuses on complex cellular biochemical processes; philosopher of science and CSC Director Stephen Meyer, whose specialty is origin of life studies; and astronomer Guillermo Gonzalez, who looks to the large-scale universe. Gonzalez and others fully embrace all the anthropic coincidences as clear evidence of design, but he takes them one step further—noting that the unique set of conditions for habitability also renders the Earth a unique site for scientific discovery. For him, this correlation is the "strongest evidence for purpose in the universe to date" (Witham 2003, 139).

Although critics associate ID advocates with the now disfavored natural theology of the past and label them creationists in disguise with a covert religious agenda, ID proponents are eager to distinguish themselves from both. In *Intelligent Design* Dembski claims, "ID is at once more modest and more powerful than natural theology . . . [which] reasons from the data of nature directly to the existence and attributes of God" (Dembski 1999, 107–8). ID is more powerful in providing a more precise detection method than the intuitive analogical reasoning of earlier design advocates such as Paley. It is more modest in making no claims as to who the designer is and for what purpose or end something was designed. "To connect the intelligence inferred by the design theorist with the God of Scripture,"

Dembski says, "is a task for the theologian" (Dembski 1999, 107). In this way, ID proponents claim to differ considerably from the biblically oriented creationists, although they certainly share opposition to naturalistic, Darwinian thought.

A major goal underlying their challenge to Darwinism and their proposal of design detection methodology is to topple a foundational principle of modern science—naturalism, methodological as well as philosophical. Not only do they seek to confront materialist interpretations of science; they also want to change the rules of science itself by gaining acceptance for causal agents that go beyond natural laws and processes, in particular intelligent ones. They often speak of this goal in militaristic, political or even medical terms, as the "impending demolition" or "unseating" of scientific naturalism or "finding a cure" for it. To this end, CSC also sponsors social science and humanities research on the cultural effect of scientific materialism and initiatives to improve science education by presenting weaknesses as well as support for Darwinian theory. Opposing Darwinism and the whole of naturalism in academia has been the focus of another leading light of the movement—CSC Advisor Philip Johnson, a UC Berkeley emeritus law professor. In *Darwin on Trial* (1991) he not only critiques evolution science and reviews the educational legal battles but also criticizes the biased mind set of a scientific establishment too deeply entrenched in the paradigm of naturalism even to consider evidence for divine action (Peters and Hewlett 2003, 105–6). He explores this position more fully in *The Wedge of Truth* (2000).

Although Dembski and Meyer believe that Intelligent Design can form a helpful bridge between science and theology, it meets with fervent criticism from both sides. Critics of ID focus on numerous issues: the failure to distinguish science's naturalistic method from belief in materialism, violation of the "rules" of science, use of "God of the gaps" reasoning, insufficient accounting for the flexibility of natural selection, and failure to produce a testable alternative hypothesis for life's origin and development. There is, for example, a very wide range of views among ID advocates as to how and when the designed feature—complex specified information (CSI)—is introduced and works in nature. Some design advocates propose that it was "front-loaded" at the very beginning and gradually becomes active over time—a position that is vulnerable to being labeled deism. Others envision that it emerged by intelligent activity in discrete interventions over time. With such a prominent role for information, however, all agree that Darwinian naturalism is not an option.

THEISTIC EVOLUTION

Moderate and liberal Christians generally embrace both cosmic evolution and belief in a supreme God who created, sustains and continues to interact with the world. In the last twenty-five years 35–40 percent of the population as a whole, as revealed by Gallup Polls (2007), have consistently held this view, expressed as follows: "human beings have developed over millions of years from less advanced forms of life, but God guided this process." Theistic evolution is supported by the Roman Catholic Church, mainline and liberal Protestant denominations, and a significant body of conservative, evangelical Christian scientists—members of the American Scientific Affiliation (as declared in official statements printed in *Voices for Evolution* [Matsumura, 1995]). In 1996 Pope John Paul II articulated his own particular version of theistic evolution. He endorsed evolution as being "more than a hypothesis," a theory well supported by independent lines of research and capable of explaining the time line and mechanism for physical development of living creatures, including humans. However, science can never explain how our spiritual nature arises. He endorsed Pope Pius XII's position that "if the human body takes its origin from pre-existent living matter, the spiritual soul is immediately created by God" (Peters 1998, 150–51).

The Pope's statement well illustrates the kind of challenge religious leaders and scholars face in forging a sound position of theistic evolution. The task is to reconcile two key positions. One is the seemingly self-sufficient evolutionary process driven by the blind, apparently purposeless operations of chance and natural selection. The other is belief in the special dignity of the human being and in the action of a loving, purposive God who designed and still guides the whole process to some consummate end. All who approach the dilemma must answer certain basic questions. If God created and is still guiding the process, how exactly does he interact with nature to accomplish his ends without disrupting natural laws and how can its chancy, purposeless appearance be understood? What conceptions of God are consistent with a full acceptance of both evolution and Christian theology? Can design by God now be detected in the natural world in the fine-tuning of physical laws for life and the progress of living forms toward greater complexity and consciousness? Does this call for a teleological explanation and provide the basis for a new natural theology and perception of purpose in the cosmos? If all is seen as part of God's purpose for creation, how can one accept as just and God-ordained a process full of pain, struggle and evil—the problem of theodicy?

Needless to say, these are each enormous and important theological subjects, and they have been explored in a vast literature by scholars who have grappled with them in a great diversity of ways. Some focus on specific issues, such as divine action, teleology, or theodicy, while others address all the questions together within comprehensive metaphysical systems, such as neo-Thomism, process theology, Trinitarian theology, or the thought of Teilhard de Chardin. A number of helpful sources review various positions through a variety of organizational rubrics—among them

Barbour's typologies for ways of relating (1997, 2000), Peters and Hewlett's spectrum of positions regarding divine agency and divine purposiveness in nature (2003), Russell's typologies of theological views of divine action (1995) and his categorization of noninterventionist models of God's agency (2000), and the classification of models of God and views of divine agency in Southgate et al. (1999), to name but a few. Much excellent discussion of divine action can be found in Keith Ward's 1990 study *Divine Action* and three volumes of scholarly essays from conferences about scientific perspectives on divine action convened in the 1990s by the Vatican Observatory and the Graduate Theological Union's Center for Theology and Natural Science (CTNS) in Berkeley, California (Russell et al. 1993, 1995).

Cosmic Evolution and Christian Theism as Independent

One approach to reconciling Christian faith and cosmic evolution is simply to accept both fully but regard them as separate and independent from one another, a position held by many conservative and evangelical scholars, neo-orthodox theologians, and Christian existentialists. Conservative scholars who hold to independence are not necessarily biblical literalists, but they do give central importance to scripture, the message of Christ's atonement and the personal experience of conversion and transformation through Christ. They also take very seriously the findings of modern science.

Evangelical Christian physicist and astronomer van Till of the Reformed tradition exemplifies this position. In *The Fourth Day* he argues for a clear separation of scientific questions about the internal workings of the cosmos (material properties, behavior and cosmic history) from religious ones about the external relationships (questions about status of nonmaterial being, origin, governance, value and purpose). These two types of questions have "categorical complementarity"; that is, they seek knowledge of different aspects of the cosmos. By keeping them separate, it remains very clear when one is taking a scientific versus a theological perspective. Van Till is critical of both scientific creationism and philosophical naturalism for their disregard of this important distinction.

In van Till's "creationomic perspective" the findings of science are interpreted theologically in the wider framework of Biblical truth, but both are highly respected. Thus the scientific picture is fully appreciated for the "dynamic order" it reveals—a "magnificent tapestry woven from the different strands of temporal development to form the intricately designed pattern of cosmic evolution." The same thing, when viewed in Biblical theological terms, is "an inexhaustible tribute to the boundless vastness of divine creativity"—a far greater creativity than creationism's sudden making of a finished product (van Till 1986, chapters 10 and 12). In similar pairings, the "integrity of the created order" is seen theologically as expressing "the unity of God," and science's "natural law" reveals God's "faithfulness." It is noteworthy that while van Till's view exemplifies independence in clearly separating scientific and religious questions, his theological exploration of cosmic evolution and other scientific findings seems to move him beyond this label.

Regarding divine action and teleology, Van Till follows Augustine in holding that all the potentialities for this process were present at the beginning and then gradually realized over time without any special divine intervention. There are no gaps which God must fill. Thus in biological development God provided all the possibilities for workable living forms and the process that led to them—mechanisms which the scientists can study. However, any consideration of the governance or purpose of the unfolding process can only be considered from a wider religious framework and not reasoned directly, as in natural theology. While divine action in nature appears to have been relegated to the beginning, as in deism, van Till holds that God still acts directly through "special revelatory and redemptive acts" (Barbour 2000, 103).

Protestant neo-orthodoxy, exemplified by Swiss Reformed theologian Karl Barth and his followers, and Christian existentialism emphasize an even more distinct split between scientific and religious spheres. Neo-orthodoxy sought to return to the Reformation principle of the central importance of Christ and revelation but at the same time fully accept science's evolutionary findings as belonging to a different realm. Genesis I carries only a theological message about the goodness of creation and its complete dependence on God and not a literal message describing physical beginnings. Knowledge of transcendent God can only come through his direct revelation to the human person and never through reason via arguments for design or natural theology. Barth also uses the concept of primary and secondary causality to express his belief that God's supreme rulership over nature, which is always foreordained to obey his will, exists on an entirely different level from human activity. On the secondary level, natural laws operate and humans have a certain freedom, but all causation derives from God, and creatures find that in truth they can only submit to his will, as a pen must submit to the hand that writes with it. Christian existentialists, such as German theologian Rudolf Bultmann, likewise split the subjective and objectives realms, concerning themselves only with the personal, direct relationship with God (Barbour 1997).

Theistic Evolution: Dialogue and Integration

A large number of theistic evolutionists seek a closer relationship than independence between scientific and theological understanding of cosmic evolution. They typically begin this task by assuming the basic Christian belief: God is both the transcendent Supreme Being who has brought the whole cosmos and its governing laws into being out of nothing, and the active continuing Creator, present within the universe and using natural laws and mechanisms to increase the complexity of both nonliving and living forms. They then ask how this accepted truth can be interpreted in the light of science. For example, the theological notion of God's "creative and providential action in the world" is seen as being accomplished *through* neo-Darwinian evolution—"Evolution is thus the way God creates life" (Russell 2000). All happens as science describes, but God is in some way directing and guiding and/or immanent within the process. Exactly how he accomplishes this without "intervening" and disturbing the natural causal order is the central subject of much current research in theology and science.

Many scholars are attempting to find a middle path between two extremes—one being a God who is removed to the beginning or acts only on the whole and never in particular circumstances, as in deism and uniformitarianism, and the other being a God who is continuously active but intervenes supernaturally in the causal order, as in traditional theism. The goal is to understand how God might act in a manner that does not interrupt the causal flow, in a real "objective" way (not just subjectively perceived) and in special or particular situations—a view that has been called "non-interventionist, objective, special divine action" (Russell 1995). In the following review (which follows the classification of Southgate et al. 1999), the first three approaches are conceptions in which God acts on the world as a whole—at the macro level—without using any natural gaps in the causal order. The second three exemplify approaches where God acts in many specific instances, utilizing gaps or indeterminacies which may exist in the physical world. Additional approaches explore the relation of theology of the Holy Trinity to evolution and holistic syntheses of evolution and Christology.

Neo-Thomism   One significant approach uses the principle of primary and secondary causality developed by St. Thomas of Aquinas and later reformulated and utilized by many Catholic and some Protestant scholars. God is seen as the primary cause who acts in nature through secondary causes—the laws of nature that science studies and the actions of human agents. By his primary action God gives existence to and sustains the entire natural world and endows secondary causes with the power to operate. The level at which they operate is complete unto itself, requiring no direct intervention by God. The entire process of cosmic evolution can be viewed as God's "continuing creative action" in the universe, as Jesuit astronomer William Stoeger describes:

If we put this in an evolutionary context, then, and consider what we know of the complexification of structure and the diversification of physical, chemical, and biological processes from a time shortly after the Big Bang, we see that we can conceive of God's continuing creative action as being realized through the natural unfolding of nature's potentialities and the continuing emergence of novelty, of self-organization, of life, of mind and spirit. (Stoeger 1995, 248–49)

In Stoeger's view one must fully accept the wholeness and integrity of both the created world and science. Neither life's emergence nor the rise of consciousness necessitated direct divine action. He also emphasizes the difficulty of truly understanding divine action. His direct or primary action of creating and preserving, for instance, "occurs at the very core of our beings and is hidden from our eyes." It is equally difficult to understand the "causal nexus" between God and secondary causes—whether God imbues those causes with his will using natural laws, or uses laws relating to the personal and to consciousness that go beyond the ones we know (Stoeger 1995, 252–53).

A scholar who contributed enormously to forging a connection between Thomistic divine creative action and scientific evolution was the influential Roman Catholic theologian Karl Rahner. For him, the whole evolutionary process is driven by the indwelling of God's power of transcendence in matter, which creates an urge to self-transcend and develop into something truly new. The process itself, as a secondary cause, requires no divine intervention; it emerges from a primary cause—the presence of God's own transcendence within the material world (Mooney 1996, 158–59).

Peters and Hewlett also discuss divine action in evolution in terms of primary and secondary causation, emphasizing how the scientific method, limited as it is to physical measurement and to secondary causes or their effects, is "blind to primary causation." Since Darwinian theory developed solely through the scientific method, it can only involve secondary causes and make no statement about God as a primary cause, a task which must be left for the theologian. They emphasize how God's primary action operated not only at the beginning but "over the entire historical sweep of the created order, even to its eschatological limit" constantly drawing novelty from all its potentialities (Peters and Hewlett 2003, 170–71).

In its basic formulation neo-Thomism separates scientific and theological descriptions, and in that sense it illustrates a relationship of independence. However, many of the scholars who use neo-Thomistic concepts seem to advocate a closer relationship than total independence (Stoeger, Rahner, Hewlett), while others do not (Barth).

God's Creativity and Top–Down/Whole–Part Causation   One prominent scholar who has considered and written extensively about divine action in the light of cosmic evolution is the late physical biochemist and

Anglican theologian Arthur Peacocke. He notes how the new dynamic, evolving cosmos necessitates a change in our concept of God—from one who just creates and maintains a static order to one who sustains a continual process of creativity. At every instant of time God is immanent in nature "creating . . . in and through the perpetually endowed creativity of the very stuff of the world" and making "things make themselves," as expressed by a Victorian novelist he quotes. "The processes themselves . . . *are* God-acting-as-Creator" (Peacocke 1998, 358–59). There are no gaps where God acts over and above the normal processes. The metaphor of a composer illustrates the relation he sees between transcendent and immanent God: a composer transcends his music, but deeply absorbed listeners experience his creative nature (his "inner musical thought") as immanent within it (Peacocke 1993, 176).

Peacocke fully embraces chance as an integral part of the creative process of God, who is the "ground and source" of both chance and law. Chance allows for the maximum exploration of potentialities of animate and nonanimate forms, and the constant interplay of chance and law determines the course of evolution. He sees it as "the only way in which all potentialities might eventually, given enough time and space, be actualized . . . it is as if chance is the search radar of God, sweeping through all the possible targets available to its probing" (Peacocke 1993, 120). While the outcome is open-ended and not predetermined at the microlevel, there are inherent tendencies for life and complexity and consciousness to develop, so that there could be "determinate ends" at the macrolevel—such as the evolution of intelligent humans who could interact with God. A number of other theologians have also discussed a positive role or interpretation for chance in a theistic world, among them Haught (1984, chapter 6; 1995, chapter 3), Barbour (1997, 239–40), and Ward (1996).

Peacocke conceives of God as acting on the world-as-a-whole perhaps through "top-down" or "whole-part" influence. In the former, processes at higher levels affect those at lower levels; an example would be the way a mind operates on a body, an idea incorporated in his God-as-the-world's-mind model described below. In whole-part influence conditions at the boundary or in the environment of a whole system—a biological ecosystem or a nonequilibrium chemical system, or, again, a brain in a body—constrain or direct the individual behavior of component parts. Such nonlinear systems can spontaneously become more ordered without violating the law of entropy—producing "order out of chaos," as Ilya Prigogine first called it. In the same way, novelty and higher level properties might emerge in nature. God, then, as the whole "environment of the cosmos," could bring about the emergence of novelty in particular parts (Southgate et al. 1999, 257).

Peacocke uses a number of rich metaphors to describe his panentheistic conception that God is both transcendent to and immanent within the

world. One image which underscores the combined importance of those
two qualities is that of the mother who gives birth to the world within
her own body. Another fruitful model is that of the composer who is
still improving and creating his symphony or a choreographer who allows for individual decision on the part of dancers. Theologian Conrad
Hyers similarly suggests that the interplay of lawful order and chance
is analogous to the way an artist works with a particular medium. Just
as a poet or dramatist or novelist intersperses the unexpected spontaneous idea with a overall direction or plan, so God might work with
order and chance in the universe (Barbour 1997, 241–42). It is interesting
to note the similarity of these images with ones some scientists envisioned
(Chapter 4).

Embodiment Models   A number of theologians, including Peacocke,
envision God's action on the whole of creation as analogous to the way
in which a mind works within a body. Such models make the strongest
use of analogies to human action, while also stressing the immanence
of God in creation. In this analogy, evolutionary history is the result of
the intentions of God's "mind" being acted out upon God's "body"—
the physical world. Theologian Sallie McFague also suggests this model
in *The Body of God*, as well as conceptions of God as mother, lover, and
friend. The interactions important in such human relationships—empathy
and mutual interdependence—are stressed over the dominance of an all-
powerful creator as in the monarchical model of God. In *God's World, God's
Body* Grace Jentzen goes further to propose that the world is the "medium
of God's life and action." The relationship is different from that of humans
to their bodies: God, as perfectly embodied, is more omniscient about
the world than humans are of their bodies and has total awareness of all
events, acting both universally and in special events (Peacocke 1993, 168;
Barbour 1997, 320; Russell 2000). Other feminist theologians who elaborate
on the model of nature as mother are Elizabeth Johnson, who celebrates
the merging of God's transcendent and immanent aspects in the mother-
creator image, and Anne Clifford, who calls for a similar merging of the
biblical creator God with dynamic, evolving nature in the image of "nature
as a mother giving birth" (Russell 2000).

   Philosopher Philip Clayton has expanded richly on the "panentheistic
analogy," arguing that "God's action can be much more coherently conceived if the world bears a relationship to God analogous to the body's
relationship to the mind or soul" (Clayton 1997, 100–101). For example,
by analogy with automatic functions within a human body, the action
of natural laws can be interpreted theologically as automatic actions taking place within the body of God. Intentional action, however, must be
understood using our most solid ideas about mind–body relations, and
a view he favors sees "mind as an emergent property of a particularly

complex physical system—a property which can then be causative in that system" (Southgate et al. 1999, 252–54). He shares Peacocke's view that such emergence is God's immanent creative activity at work. Panentheism, he thinks, develops naturally from reexamining theism in the light of science and even offers a closer relationship between the being of God and humanity than classical theism. Whereas traditional Christianity would speak of sensing the divine as our creator or sustainer or perhaps through direct communication, panentheism offers a fourth mode of being "aware of God because we are within God" (Clayton 1997, 102).

Chaos Theory   Modern science suggests that there are intrinsic gaps or unpredictable, spontaneous occurrences that leave openings or opportunities in the causal flow. Theologians have been exploring such gaps as ways God might bring about desired results in the world without directly intervening. Both chaos theory and quantum mechanics appear to offer such possibilities.

Physicist and Anglican theologian John Polkinghorne and others have advanced the idea that an arena for divine action might be chaotic systems, in which very small triggers can amplify into large-scale effects. This is the well known "butterfly effect," by which the fluttering of a butterfly's wings in one part of the world causes a storm half the world away a month later. Polkinghorne has argued that this extreme sensitivity to initial conditions can produce very unpredictable behavior—not just because we can't know them but, in his view, because they are genuinely open. God can therefore act by injecting not energy, which could be observed, but "pure information." No input of energy is needed because God exists everywhere, even at the microlevel. However, the "active information" will not enter via "localized mechanism" but will rather have a "holistic top–down character" involving the "the formation of a dynamic pattern" or an overall context. He relates this combination of non-material "information" and a physical matter to the metaphysic of dual-aspect monism in which the physical and mental or spiritual aspects are complementary attributes of the same stuff of the created world. This mental or spiritual world might be the means whereby God's information comes into the physical realm (Polkinghorne 1995, 154–55). Use of current chaos theory to explain divine action has been criticized on the grounds that it is still basically deterministic, but it is hoped that development of more complex "holistic chaos" and "quantum chaology" theories may be productive (Russell 2000).

Quantum Indeterminacy: Bottom–Up Causality   According to the most favored interpretation of quantum mechanics, events at the subatomic level are truly open and their future is unpredictable. This indeterminism affords an opening in which God could act in the universe within the bounds of natural law to affect outcomes in a way that would be neither interventionist nor perceivable. God would become the "hidden variable," as it were, which determines the actual path of an electron or other subatomic particle. The course of evolution could be affected by God's action at the quantum level on genetic mutations. As philosophical theologian Nancey Murphy expressed it, "The apparently random events at the quantum level *all* involve (but are not exhausted by) specific, intentional acts of God" (Murphy 1995, 339). This approach has also been explored recently by Thomas Tracy, physicist-theologian Robert Russell, and cosmologist George Ellis, all of whom find the approach promising but acknowledge a number of unsolved problems. There is still dispute about whether quantum uncertainty really represents true indeterminism, and numerous philosophical issues still remain with quantum mechanics itself. Both Peacocke and Polkinghorne are critical of quantum indeterminacy as a locus of divine action, as they both seem to favor a more whole/part kind of influence.

Process Theology   The metaphysical system of philosopher Alfred North Whitehead, developed as it was in the twentieth century in the light of relativity theory, quantum mechanics, and evolutionary theory, has an inherent consonance with modern science. Process philosophy posits that series of events and "interpenetrating fields" (as in quantum theory) and processes of change (as in evolutionary theory) are more real and fundamental than separate material objects. No hard and fast line exists between nonhuman and human life, now or in the past. Every organism is a coordinated network of mutually dependent events, and every occurrence happens in an environment that affects it. Thus the overall view of process philosophy is ecological, seeing the world as comprised of a whole integrated network of relationships (Barbour 2000, 115–17; Russell 2000).

Every event has a triune set of causes: past causes or occasions, divine purposes or the "divine subjective lure," and the entity's own response which might bring spontaneous, intrinsic novelty. All events also have two aspects—the view from within and the view from outside—and the "interiority" develops in evolution along with physical form. God is present in the interiority of every event, offering new possibilities in an inviting and persuasive way rather than coercing an outcome. He orders and structures the potentialities and new possibilities, thus acting as both the *"primordial ground of order"* and the *"ground of novelty."* Just as God is present in every event, he is also affected by events. God's nature as creator or ground of order and novelty never change, but his interactive nature and knowledge of the world do change. In John Cobb and David Griffin's reformulation of Christian belief in the light of process thought, this "dipolar character" of process theism becomes God's *"creative-responsive love,"* where the "ground of order" is identified with the biblical eternal divine Word and the responsive God with the aspect that changes and is affected by the world (Barbour 2000, 174–76; Cobb and Griffin 1976).

Process thought clearly allows for both general divine action—in the primordial creation of order—and special divine action—in particular events. Since God supplies new potentialities at every juncture, "no event is wholly an act of God, but every event is an act of God to some extent" (Barbour 2000, 176). God acts in essentially the same mode for all entities, although what results from his action varies greatly according to the level of sentience of the recipient.

A number of scholars have applied process concepts to a theological understanding of cosmic evolution and the role of chance, among them Barbour, Cobb and Birch, Hartshorne, and Haught. Barbour incorporates a panentheistic view within process thought. Nature is a "dynamic process of becoming . . . an incomplete cosmos still coming into being," where God is constantly active and affecting events through his persuasive but not coercive love. The long, slow process is understandable when one considers that what drives it is a luring rather controlling force. God is "pre-eminent but not all powerful" (Russell 2000), a creative, responsive player in the ever evolving community of living beings. With compassion and gentleness, God encourages all of creation toward his goals, never forcing the outcome. An ecological perspective is central and no soul/body or human/animal separations are emphasized. Humanity's commonality and community with other creatures are stressed. The panentheistic synthesis of God's transcendent and immanent natures fosters dignity and respect for all of nature. Charles Birch and John Cobb likewise de-emphasize the life/non-life split without disregarding clear-cut levels of complexity and capacity for conscious experience. God's immanence expresses itself as the "life-giving principle" and "the supreme and perfect exemplification of the ecological model of life." Thus, life is in no way purposeless and governed by blind chance; it is "suffused with 'the cosmic aim for value'" (Russell 2000). Roman Catholic theologian John Haught focuses on God as the originator of order, novelty and creativity and suggests that it is "God's will . . . to maximize evolutionary novelty and diversity" (Haught 1995, 68–69).

Trinitarian Theology   A number of Christian scholars have worked to understand how a trinitarian God acts in relation to the evolutionary process and especially what the incarnation and death of Christ on the cross signify in relation to evolution. Australian Roman Catholic theologian Denis Edwards believes that the process of natural evolution is the means by which God works out his goals in nature. In *The God of Evolution,* he grounds this view in a "trinitarian vision of God as a God of mutual relations, a God who is communion in love, a God who is friendship beyond all comprehension" (Peters and Hewlett 2003, 141). To make creation possible, he has freely limited himself. "The divine act of creation can be understood as an act of love, by which the trinitarian Persons freely make space for creation and freely accept the limits of the process . . . the limits of physical processes and of human freedom" (Peters and Hewlett 2003, 142). Nonetheless, God acts through processes, such as natural selection, to achieve his goals.

Jurgen Moltmann is another theologian who advocates the theology of God's "self-emptying" or "kenotic" love as the basis for creation and evolution. In this conception God self-limits his own being to allow space for creation to be and for freedom to exist within it. He situates all of past and future cosmic evolution within a trinitarian framework. The entire process of creation and evolution is a long series of self-limitations by God starting with the first creative act of the Trinity—"a community of love"—to withdraw into itself to allow the space for manifestation. The idea of divine withdrawal and "letting be" accords well with the scientific picture of the interwoven dynamic of chance and law that drives evolution. The long sequence of self-limitations took its ultimate form in Christ's crucifixion, and his resurrection marked the turning point after which evolution began to be redeemed (Southgate et al. 1999, 219–20).

Rahner also addresses the question of the significance of Christ's life, death and resurrection in relation to evolution. He notes that the self-transcendence that God bestows on all matter and that drives its forward progress takes a different form in the human being, who has the unique capacity to receive and give God's gift consciously. In Jesus the receiving and self-transcendence was total; he was the "unsurpassable climax of God's creative immanence in the world" (Mooney 1996, 165). Having a physical form like all humans, Jesus showed us God's eventual aim for the material world and for all incarnate beings; hence he is seen by Christians as the goal of all of creation. Rahner writes, "The Incarnation [is] . . . the unambiguous goal of the movement of creation as a whole, in relation to which everything prior is merely a preparation of the scene . . . [one can] conceive the evolution of the world as an orientation *towards* Christ, and to represent the various stages of this ascending movement as culminating in him as their apex" (Mooney 1996, 166–67). These words have striking consonance with ones written around the same time by Meher Baba: "The *Avatar* [Christ] awakens . . . humanity to a realization of its true spiritual nature . . . For posterity is left the stimulating power of his divinely human example, the nobility of a life supremely lived . . . He has demonstrated the possibility of a divine life for all humanity, of a heavenly life on Earth" (Baba 1967, vol. III, 16).

The Theistic Evolutionary System of Teilhard de Chardin   An evolutionary process culminating in the cosmic Christ also characterized the visionary thought of French Jesuit priest and paleontologist Pierre Teilhard de Chardin. His grand and sweeping synthesis of evolution and Christian belief stretched from the earliest beginnings of inanimate matter to a final culmination in Christ. He occupies a unique place in the thought of the twentieth century, finding no ready niche in academic disciplines. His "theology" was too unorthodox for the Catholic hierarchy, and his works were banned by the Church during his lifetime. He considered his major work, *The Phenomenon of Man* (more recently retranslated in *The Human Phenomenon*), a scientific study, but few scientists agreed. Neither was it a philosophical treatise—a coherent metaphysical system with well-defined categories—although not all scholars might agree with this assessment. Despite all the critiques and probably because of its deeply inspired and visionary quality, Teilhard's work became a source of inspiration to many, led to the formation of the American Teilhard Association in the 1960s, and has had an enduring influence among scholars ever since. A fine collection of essays inspired by Teilhardian thought over the last few decades can be found in *Teilhard in the 21st Century* (Fabel and St. John 2003), whose introduction has guided the description below.

In Teilhard's vision, consciousness is central to the developing universe. In one grand trajectory, a spirit-infused matter develops more complex outward form and greater inner capacity for experience and consciousness through the stages of prelife, life, and thought. The human is doubly the center as both the final object and the perceiving subject—the aspect of the cosmos reflecting upon itself and able to perceive its unity. He writes, "The human is not the static center of the world, as was thought for so long; but the axis and the arrow of evolution—which is much more beautiful" (Teilhard 1999, 7).

In *The Human Phenomenon* he traces this great arc of evolution from its primordial beginnings to its final spiritual culmination, always emphasizing the unity of spirit and matter. Matter derives from some unitary state, as described by cosmologists, and has three properties—plurality, unity, and energy—revealing itself to us as "radically particulate, yet basically connected, and finally, prodigiously active" (Teilhard 1999, 12–14). He sees it as being drawn forward, the forward movement balanced by the dissipating activity of the second law of thermodynamics.

More radically, he asserts that matter has an interior dimension:

Indisputably, deep within ourselves, through a rent or tear, an "interior" appears at the heart of beings. This is enough to establish the existence of this interior in some degree or other everywhere forever in nature. Since the stuff of the universe has an internal face at one point in itself, its structure is necessarily *bifacial*; that is, in every region of time and space, as well, for example, as being granular, *coextensive with its outside, everything has an* inside. (Teilhard 1999, 24)

Evolution is the on-going development of both of these dimensions, the physical and the psychic, through the activity of two different kinds of energy, which are but two aspects of a unitary process. The two kinds of energy involved are: "tangential" energy, which operates between elements "of the same order in the universe as itself (that is of the same complexity)," and "radial" energy, which draws the "element in the direction of an ever more complex and centered state, toward what is ahead" (Teilhard 1999, 30). In this interwoven process complexity of form and ever deepening consciousness (or "centricity") grow.

The evolutionary process crosses significant thresholds as it moves first from the development of inanimate matter in cosmogenesis to the rise of living forms in biogenesis and then to the emergence of the apparatus for thought in anthropogenesis. With the birth of thought the sense of self intensifies—there is greater personalization or "hominization." With man the cosmos becomes fully conscious of itself, reflecting back on but also now affecting the course of evolution. Humans can consciously join the process that inwardly draws all creation forward, helping to spiritualize and transform it through love. For Teilhard, becoming more spiritual meant joining the evolutionary flow and directing human energy towards all activity that fostered unity, increased consciousness, and enlivened the spirit of upward growth. As this process intensifies, noogenesis occurs and a collective global human consciousness emerges, eventually attaining its highest spiritual state in the Cosmic Christ of the universe or "Omega point." God is drawing the whole cosmos toward this state from the beginning and throughout the entire span of creation and evolution. Teilhard saw the world "as a mysterious product of completion and fulfillment for the Absolute Being himself." Divinity becomes immersed in matter at the very instant of creation and is its immanent driving force throughout the evolutionary process; there is "no creation without incarnational immersion" and one can achieve "communion with God through the Earth" (Grim and Tucker 2003).

Just how deeply his own spirituality was grounded in matter and belief in the forceful directedness of the natural world can be seen in a remarkable statement he made about his faith. He wrote:

If, as the result of some interior evolution, I were to lose in succession my faith in Christ, my faith in a personal God, and my faith in spirit, I feel that I should continue *to believe* invincibly *in the world*. The world (its value, its infallibility and its goodness)—that, when all is said and done, is the first, the last, and the only thing in which I believe. It is by this faith that I live. And it is to this faith, I feel, that at the moment of death, rising above all doubts, I shall surrender myself. (Teilhard 1971, 99)

Divine Action in Theistic Evolution: A Summary and Evaluation   The various approaches to divine action outlined above differ widely in approach, some using science as a springboard, some employing models of human agency and mind/body analogies, and others working from an existing metaphysical and theological system. All appear to accommodate the workings of chance, albeit in different ways. God works through quantum events, or God designed chance as part of his own creativity immanent in nature, or God works persuasively with law and chance in every event, or chance operates because God has self-limited his own power to allow humans and the laws of nature to operate freely. Reflection on divine action has also produced rich new conceptions of God and nature that go beyond the traditional monarch or craftsman to more organic and relational models: God as mind to the world's body, God as Mother giving birth to the cosmos, God as communicator of information, God as artist choosing among possibilities (determiner of indeterminacies), God as a parent limiting his power to participate creatively in a child's growth.

Various scholars have evaluated the efforts to understand divine action in relation to the natural world and noted that a coherent theory is far from achieved. The approaches through science seem too stretched or full of guesswork or bold assertions. Process philosophy and neo-Thomism, which both suggest a kind of double agency (part from God, part from nature) both have the problem that they lack grounding in what can be described by science or observed in everyday experience—they do not provide a mechanism (Southgate et al. 1999, 266–67). Boston University theologian Wesley Wildman puts it bluntly: "After all this time...the theological problem of divine action retains much the same shape and sharpness as it has had for 2,500 years, except that the most audacious account of all—the traditional Jewish–Christian–Muslim insistence that God acts in history and nature—has become quite obscure." He asserts that this failure of modern theology is a compelling reason for the new "discipline of science and religion" to exist (Wildman 1996, 57). Such collaboration is indeed proving to be promising, as scholars are researching how the different possible accounts of divine action may not be mutually exclusive but might be working in some combinatory fashion (Murphy 1995; Clayton 1997).

The Anthropic Principle: Design Arguments for God

An important issue for theistic evolutionists is how to interpret features in the universe suggestive of design. As discussed in Chapter 4, findings in physics and astrophysics have shown that natural laws and other conditions appear remarkably attuned to allow for a long preparatory growth of the cosmos and its eventual flowering into life and intelligence. The plethora of such "coincidences" are happily received by creationists as obvious evidence for a God who has perfectly designed natural laws and other conditions to accomplish his aim of producing conscious humans beings capable of communication with their Creator. Theistic evolutionists, however, are much more cautious for several reasons. For one thing, reasoning directly toward divine design by observing nature—natural theology—has generally no longer been considered wise or viable. The practice of ascribing features unexplained by science to the action of God— "God of the gaps" thinking—failed so consistently as science advanced that theologians now are eager to avoid it. Anthropic coincidences may in the future be explained naturally or shown to be necessary by a unified theory and not at all improbable. Even now, one side of the scientific anthropic debate argues that fine-tuning is inevitable and not surprising, if multiple universes exist—ours is just the expected one among an infinitude where the laws were just right for life. In addition, reasoning to the existence of a metaphysical entity such as God from empirical evidence in nature has been criticized by philosophers since Hume and Kant. Furthermore, the design argument would not uniquely specify the biblical Creator God, a problem for Christian theists. For theologians who explore new ecologically oriented creation theologies that avoid anthropocentrism, anthropic design arguments are also problematic in assuming "the existence of rational carbon-based life forms (i.e. humankind)" as the "ultimate goal of creation" (Southgate et al. 1999, 130).

Roman Catholic theologian Ernan McMullin admits that the design argument based on anthropic features is an "obvious" and "attractive" possibility to many, especially when it avoids direct intervention by God by focusing only his initial fixing of the laws. This is a broader type of natural theology in which the laws of nature themselves, rather than specific features, are evidence of divine design. McMullin nevertheless warns of the above weaknesses in this approach—its reliance on current knowledge gaps and its fragility in the face of the scientific controversy and possible changes in theory (1981, 45). It also has the philosophical weakness of arguing for a metaphysical reality from empirical evidence. This concern is shared by Stoeger, who writes, "The more profound grounds of explanation, necessity and possibility remain forever veiled" (Stoeger 1993, 222). McMullin concludes that "it ought to be clear that 'anthropic' features . . . cannot properly be used as *argument* for the Christian doctrine of a Creator" (1981, 45).

Several scholars emphasize that anthropic design arguments more logically lead to a designer operating within the universe itself and not the transcendent, biblical creator God. Possibilities identified by theologian Mark Worthing include a cosmic architect or demiurge or even the universe itself, as atheistic biologist Richard Dawkins has pointed out (Worthing 1996, 46–47). Philosopher John Leslie conceives of the God who might design anthropic features in neo-Platonic terms—one not beyond reason but a "creatively effective ethical requirement for the existence of a (good) universe or universes" (Leslie 1989, 186). This possibility of "designers" different from the biblical Creator God clearly limits the implications that anthropic design arguments have for Christian theology, according to Barbour (1997) and Russell (1989).

Despite the drawbacks and limitations, many scholars nonetheless think that there is a significant relationship to be explored between the anthropic principle and Christian theology. They would agree with McMullin that "a Being who 'fine-tunes' the universe... is *consonant* with the Creator God of the Christian tradition... consonance [being]... more than logical consistency, but much less than proof" (McMullin 1988, 70–71). Several lines of thought are followed in exploring this consonance. One is to re-examine theologically the current status of the scientific debate as a choice between design and multiple universes. Another is to work from what science says about cosmic evolution and life-tuned features and note that a metaphysical being like God is a plausible or probable hypothesis. Yet another is to work from the Christian belief in a creator God and identify what consequences might follow and what qualities a divine being would have who fine-tuned cosmic evolution to lead to life and intelligence.

Several authors disagree with some scientists' argument that we must choose between design and multiple universes. In *Universes* (1989), Leslie declares that both the many worlds and God hypotheses are strong, but they are not mutually exclusive. He questions why it is assumed that God made only one universe. Russell insightfully analyzes how different levels of multiple universes each have their own form of contingency, which could lead to design, so that "science will never eliminate the meaningfulness of contingency in the creation tradition" (Russell 1989, 198–201). For Peacocke, the possible existence of multiple universes has little impact on the remarkable fact that life has evolved to complexity and consciousness here. It just multiplies the range over which many possibilities were explored. He concludes, "Hence any argument for theism based on 'anthropic' considerations may be conducted independently of the question of whether or not this is the only universe" (Peacocke 1993, 106–8). John Haught agrees that for a theist a "plurality of 'worlds' is quite compatible with the idea of God." The infinitude of other universes is a metaphor for God's awe-inspiring "divine infinitude" (Haught 1995, 139–40).

Some scholars work from science towards God in a manner that steers away from strict logical proof and toward a hypothesis of probability.

Philosopher of religion Richard Swinburne follows this approach in *The Existence of God* and the shorter *Is There a God?*, where he used principles of confirmation theory in philosophy of science to argue for the probability of God's existence. A beginning plausibility of God's existence, grounded in simplicity and personal agency, grows in probability with additional evidence, such as order in nature, the need for external cause for the appearance of consciousness, and individual religious experience (Barbour 1997, 99). Peacocke makes a similar argument about the rise of consciousness, noting that the strong anthropic principle strongly ties the human being to the cosmos. Cosmic evolution, which began as a random jumbling of insentient atoms, has now produced a part of itself which science cannot explain—persons who experience subjectivity, have consciousness, are purposiveness, and hold values. Whatever produced all these disparate material and nonmaterial things must include both "personhood" and "matter" but as their source, must transcend and differ from both. The best explanation of such an entity is a personal transcendent God who acts with purpose and has dynamic qualities as an immanent, continuously creating Creator (Peacocke 1993, 106–12). For Polkinghorne, also, "the fine-tuning of a potent universe" is a meaningful finding that calls for a metaphysical explanation, since it must account for the natural laws themselves. Admitting that there is no sure argument for religious belief, he feels that the special life-tuned quality of our "potent universe finds deeply satisfying understanding within the intellectual setting of theism" (Polkinghorne 1994, 114–15).

In a new version of the design argument applied to the dynamic universe, Jesuit theologian W. Norris Clarke argues the intricate workings of the basic, active elements in the "cosmic-wide order" can only be explained by the existence of a "World-Ordering Mind," without whose "primal ordering . . . nothing could happen at all, not even by chance" (Clarke 1988, 119). Like Peacocke, he speculates imaginatively about the qualities of the Creator of our amazing universe:

Must not the "personality" of such a Creator be one charged, not only with unfathomable power and energy, but also with dazzling imaginative creativity? Such a creator must be a kind of daring Cosmic Gambler who loves to work with both law and chance, a synthesis of apparent opposites—of power and gentleness, a lover of both law and order and of challenge and spontaneity. (Clarke 1988, 121)

The design hypothesis is clearly very appealing for those already committed to theistic belief, and some scholars begin with this belief and ask what observable consequences flow from it, thus interpreting science in the light of theology. Ellis follows this approach in developing his "Christian Anthropic Principle"—a synthesis of science's anthropic principle and the theology of religious leader William Temple and his own Quaker perspective (Ellis 1993b). He begins with the "essential core" of Christian teaching, that God is the transcendent creator and immanent sustainer, the embodiment of justice and holiness, and a personal, active God who loves each being and whose divinity is perfectly revealed in Jesus. Forgiveness and self-sacrificing love characterize both the aim of the Christian follower and the nature of God's action. His goal of loving action "shapes the nature of creation," which he designed with just the characteristics necessary "to attain the goal of eliciting a free response of love and sacrifice from free individuals." Thus orderly laws, provision of physical needs, hidden action by God all allow for free action, while direct revelation provides encouraging glimpses of ultimate reality to those who are open. In this context, anthropic features receive a more profound explanation than science alone can give—they are part of the original design that allowed the spiritual goal of the universe to unfold. Christian theology thus gives the anthropic principle a more profound explanation than science alone can give.

Theological consideration of the anthropic principle and design arguments is an illustrative example of the rich new ways scholars are advancing understanding and methodology at the interface of science and religion. It has led some to formulate a new natural theology based on the dynamic cosmos or plausibility arguments, as with Clarke and Swinburne, and others to advance and use new methodology for studying science and theology together, as with theologian and philosopher Nancey Murphy. For her, Ellis' Christian Anthropic Principle helped to frame and elaborate her broader "science-like" theological research method, based on Imre Lakatos' theory of knowledge. Cosmological fine-tuning in this light argues for Christian theism only as one of a whole network of "auxiliary hypotheses" which surround the core theistic belief and which must predict new facts and seek confirmation.

In the approaches described above, the understanding of God can be aided by science, and intractable scientific problems such as the anthropic principle can be explained more profoundly. Scientific and religious knowledge can be studied together, in what theologian Mooney calls "a completely different kind of epistemological project, one with hitherto unexpected illuminative power." The two lines of knowledge run side by side, like "two meridians on the sphere of the Christian mind" (as Teilhard de Chardin suggested). At the equator, or "present time," they run parallel and each separately gives "signs of both their present consonance and their possible future convergence at some pole of common vision" (Mooney 1996, 62–63). On the religious line, Christian "data" reveals the supreme importance of human persons as the ultimate product of God's creative activity over all of cosmic history culminating in Christ's incarnation into material form. Scientific data about cosmic evolution and fine-tuning gives the Christian specific information about how God has undertaken the design activity, or, as Russell expressed it, "concrete language for our deepest insights about God's relation to creation" (1989, 201). Thus the two combined "sets of data" nourish deeper, more expansive understanding of humanity in relation to the Christian God and to the cosmos.

The merging of scientific and some theistic perspectives can also correct what many regard as the overly anthropocentric character of the Christian view. Cosmos-wide fine tuning reveals both the immense canvas on which God's creative activity works and also his concern with "the potential riches of all matter, all energy, all forms of life" (Mooney 1996, 650). Thus, certain theistic perspectives urge enlargement of the anthropic principle in recognition that another scientifically verifiable process has been at work—to experiment with as many different forms as possible—what Freeman Dyson called the "principle of maximum diversity." Haught characterizes this adventurous drive toward ever greater diversity and beauty as the "aesthetic cosmological principle" with God as the "One who wills the maximization of cosmic beauty." The emergence of living forms and intelligence, important as it is, is thus just part of a "more encompassing cosmic adventure toward an ever greater breadth of beauty" (Haught 1995, 140).

Directionality, Divine Purpose, and the Problem of Pain, Suffering, and Evil

An all-important question lying at the heart of all theistic evolution discussion is whether the universe can be inferred to have directionality, and if so, a divine cosmic purpose. Such a purpose is assumed in virtually all traditional religions, even if knowledge of it might remain inaccessible as part of God's mystery. If there is such a purpose, and if God is acting in the cosmos to fulfill his goals, two problems must be addressed. First, how does it accord with the seemingly purposeless activity of randomness and natural selection in cosmic evolution, and second, how can one explain the existence of evil, pain, and suffering? This latter problem, theodicy, is an on-going problem for theology, which has addressed the issue of moral evil and suffering at the hand of nature for centuries. What cosmic evolution adds to the picture is the idea of evolutionary "evil"—that progressive development in the created world involves so much struggle, suffering, and death on the part of individuals and species. These are each vast topics within theology, and what follows is but a sample of representative thinking on the subject.

Most, but not all, evolutionary biologists firmly deny the existence of any directionality in evolution, because of the random, unpredictable nature of change, but Jesuit astronomer-theologian Stoeger (1998) makes a strong argument for it. He surveys every phase of cosmic evolution, beginning with the Big Bang, and examines how the natural laws move material reality in the direction of developing more structure. Beyond the Planck Era and the Inflationary Epoch within the first microsecond, the expanding, cooling universe determines a global cosmic directionality. Gradually original quantum fluctuations grow into the macroscopic seeds of later galaxy groups, where stars form and begin to manufacture and distribute heavy elements. The same directionality gets more focused as planets form and provide environments for development of chemical and biological complexity. Just the simple laws, chance, and changing conditions functioning together at each step produce order and "directedness" and "orientation towards complexity" (Stoeger 1998, 169). Science cannot say what causes this inherent directionality, for it is not perceivable at the level of individual interactions, but only in the progression of the system as a whole.

A kind of teleology is definitely at work here—not a fixed, dictated plan for a specific end form but a movement toward realizing possibilities in a systematic manner. Stoeger describes it as a "very rich notion of directionality and teleology, which gives freedom and autonomy to the laws and processes of nature and encourages them to explore and realize the full range of . . . potentialities of the universe" (Stoeger 1998, 185). Whether such a process is intentionally directed for a purpose by divinity can neither be known nor disproved by science but is knowable through revelation for the Christian, for whom divine purposiveness is an inevitable conclusion.

Peacocke believes that directionality can be seen in nature's inherent propensities to develop complexity, ability to process and store information, and language. Divine purposiveness can be inferred, since God is working in and through all processes in creation. In this light, the enormous time humans took to evolve and the enormous number of forms that developed and died out must have been part of God's intention. Hints exist that God must have taken great delight and joy in this rich proliferation, as in the Genesis I statement that God "saw . . . that it was very good" and in the Hindu concept of God's divine play (*lila*) in creation.

The increase in information-processing ability brings expanded consciousness but also inevitably an increase in pain and suffering. As a warning of danger and disease, pain is necessary to survival and thus to the ongoing forward movement of evolution; such sensitivity is selected by nature to prevail. Death is necessary to make room for the new and more complex forms, and the assimilation of old forms in evolution and feeding on other species is necessary and efficient for timely progress. Peacocke suggests that all of these workings of nature "are . . . the very action of God" and that "God *suffers in, with and under the creative processes of the world* with their costly unfolding in time" (Peacocke 2001, 86). Birthing the new—a child or work of art—always involves painful and costly struggle. In an act of loving self-sacrifice, all-powerful God limits his power—emptying himself—to participate as a fellow sufferer in the natural evils of the world for an important divine purpose:

to bring about a greater good thereby, that is, the kaleidoscope of living creatures, delighting their Creator, and eventually free-willing, loving persons who also have the possibility of communion with God and with each other. (Peacocke 2001, 88)

God's suffering is active and, as part of his divine love, has creative power to bring about the new. It is a risky process for God to create free beings who can rebel against his process, but one totally necessary. His goal of instilling values of "truth, beauty and goodness" in the world can only be achieved by persons freely choosing to work toward them and hold them here in creation.

In Peacocke's view God's immanent role within cosmic evolution necessitates a revision of certain older Christian doctrines, especially about original sin, redemption and the meaning of Christ's incarnation and suffering on the cross. First, biological death can no longer be viewed as the "wages of sin" but is a necessary part of God's ongoing process of creating new forms. Second, there could not have been a first man and first woman who enjoyed a state of perfect union with God and whose rebellious acts introduced pain and suffering into creation. Third, Jesus' suffering and death were not meant to save humanity from an original sin but to reveal God's suffering with us. He writes: "The suffering of God, which we could glimpse only tentatively in the processes of creation, is in Jesus the Christ concentrated into a point of intensity and transparency which reveals it to all who focus on him" (Peacocke 1998, 372). Finally, his incarnation and resurrection revealed to humanity what is possible for each one of us, "the paradigm of what God intends for all human beings, now revealed as having the potentiality of responding to, of being open to, of becoming united with, God" (Peacocke 1998, 375).

A theologian who has given much thought to the issue of cosmic purpose and theodicy is Haught. In 1997 he brought together an international group of scientists and religious scholars for a conference on "Cosmology and Teleology" cosponsored by the Georgetown Center for the Study of Science and Religion and the American Association for the Advancement of Science's Program on Dialogue between Science and Religion. The presentations are summarized in the collection of essays, *Science and Religion in Search of Cosmic Purpose*, which Haught edited. In his own paper, he notes how cosmic meaning in the physical, temporal world was always understood in traditional religions as flowing down from higher nonmaterial realms in a hierarchy or "Great Chain of Being." Modern science's attempt to explain all reality in terms of material energy and particles and the linear story of cosmic evolution has collapsed this hierarchy, and theology must now seek to recover it or find a way to synthesize it with science's new horizontal evolutionary picture. Additional levels of value and meaning, whether they be conceived in a vertical hierarchy or in nested circles or somehow infused into linear progression, are indispensable to understanding any cosmic purpose in the material world.

Haught believes that a promising path lies in studying science's own hierarchical structures and the elusive concept of information. Science has seen that new properties emerge at higher levels of complexity that are unexplainable in terms of elements of lower levels. An example from Chapter 5 was the "information-rich" sequence of base pairs in the first molecules of life, whose origin cannot be determined by physical processes alone. Haught believes that information can flow from higher to lower levels of existence without disrupting the normal causal flow of matter and energy, an idea he illustrates with an example drawn from Polanyi's writing. Imagine one draws random markings with a pen but then without lifting the pen begins to form letters and meaningful words. Information was "injected" into the system without interrupting the material flow. He likens the elusive "information" to the mysterious "Tao" of Taoism—"the unnamable Way or Truth" behind all life, self-concealed and yet powerful, and knowable only to the spiritually aware. Perhaps one becomes conscious of the "noninterfering effectiveness of information" and the nature of cosmic meaning "only after we have ourselves undergone a personal transformation in which the Taoist humility and sensitivity to the power of non-being has begun to reshape the center of our own lives" (Haught 2000b, 116). Such a source of meaning and purpose lies beyond the limited knowing of material proof but is knowable for those with inner vision. As in the pen and writing example, there is no material discontinuity between the random scribbles and the meaningful words, but for one who knows the words there is an abrupt injection of something new (Haught 2000b, 112–19).

In other writings Haught views cosmic purpose from the vantage point of process thought, where it is seen as an aim toward beauty. He contrasts this with the view of Teilhard de Chardin, for whom divine cosmic purpose lay in the development of human consciousness, first in individual and then in collective manifestations. The aim toward beauty, Haught contends, provides a more encompassing and less anthropocentric notion. Beauty is the focus because it involves a balancing or synthesis of contrasting or opposite elements—"harmony and complexity, order and novelty, stability and motion." Too much of one leads to chaos, while too much of the other leads to boring sameness. For Whitehead the cosmos is an "aesthetic reality," much like a work of art, music or writing, whose value derives from the extent to which unity and harmony of polar opposites are found. Thus "we might value a universe in which contradictions are constantly being unified into an aesthetic whole: entropy and evolution; order and chaos; novelty and continuity; permanence and perishing." An individual perspective may see only discordance, but such a view lacks a "wider angle of vision" necessary to see the whole (Haught 1984, chapter 8).

From a Christian perspective, Haught describes God as carrying out his purpose through "kenotic love" and the "power of the future" (2000a, 110). By his self-giving love God lets the world undergo its evolutionary unfolding, yet he remains intimately involved with the evolving world. Christian faith in the resurrection also sees God as opening up a bright new future for humanity. Bringing about novelty in evolution points the way to this grand future renewal. Theological openness to spontaneous new future outcomes resonates with the unpredictability in science's chaos and complexity theory. Furthermore, cosmic evolution, as a still unfolding story with an open future, can become part of the religious story of promise of hope for an ultimate "cosmic fulfillment." From its earliest moments the universe had the fine-tuned features to lead to life—it held the "*promise* of emerging into life." Why should we not then "claim confidently that the *present* state of the cosmic story is not also pregnant with potential for blossoming into still more abundant new creation?" (Haught 2000a, 118).

As with most theologians, Haught struggles with the issue of suffering, which he calls "an open sore that theology can never pretend to heal" (Haught 2000a, 55). One perspective that evolution offers is the idea that the universe is not yet finished but is still imperfect, and part of that imperfection is the existence of pain. Haught also explores the meaning of suffering in terms of the aesthetic teleology discussed above, where suffering has meaning in relation to the goal of maximizing cosmic beauty and our intense experience of it (1984, 128). While God never causes or wills for suffering to occur, entities are drawn by God toward ever more intense "aesthetic enjoyment," which involves risk. To advance and gain more capacity, an entity must be open to what is new, but it can fail and fall apart into disorder and "evil." When such failure does occur, it is saved by "God's aesthetic care . . . which is infinitely sensitive to particular sufferings, identifies with them, takes them into the divine life and transforms them into an aspect of the beauty of the cosmos in order that they never be forgotten or lost" (Haught 1984, 129). In this way, God's "compassionate embrace enfolds redemptively and preserves everlastingly each moment of the cosmic evolutionary story" (Haught 2000a, 119). Even though God is in some way responsible by having created the process wherein evil and suffering occur, he is faithfully present as a fellow sufferer at every instant of pain.

Throughout this sample of representative views about cosmic purpose and the problem of evil and suffering, certain themes emerge. One is that science and religion both agree that neither cosmic directionality nor cosmic purpose can be discerned at the material level. Directionality can be seen only when human consciousness observes the whole history and notes the obvious changes from simple to complex structure and the emergence of mind and spirituality. Theistic evolutionists all agree that God's purpose is definitely at work in the whole process of cosmic evolution, but it can only be discerned by a different kind of knowing than direct proof. God's involvement is never in a direct manner, but is through secondary causes or in a very self-restrained, kenotic way or from the perspective of the future, in all cases nurturing the whole enterprise along and suffering with each being in it. In a loose teleological way, the cosmos is growing and developing, not so much according to a preconceived blueprint but following propensities and exploring a maximum of possibilities. Finally, with cosmic evolution pain and suffering are inevitable in a still growing, still imperfect universe, but are also the natural accompaniment of increasing sensitivity and consciousness. For humans evil and suffering naturally result from God's gift of freedom and from our frequent failures as we struggle to develop our full potential.

Chapter 7

The Cosmology of Meher Baba

No description of cosmic evolutionary systems would be complete without including the perspective of a unique modern spiritual figure from India who has written extensively on the subject of creation and its purpose—Meher Baba (Figure 7.1). Born Merwan Sheriar Irani to Persian Zoroastrian parents in Pune, India, in 1894, Meher Baba as a young man is said to have achieved a full and complete experience of divinity and its workings through association with respected spiritual masters of diverse faiths working in central India. Disciples began to gather around him in the early 1920s and gave him his name, "Meher Baba," meaning "Compassionate Father."

Following a period of training disciples and traveling in India and Persia (Iran), Meher Baba settled into an abandoned British military camp in central India, which came to be called Meherabad. There he established an ashram, or spiritual community, which undertook a number of service projects—a free hospital and dispensary, shelters for the poverty stricken, and a school that welcomed children of all castes and creeds and emphasized spiritual learning. These activities were all undertaken while Meher Baba maintained strict silence, which he began in 1925 and maintained for the remainder of his life.

Meher Baba's silence seemed to reflect his belief that mankind had been given enough words about how to live a spiritual life; what was needed now was a reawakening to the truth at the heart of all religions. While he encouraged some disciples to continue to practice the rites and rituals of their own religion, he emphasized a common core belief of all religions—that true spirituality could be sought only by pure love for God and by selfless service to others. His views about the religions of the world and the true path to God are represented on his "Mastery in Servitude" emblem which displays the symbols of all the major religions of the world (Figure 7.2).

Figure 7.1 Meher Baba. This unique spiritual figure of modern India offered detailed explanations of the workings of the cosmos and the nature of the spiritual path in his major works God Speaks and The Discourses. His cosmology blends ancient schools of Hindu thought with Christian emphasis on the important role of the Christ in the ongoing progress of creation. The scientific picture of physical and biological evolution is also fully integrated into his system. (© Lawrence Reiter. Used by permission)

On one of the first of many travels to Western countries in the early 1930s, he delivered the following message regarding his views on religion:

My coming to the West is not with the object of establishing new creeds and spiritual societies and organizations, but it is intended to make people understand religion in its true sense. True religion consists in developing that attitude of mind, which would ultimately result in seeing one Infinite Existence prevailing throughout the universe; when one could live in the world and yet be not of it, and at the same time, be in harmony with everyone and everything; when one could attend to all worldly duties and affairs, and yet feel completely detached from all their

Figure 7.2 Emblem of Meher Baba. A central aim of Meher Baba's work was to foster unity among the world's major religions—to bring them together like "beads on one string," as he expressed it. This goal is illustrated on his emblem, or colophon, where symbols of the world's major religions surround a unifying theme of gaining mastery of the self by loving and serving others. (Reprinted by permission of the Avatar Meher Baba Perpetual Public Charitable Trust)

results; when one could see the same divinity in art and science, and experience the highest consciousness and indivisible bliss in everyday life . . . I intend bringing together all religions and cults like beads on one string and revitalize them for individual and collective needs. This is my mission to the West. (Kalchuri 1979, 1554)

In such statements, Meher Baba defined spirituality for the modern age and also emphasized the underlying unity of all religions and the possibility of unifying science and divinity. In *The Discourses* he expanded on this latter point:

It is a mistake to look upon science as anti-spiritual. *Science is a help or hindrance to spirituality according to the use to which it is put.* Just as true art expresses spirituality, so science, when properly handled, can be the expression and fulfillment of the spirit. Scientific truths concerning the physical body and its life in the gross world can become a medium for the soul to know itself; but to serve this purpose they must be properly fitted into the larger spiritual understanding. (Baba 1967, vol. I, 19)

THE MESSAGE OF *GOD SPEAKS*

Meher Baba laid out the "larger spiritual understanding" in his major work, *God Speaks: The Theme of Creation and Its Purpose,* probably one of the most comprehensive descriptions of the fundamental workings of the cosmos and life ever written. It strongly echoes certain ancient Hindu beliefs but reformulates them for the modern age with much new elaboration about the growth of consciousness, its evolutionary framework, and the higher states of consciousness as understood in several religions. *God Speaks* traces in careful detail the journey of the soul from the moment of its manifestation from within God's being through its evolution in inanimate and animate forms and its reincarnation in human forms to its final involution and return to God. In many ways the progress of the forms through evolution closely mirrors the picture given by evolutionary biology but with two crucial distinctions. First, in evolutionary biology consciousness emerges as a by-product of evolution, whereas in Meher Baba's system the development of consciousness is of central importance from the very beginning of the process. An inner urge within every being to seek greater consciousness drives the process along, giving it directionality and purpose. This drive is really God's own initial urge to know himself. The embodied soul, ever one with God, unconsciously feels and responds to this force, seeking wider and wider consciousness of the world and eventually of its own being as God. A second major difference is that the entire process of evolution is, in Meher Baba's system, embedded within a larger spiritual whole. It is but one of five major stages of the journey of consciousness: Creation, Evolution, Reincarnation, Involution, and Realization. The developing consciousness emerges from within the being of God at the moment of creation, grows and evolves through experience in multifarious forms, fully experiences the created world through human reincarnations and eventually turns inward in involution to realize its oneness with God. It is a journey from "eternity to eternity," as one essay expressed it (Baba 1958, 77).

Meher Baba's explanation of creation and evolution begins in *God Speaks* with an affirmation of the absolute unity of all beings: "All souls (*atmas*) were, are and will be in the Over-Soul (*Paramatma*). Souls (*atmas*) are all One. All souls are infinite and eternal. They are formless. All souls are One; there is no difference in souls or in their being and existence as souls" (Baba 1997, 1). What distinguishes them is their level of consciousness—gross (or physical), subtle, or mental. As the soul proceeds along its journey, it gathers and stores these three types of impressions as a by-product of living. They are an invaluable tool by which consciousness expands, but in the end they must be eliminated for the soul to become aware of its deepest self and realize its oneness with all other beings.

CREATION AND EVOLUTION

The Beginning of Creation—the Initial Urge

In Part 2 of *God Speaks*, Meher Baba traces the development of one "unconscious soul" from the "initial urge" through the "journey of evolving consciousness" (1997, 8), which is depicted visually in a painting by an American woman artist, Rano Gayley (Figure 7.3). The soul at the beginning is in an "infinite, impressionless, unconscious tranquil state," at one with God in his Beyond-Beyond state, where God purely exists. All opposites are merged within God's being: light and darkness, everything and nothing, consciousness and unconsciousness. It is like a state of deep sleep, where everything is at rest, and all things are latent but absolutely still and unmoving, neither conscious nor unconscious.

In this original, "deep-sleep" state of God there once surged a whim or an urge to know himself. Since no movement or thought can ever occur in the Beyond-Beyond state, this urge actually occurred in a slightly different aspect of God's transcendent being—the Beyond State. There the urge to know first manifested itself as a most finite point called the "*Om* Point or Creation Point." The "shadow of the Infinite... gradually appeared... seeped through or oozed out of the most finite point... and went on expanding."

With the surging of the whim, God thus began to wake himself up, and he experienced simultaneously both infinite consciousness and infinite unconsciousness. Such a strange paradox is impossible for the human mind to comprehend, according to Baba, but it will be known eventually to everyone through direct experience. The infinitely conscious aspect of the impersonal God, unaware of anything but himself, enjoys and delights in his Godhood. God's infinitely unconscious aspect is driven to become conscious and yearns to reestablish the oneness that was lost. Thus is initiated the divine drama of creation. The reunion cannot happen instantaneously, since there is an "infinite disparity between the two." Infinite unconsciousness has now received the tiniest, most finite impression of its opposite and descended into a "primal duality," which it cannot extricate itself from or resolve immediately. It must reach out towards full consciousness "through a long-drawn-out temporal process of evolution"... first trying to "fathom its own depths, then by backward treads [seeking] and ultimately [finding] the infinite consciousness through numberless steps, thus fulfilling the whim from the Beyond" (Baba 1958, 8–10; 1997, 8–9, 164).

Additional details about the descent of divinity into material form are given in another work, *The Nothing and the Everything*, composed under Meher Baba's direction by his close disciple Bhau Kalchuri. There Kalchuri explains the phases of existence between the initial urge and the "creation" of the physical world. He describes how the "Ocean of Everything" (Infinite Consciousness) and the "Ocean of Nothing" (Infinite Unconsciousness) emerged from the Beyond-Beyond "Ocean" of God's Being. When the whim arose in the Ocean of Nothing, it was also felt in the Ocean of Everything, which awoke slowly over ages and ages to ask the eternal question "Who Am I?" When it answered "I am God," the Infinite Consciousness "was established in the Ocean of Everything" as the

Figure 7.3 "Creation, Evolution, Reincarnation, Involution and Realization, according to Meher Baba." The chart depicted here is a pictorial version of Meher Baba's major work, *God Speaks: The Theme of Creation and Its Purpose.* This version is a new rendering by artist Norman Remer of the original, which was painted by Rano Gayley under the supervision of Meher Baba and reproduced in *God Speaks* with the caption below.

God in the Beyond-Beyond state represents God as pure essence, infinite, original and eternal, unaware of anything, even of Himself. God in the Beyond state represents the Over-Soul (*Paramatma*), essentially the same as God in the Beyond-Beyond state except that here surged the whim to know Himself and He became conscious of infinite power, knowledge and bliss, and simultaneously conscious of Illusion which manifested as the Creation. By completing His journey through the worlds of forms He sheds the illusion of their apparent reality. Reading counter-clockwise, the first forms taken by souls emanating from the Creation point are gaseous. As consciousness evolves, souls take the innumerable forms indicated, experiencing increasing impressions (*sanskaras*). Arriving at the state of man, the soul has achieved complete consciousness and reincarnates innumerable times until it is ready to experience involution, all of which takes place while embodied in the gross world. While getting free of *sanskaras*, the ascending soul gradually becomes aware of the seven planes and higher spheres until it is liberated from all bindings and becomes one with God (God-realized). The first three planes depict subtle awareness; the fourth portrays the vast powers and energies encountered there; the fifth is the plane of sainthood and is in the mental sphere; the sixth is the plane of illumination and the seventh is the plane of God-realization, i.e., unity with God. (Rendering of the original chart and reproduction of the caption to Chart VIIIA in *God Speaks* [between pages 190 and 191], by permission of Sufism Reoriented) eternal God who knows Everything. This transcendent state, in which God is conscious only of being God, is known as "*Parabrahma-Paramatma*" in Hinduism, "*Allah*" in Islam, "*Yezdan*" in Zoroastrianism, and "God the Father" in Christianity.

"But the poor Ocean of Nothing!" Kalchuri writes. When it heard the question "Who Am I?" it had no response, being unconscious. At the *Om* Point where it heard the question, numberless queries and responses poured forth into its being, and the Infinite Unconscious became established in the Ocean of Nothing. This state of God's being is named in Hinduism "*Ishwar*," better known by its three aspects—*Brahma* (the Creator), *Vishnu* (the Preserver), and *Mahesh* or *Shiva* (the Dissolver), who originate, sustain and transform manifested creation.

Thus, between the whim and the beginning of the created world, there was an interlude during which apparent duality had to become established. This process is described further in *The Nothing and the Everything*:

In this time before creation began original fire (Tej) in infinitely finite form manifested in the Ocean of Nothing...near the Om Point...The Whim gave rise to this fire, [which] gave rise to infinitely finite energy (Pran),...which required space (Akash) to manifest...and instantly there arose conflict between these two...a powerful clash between space and energy [which]...over aeons of time...awakened the ocean of Everything to ask and answer the eternal question. (Kalchuri 1981, 32)

The desire or urge to know that sprang forth from the whim gave rise to seven major desires, which flooded the Ocean of Nothing before creation and established themselves afterwards in seed form in the mental world, in a "germination" state in the subtle world, and in the physical form of "action" in the gross world. Humans experience these desires as "lust, anger, greed, hatred, pride, selfishness, and jealousy."

In a long slow progression over hundreds of thousands of years, the original fire of God became transmuted in its descent down through the world of mind (mental sphere) and the world of energy (subtle sphere) finally to take on a gross or physical form. Using the ocean analogy, the energy or movement created large "wave-bubbles" in the Ocean of Nothing, which were the universes and physical worlds, and "drop-bubbles," which were the individual forms that souls acquire. As these drops individuated, they passed down through the six planes of existence, two mental and four of subtle energy, before they acquired a physical form in the world of gross energy. The original energy or movement contained these seven states, each of which imbued the drop-soul with a "bubble of energy," or "body," as it descended. In the case of the first six—the mental and subtle states—"the stir and the bubble are so subtle, and the movement is so exceptionally rapid, that both are absolutely unseen . . . The seventh state of movement created the bubbles that are seen. (These bubbles are the gross forms of evolution.)" These bubbles that surround the drop are the mental, subtle, and physical bodies.

The drop-soul, or "*jeevatma*," thus descended through levels of highly rarefied matter or gases on its way into physical form. It "[passed] unconsciously through two planes of fire and light in the mental world . . . and four planes of 276 gases in the subtle world." These subtle gaseous states hold subtle energy (*Pran*) and express themselves fully throughout the infinite space (*Akash*) of the subtle world. The activation of matter through subtle energy eventually brought about its fullest manifestation—in the physical realm—as the 277th state of subtle gas, the gross gas hydrogen. In succession, the proton and electron formed, and then the atom, in a process that required some millions of separate impressions or "sanskaras." With the coalescence of roughly 10 million atoms, a particle of dust or "stone" was formed. The journey of the drop-soul through the subtle and mental states of fire and gases, through the form of atoms and into the physical form of stone required hundreds of thousands of years (Kalchuri 1981, 26–42)

Between the last form of subtle gas and the appearance of the first gross gas, hydrogen, however, there was an even more insubstantial first physical form, one so "inconceivably, infinitely finite . . . so very infinitely shapeless and substanceless, matterless and formless, that it cannot lead one even to imagine that it is gross." It appeared "simultaneously as if in three prongs, as the first three of the foremost seven 'gas-like' forms," indescribable except in terms of having in turn "infinitely negligible density . . . negligible density . . . and the first traces of density." The next three, which are conceivable to our minds, can be called "semigaseous and semi-material." Evolution starts with the first of these: the fourth gas-like form. Hydrogen is the seventh and final of the "gas-like" forms (Baba 1997, 174).

Evolution of Worlds

The passage of individualized forms, or "drop-bubbles," through all the subtle and gross gaseous states occurred throughout a vast universe, containing "billions and billions of nebulae [galaxies]," each containing "millions of worlds," the "wave-bubbles" stirred up by the original whim. These too evolved. Planets formed from gaseous mixtures which cooled and hardened into irregular rocky crusts on the surface, remaining hot and liquid within. With further cooling and liquefying of gases, water filled the uneven spaces between large rocky structures, the newly forming mountains, and became seas, lakes and oceans. There, eventually, evo-lution of living form began with simplest of plant forms (Kalchuri 1979, 1875).

Evolution Begins in the Physical World

The long evolutionary process thus takes the individual drop-soul through many earlier subtle and gross gaseous states over a long period of time before it reaches even the most rudimentary of visible forms, the stone. Even a stone, according to Baba, has rudimentary consciousness, though it is "most, most-finite" (1997, 10). What happens from then on is a pattern of development that repeats itself over and over, a process by which consciousness evolves bit by bit through innumerable forms in each of the kingdoms: stone, metal, vegetable, worm-insect-reptile, fish, bird, animal, and finally human. In each form, the conscious soul receives impressions and it "must necessarily experience these impressions, and in order to experience the impressions, the consciousness of the soul must experience them through proper media [or bodies]" (Baba 1997, 10). He emphasizes again and again how the consciousness gained is not of reality but of the illusory world of separate form.

Within each kingdom there are numberless "varied species," each of which is a vehicle for the soul's consciousness to learn certain qualities and gather certain impressions. With each form, the soul reaches the end of what it can learn from that species and ceases to identify itself with that species. It realizes it is not that form. That form "is dropped" or dies, and for a while the soul has no form or "medium," but the impressions remain. The consciousness, while still focused on the impressions left by the last species, begins to "associate with" the next species of that kingdom through a new gross form or medium. The new form "is always created and moulded of the consolidated impressions of the last species of form with which the soul associated and identified itself." In the new form, it experiences the impressions from the last form, even as new impres-sions are gathered. Thus the soul develops its next body in seed form by gathering impressions in its present body, even as it is living out and ex-periencing the impressions from the previous body. In a single kingdom then, endless different experiences of numberless impressions gathered by the soul's consciousness through different species produce the growth and development of consciousness of the physical world (Baba 1997, 12–13). When a new form is taken on, the soul perceives more of the created world, as its "'angle of vision' widens and the consciousness of 'knowing' increases in proportion to the simultaneous evolution of form."

When a physical form dies, the soul never loses its subtle and mental forms, which serve the crucial function of continuity. The subtle body provides the energy for further advancement, while the mental body re-tains and stores all the impressions gathered throughout its life in the limited gross form. Baba explains, "it is the unconscious association of the soul with its subtle form that fortifies the soul, then without any gross medium, with finite energy—the driving force—to tend the consciousness of the soul towards identifying itself with yet another form ... to expe-rience the impressions of the last dissociated finite gross form, retained and reflected by the finite mental form of the soul" (1997, 178). Thus, the mental and subtle bodies must always evolve alongside physical forms to be able to hold the growing number and diversity of impressions and to provide impetus for their experience in new, more developed physical forms.

Gradually the consciousness of the soul assumes more and more com-plex forms, and slowly develops its subtle and mental bodies, until a final "highest and most sublime form" is attained—the human form. In this form the soul is fully conscious of the physical world, and its subtle and mental bodies are fully developed, though not yet perceived. As a human being, the soul will be able to live out and "completely exhaust all im-pressions," eventually eliminating them all to realize its "real, eternal and infinite state in the Over-Soul" (Baba 1997, 26–27).

The Leaps and Turns of Evolution

On its way to full consciousness in the human form, evolution makes seven jumps or "leaps"—from stone to metal, metal to plant, plant to worm, worm to fish, fish to bird, bird to animal, and finally from animal to man (Baba 1973, 28, 176). Countless are the number of prehuman forms the soul must take on—8,400,000 in each of the major kingdoms (Kalchuri 1981, 45–48). Within each one of these forms, however, buried in seed form, is the latent human form, so that "strictly speaking there is only one form—the human form". As consciousness develops, this latent hu-man form makes a "series of partial turns," viewing creation from a dif-ferent angle of vision and associating with different elements before it manifests fully as a human being. In the stone form, it is the most highly compacted and extremely latent, and it experiences primarily the earth element. There is no voluntary movement, and the consciousness experi-ences a reclining, horizontal posture, although in the rock form of crys-talline granite, the human form becomes inverted with its head pointing downward. This is also the posture of plants and trees, with head under-ground where nutrients are absorbed, the roots like hairs, the trunk like the torso, and the branches like limbs. Plants have consciousness that is partly inanimate and partly animate, remaining erect only with the help of earth and rock and unable to move themselves at will. They use energy through breathing and taking in nutrients and water; thus their subtle or "energy" body begins to evolve.

In the further progression of forms, the position of the latent human body slowly rotates, becoming horizontal in the first animate forms of worms, insects, reptiles and amphibians, and gradually turning upward in higher and higher animal forms, until it becomes totally erect and manifest as a human being. In the worm-insect-reptile kingdom, thinking begins in a most elementary way. Moving voluntarily by crawling to find food, engaging in self-protection through instinct, experiencing sensations of pain and pleasure to a small degree, and seeking to procreate are all actions that require a modicum of thinking and thus begin to develop the mental body. While most forms in this kingdom experience primarily the earth element, some associate also with water, an element experienced most fully in the next kingdom of fish forms, where the posture continues in the recumbent state. Consciousness takes to the air in the form of birds, whose ability to fly and stand erect "enriches (enlightens) consciousness with new experiences."

In animal forms consciousness has a rich variety of experiences but must face a struggle for survival. Fire is the element most associated with animals. While earth, water and air are easily visible, the fire in the world, "a kind of blaze, or "*tej*," lies hidden but is felt intensely by animals as a "hunger-heat" driving them to seek food constantly, "as if they were born for the sole purpose of eating" (Abdulla 1954, 20). In animals, instinct, which gradually developed in earlier animal forms, evolves fully as an aspect of the mental body. In humans, this aspect will develop fully as it transforms into intellect (Baba 1997, 28–30, 175–78; Kalchuri 1981, 42–49; Abdulla 1954, 19–20).

THE PROCESS OF REINCARNATION

Evolution comes to an end in the human form, where consciousness of the world and the development of the subtle and mental body are "full and complete." Having moved through creation and evolution the soul now embarks upon the third major phase of its grand journey: reincarnation in human form. In this stage, the soul in the human form have full and complete consciousness only of the gross world, not of the subtle world, nor the mental world, nor of its own "indivisible, eternal and infinite" Self. Much the same process occurs in human form as occurred in all the subhuman phases. In any one life, the soul is living out and "exhausting" the accumulated sanskaras gathered in the previous life. The new physical body is the "consolidated mould" of those impressions from the past. When all the impressions are used up, the soul leaves that form—a process called the death of the human being. But in human form, which is associated with a fully developed subtle and mental body, there follows a period of time when the soul is said to be "in heaven" or "in hell," states of mind determined by the quality of experiences gathered in the lifetime just ended. As Baba explains in *God Speaks*:

[If the] predominant counterpart of the impressions of opposites (such as virtue and vice, good and evil, male and female, etc.) . . . is of virtue and goodness (*i.e.,* the positive aspect of the opposite impressions), then the soul is said to be in heaven. If it is of vice or evil (*i.e.,* the negative aspect of opposite impressions), then the soul is said to be in hell. (Baba 1997, 34)

Without a physical form, such subjective states are experienced much more intensely than when the impressions were first gathered in physical form, resulting in a more heightened joy or suffering. The process constitutes a review of the lifetime just lived and ends when the residual impressions are "experienced and exhausted." Baba compares the process to the playing of an old phonograph record: "Just as the gramophone record is set aside after the needle of the sound-arm has travelled through each groove, *so the hell-state and the heaven-state terminate after consciousness has traversed the imprints left by earthly life*" (Baba 1967, vol. III, 63).

The period of review has a helpful role in the development of the soul's total understanding. The experience of reliving impressions is not just reviving the past but is a *"leisurely and effective survey of the animated record of earthly life . . . a predominantly subjective and retrospective"* view rather than the *"predominantly objective and forward-looking"* view experienced under the "pressure" of unexhausted impressions and needed actions during physical life. He summarizes: *"The snapshots of earthly life have all been taken on the cinematic film of the mind and it is now time to study the original earthly life through the magnified projection of the filmed record on the screen of subjectivised consciousness."* Such "stock-taking and reflection" enable the soul to learn the lessons of a life, and "they become, for the next incarnation, part and parcel of the intuitive make-up of active consciousness . . . and the inborn wisdom," even though the exact experiences are not consciously remembered. Thus, *"developed intuition is consolidated and compressed understanding distilled through a multitude of diverse experiences gathered in previous lives."* Clearly, the amount of experience in reincarnation that a soul has had will determine the degree of development of their intuition, but one point is crucial to understand. The increase of intuitive knowledge is not purely a distillation of past experience in which a certain amount is added each lifetime but is rather a gradual unfolding of deeper knowledge already present in the soul in latent form. This more profound perspective suggests that *"the experiences of earthly life as well as the reflective and consolidatory processes to which they are subjected in life after death are merely instrumental in gradually releasing to the surface the intuitive wisdom which is already latent in the soul from the very beginning of creation"* (Baba 1967, vol. III, 58–65).

As the soul integrates the learning of a lifetime, the predominant impressions wear out and come into near equilibrium with the opposite type of impressions. Should a perfect balance occur, the soul would be freed from all impressions, but just as this equilibrium is about to be achieved, the soul immediately takes birth in a new physical human body, "propelled by the momentum of illusion" (Kalchuri 1981, 54). This occurs because the soul becomes so intensely engaged in the process of wearing out the impressions that it overshoots and the balance swings in favor of the opposite type of impressions, which become dominant and seek expression in a new physical body. Thus the soul enters each new human body with slightly unbalanced impressions. While this unbalance may seem like a failure or glitch, it is actually a very crucial element, which insures that the soul is propelled forward to a new phase of experience and eventually attains all possible human experience. And so it goes on for millions of lifetimes—8,400,000 to be exact—in which the soul has innumerable experiences of opposite impressions. It experiences lives as a male and female, in richness and in poverty, and in a variety of social classes, religions, races, countries, jobs, and states of health. In a seemingly never-ending cycle, perfect balance is never achieved. Although the soul is trying to disburden itself of and wear out impressions, it actually becomes more entangled and "firmly centralized in the more and more concentrated impressions of human-forms" which it has assumed and dropped. "There seems to be no escape." Only through more and more frequent experience of opposites are impressions shaken loose and eventually "thinned out." The whole cycle of reincarnation is supported by "this play of balancing and counter-balancing the opposites of impressions . . . [and] on this play depends the eventual emancipation of the human-conscious soul from the chains of ignorance, and the ultimate realization of Self-consciousness" (Baba 1997, 38–39).

INVOLUTION AND REALIZATION—THE FINAL STAGES

When learning in the gross world is complete—when a soul has experienced every kind of life—it gradually becomes less interested in the outer world of human affairs and more drawn to its own inner being. As Baba expressed it:

When the consciousness of the soul is ripe for disentanglement from the gross world, it enters the spiritual path and turns inward . . . Its gross impressions now become less deep. They become fainter or more subtle, and the soul now becomes **subtle conscious** . . . [it] has started on its homeward journey. (Baba 1997, 41)

The soul has begun the new phase of involution, in which it "begins to *infold* the consciousness." Its consciousness is actively engaged only with the inner subtle world of energy, even though it still utilizes a physical body in the gross world and works indirectly with the world of mind through thinking, desiring and feeling. Gradually over hundreds and hundreds of lives working through its subtle impressions, they slowly thin out and transform into mental impressions, which themselves eventually fade and disappear. Thus, "Gross impressions become subtle impressions; subtle impressions become mental impressions and mental impressions are ultimately wiped out, leaving consciousness free to reflect the truth" (Baba 1997, 41, 57–58).

Meher Baba explains the soul's grand journey and the experience of the different worlds using several helpful analogies. One is the metaphor of the sun giving light to the Earth. Pouring out of the sun are rays of "Energy and Mind" (the subtle and mental aspects of existence) continuously bathing the earth. The gross-conscious soul on the planet absorbs and utilizes these radiations unconsciously, which are the basis for physical energy and mental processes of thinking and feeling. The subtle-conscious soul is as above the Earth, in the air, closer to the sun and absorbing its real energy to the fullest extent possible—not stepped-down to gross form. As it traverses the first three planes of consciousness in the subtle sphere, it gains more and more access to enormous power which allows it to perform what would be called miracles from the perspective of the gross plane. Finally at the end of the path through the subtle planes, the soul arrives at the threshold of the mental world. With full access to the most powerful energy of the subtle realm, but with no access yet to control of the powerful forces of thought and desire in the mental planes, the soul reaches a point of danger, where there can be tremendous temptation to use the powers inappropriately. Should it do so, the enormous release of energy disintegrates the soul's consciousness, reducing it down to the stone state. This is the only instance in which consciousness reverts to a less advanced form. If the temptation is resisted, the soul moves onwards to enter the mental planes of consciousness, where it is fully aware of the mind and able to master thought and feeling. To continue the sun analogy, the soul's position now is very near the sun where it absorbs the force of Mind to the fullest extent. When the soul has traversed both the fifth and sixth planes of the mental sphere, it becomes conscious of the Self which is like "being in the sun itself." In presenting this analogy, Baba emphasizes several times that the sun and earth suggested here are in no way to be compared to the physical sun and earth of our solar system (Baba 1997, 58–66).

In another illuminating way of describing the development of consciousness, Baba uses the concept of the "winding" and "unwinding" of "*sanskaras*" or impressions (Baba 1997, 220–24 and see Primary Source

No. 1). In the *Discourses* an image of a string is used, and impressions are described as sanskaric threads or strings (Baba 1967, vol. I, 66). These impressions are actually "recorded" on the soul's mental layer of being, much as images of action and sound are recorded on film, and capture not just an outward picture of light and sound but also all experience of action, thought, and feeling. In the path through evolution the soul gathers more and more experiences, winding them like strings around the central core of its being, weaving more and more threads into the fabric of its consciousness. In each lifetime in evolution more and more sanskaric threads are wound at the gross, subtle, and mental levels of the bubble or form. Then the form is dropped, or the "bubble bursts . . . [leading to] an increase in consciousness and . . . a twist or consolidation of impressions or *sanskaras* accumulated during the life of the previous bubble." The winding process continues and strengthens during all of evolution, as the soul undergoes "seven stages of *descent*," all the while generating more "twists," or winding more sanskaric threads, and increasing its consciousness but at the same time binding the soul. In the reincarnation process in human forms, winding stops and the soul now alternates between experiencing opposites of impressions, trying to rid itself of the bindings. It spends and exhausts sanskaras but gathers new ones, and thus remains embroiled or bound. Each alternation, however, rebalances and rearranges the sanskaric threads and slightly loosens them, in preparation for the last and final phase—unwinding. During the process of realization—involution and final realization—the sanskaric threads are slowly unwound in "seven stages of *ascent*," until they are finally fully released. Much as scaffolding is torn down when a building is completed, the sanskaras that were indispensable in the growth of consciousness nevertheless must be removed and annihilated for the soul to experience its own unbounded self as God (Baba 1997, 220–23).

   In yet another rich analogy, the stages of evolution through realization are compared to the slow awakening of man from the deep sleep state and the gradual turning of his sight inwards. During evolution the increase of God's consciousness is likened to the increasing vision of a man whose eyes are gradually opening after being aroused from a sound, sound sleep, which represents the most transcendent Beyond-Beyond state of God. With full opening of the eyes evolution ended, and during the reincarnation phase the human with fully opened eyes has many different experiences and their opposites in alternating lives. He becomes almost unaware of himself in the midst of this wealth of experiences, but the perspective changes when the soul begins to involve:

Now the urge for the involution of the consciousness of God in the man state may be compared with a man who, having been engrossed in his activities of the day, at last finds time, when the day's work is practically over, to pay attention to his own self rather than to his activities. Thus urged, man's attention shifts automatically from external activities toward paying proper attention to his own self. (Baba 1997, 114)

Just as with evolution, involution occurs in seven stages, as the soul's awareness passes through the seven "planes of consciousness." In the sleeping/waking analogy these stages are likened to "the wide-open eyes of a man, at first gazing straight ahead of him and away from him. Then, in an attempt to behold his own self, he lowers his eyes gradually, shifting them in seven stages, until eventually his range of vision includes his own self" (Baba 1997, 114). Once the mental planes are reached, all subtle impressions are worn out. After the soul establishes mastery of thought in the reflective fifth plane, in the sixth plane it gains mastery of the realm of feelings and desires. There almost all mental impressions vanish, except for a tiny residual awareness of duality. The soul sees God face-to-face, but still as separate from himself, and yearns to become one with him, eventually achieving the final goal of all life:

The loving of God and the longing for his union is really and fully demonstrated in the sixth plane of consciousness. Only when the sixth plane of the mental sphere is transcended, does Illusion vanish with the vanishing of the last trace of impressions, and Reality is realized . . . This is the final and seventh stage in the process of involution of consciousness when the full consciousness of God in the man state is now fully withdrawn inwards, so completely that it is now fixed and focussed onto Himself rather than onto the objects of His own creation. (Baba 1997, 124–25)

It is a "passing away into" God, called *fana* in Sufi terms, and has two stages: first, conscious awareness of the original "absolute vacuum state" of the Beyond-Beyond state of God and then, the full glory of the "I am God" state. This most elevated of all states of consciousness is as far removed from ordinary human consciousness as everyday consciousness is from that of a stone (Baba 1997, 114).

Of supreme importance is the very first soul who ever achieved God-Realization and became the special being called the *Avatar* or Christ. When this soul saw God face-to-face on the sixth plane, only the impersonal God of infinite consciousness existed, aware of nothing but himself. The intensity of the soul's love for God and yearning for union finally drew God's attention, who gave him God-realization. He was the first perfected being in human form. The goal of creation was accomplished, the divine drama concluded, and a new personal aspect of God became established— the Christ, a perfected being who is aware of all souls in creation and whose infinite compassion leads him to help all others attain the goal. He returns to Earth again and again in human form to fulfill this role. The divine drama is replayed each time he helps another soul achieve the goal (Baba 1958, 27–28). Meher Baba has said that he is fulfilling that role in this epoch of history, and many of his followers believe him.

In this long journey, Baba emphasizes that there is only "one real birth and one real death" and that "you are born once and you really die only once." The real birth is marked by the drop-soul's first experience of individuality, and the real death is the liberation of consciousness from all limitation. In between, the limitations very, very slowly fade away over millions of cycles of births and deaths. Baba indicates that it is not spiritually important whether one believes in reincarnation (as with the Hindus and Buddhists) or not (as with Christians, Muslims, and Zoroastrians). All are correct, since there is only one real life with multiple phases. "The so many deaths during the one whole life, from the beginning of evolution of consciousness to the end of involution of consciousness, are like so many sleeps during one lifetime" (Baba 1997, 94, 238).

Much of *God Speaks* is devoted to detailed descriptions of the seven planes of consciousness and their characteristics, including a delineation of the ten states of God, which are listed in the table below (Baba 1997, 158).

The Ten Principal States of God		
State I	God in	Beyond-beyond
State II	God in	Beyond Sub–States A, B, C
State III	God as	Emanator, Sustainer, and Dissolver
State IV	God as	Embodied Soul
State V	God as	Soul in the State of Evolution
State VI	God as	Human Soul in the State of Reincarnation
State VII	God in	The state of Spiritually Advanced Souls
State VIII	God as	The Divinely Absorbed
State IX	God as	Liberated Incarnate Soul
State X	God as	Man-God and God-man

Looking at reality from the perspective of these states emphasizes once again that all existence and all beings within it, in all states of development and self-awareness, are truly God. This perspective provides yet another angle from which to view the journey of consciousness, which is reflected in the progression through the different numbered states. In State I God purely exists and an absolute unity of all opposites, an absolute vacuum, prevails. Next is State II, which has three aspects differentiated by the extent to which God is conscious of and uses "Power, Knowledge, and Bliss." State III represents God's first descent into illusion where he acts as Creator, Preserver, and Destroyer—*Brahma, Vishnu,* and *Shiva*—within the manifested creation, and in State IV he becomes individualized as a drop-soul. The drop-bubble evolves by taking many forms in each of seven stages or kingdoms in state V and develops further in millions of human incarnations in State VI. State VII represents God in the form of souls traversing the higher planes of consciousness. States VIII through X all represent God in the state of full "God-Realization." In VIII, the soul is fully absorbed into his own Godhood, conscious of his own "Power, Knowledge, and Bliss" but not using them. If the soul is destined to remain incarnate and do further work in creation, he becomes a "Liberated Incarnate Soul" (in State IX) and eventually works in creation as Perfect Master (Man-God) or *Avatar* (God-Man) in State X, where he is conscious of "Power, Knowledge, and Bliss" and uses them for the benefit of mankind and all of creation.

Throughout the descriptions and charts in *God Speaks* outlining the planes and states of God, Baba gives terminology from a number of religious traditions, indicating that these states are well known in many spiritual systems. For example, the difficult passage across the fourth plane of consciousness is referred to in Christian mysticism as "the dark night of the soul." Another example is Baba's presentation of the different names of the perfected beings who work in creation—*Sadgurus* to the Hindus, *Qutubs* to the Muslim, and those named "Man-God"or Perfect Masters in mystic terms—and the different terms referring to the "God-Man—*Avatar* for Hindus, *Saheb-e-Zaman* in Islam and the Christ, Savior or Messiah in Christianity and Judaism. This is typical of the inter-religious perspective and inclusive spirit in which Baba frequently integrates concepts from many religions, whose practices and beliefs differ greatly on the surface. His own system most closely expresses the ideas of the Advaita Vedantic school of Hinduism, which believes in the identity of one's individual soul with the soul of all things, *Brahman*—in other words, that all is God—but he also places great emphasis on the work of the personal God or Christ, as does Christianity. Thus his theology seems to transcend any one religion, and he applies to religion itself the basic Advaitic principle of the underlying unity of all things.

DIRECTIONALITY, DIVINE PURPOSE, AND THE PROBLEM OF EVIL AND SUFFERING

It is hard to imagine a spiritual system more focused on the notion of directionality and divine purpose than that of Meher Baba. In fact, he says, "All things and beings have a purpose and must have a purpose, or else they cannot *be* in existence as what they are... Everything exists only because it has a purpose" (Baba 1963, 100). He speaks again and again of the goal of all life and the destiny of each individual to realize himself as being God. There is momentum in the forward flow of all created life toward fulfillment of its divine purpose, a momentum imparted from the first instant of God's whim to fully know himself. This momentum never ceases to draw the soul back to realization of its real nature, which is, of course, always hidden deep within his own being. Thus, it is both a push from the past, but also an attraction towards the future, when one's own real self will emerge from behind the veils of all the illusory selves one tried out and rejected as an answer to the eternal question, "Who Am I?" For ages and ages in ignorance one identifies oneself with material life, all the while moving toward one's ultimate destiny of God-Realization. In *God Speaks* this point is illustrated with a parable from the *Masnavi* of the Sufi poet Jalalludin Rumi. In the story a tiger cub that was raised among sheep took on all their behaviors and thought of itself always as a bleating, grazing sheep. But one day a tiger came from out of the jungle and met the cub and told him that he too was a tiger. He finally helped him to realize his real nature by seeing his reflection in a stream. In the same manner, Rumi says, when a human being is ripe for the spiritual path, a teacher will appear and help him find his own true self (Baba 1997, 38).

Because the experience of opposites plays such a crucial role in the forward march of the individualized soul toward its destined goal, Baba's cosmology suggests that divine purpose encompasses not only pleasure, joy, and goodness but also their opposites of pain, suffering, and evil. At first glance it appears contradictory that all is truly God and yet evil exists—how could such a thing be part of the divine being? Several aspects of Baba's system suggest a way. One is the perspective of reincarnation, which holds that a soul's knowledge is not complete until it has experienced all types of human life and that it learns something valuable from each of them. In fact, the pain and suffering during evolution and the experience of opposites in human reincarnation play indispensable roles:

It is the evolutionary struggle that enables the soul to develop full consciousness as that in the human form, and the purpose having been achieved, the side issues or by-products of evolutionary travel (the *nuqush-e-amal* or *sanskaras*) have to be done away with, while retaining the consciousness intact. The process of reincarnation therefore is to enable the soul to eliminate the *sanskaras* [impressions] by passing through the furnace of pain and pleasure. (Baba 1997, 27)

Both good and evil are produced by impressions made upon the mind during the forward march of evolution. They seem to oppose each other as totally different forces, but evil and "the Devil" are not really "active forces" opposing God. Evil is actually a "lingering relic of earlier good." Behaviors needed in earlier stages of evolution carry over to later stages where they fail to harmonize in new surroundings and seem wrong and evil. As the soul completes its journey through all the animal forms and enters human incarnations, it has more bad impressions than good, since animals are generally motivated by lust, greed and anger. In human life, such behaviors create disharmony and seem bad, and the soul tries to balance them by engaging in actions motivated by love, generosity and tolerance. "Right from the beginning of human evolution," Baba writes, "the problem of emancipation consists in cultivating and developing good sanskaras so that they may overlap and annul the accumulated sanskaras." An important point here is that what is evil and what is good are relative to the situation, a truth that holds throughout human incarnations. Even such negative actions as stealing can be good and appropriate if done for the noblest of motives—for instance, to provide food for a child when the owner of the food would not sell it.

Evil actually serves a valuable spiritual function. On one level it is clearly the opposite of good, providing a counterpoint to it, but at the same time it focuses attention on the good and can be transformed into it. In general in the long path to God, one seeks to renounce evil and embrace goodness, but in the end the soul must transcend both. Even the good binds and limits the soul, since identifying oneself as good also affirms separative existence. Bad sanskaras, deriving as they do from the most ancient experiences in subhuman forms, are the deepest and the most difficult to overcome but the easiest to recognize as limiting, while the opposite is true for good *sanskaras*. Baba writes, "The difficulty concerning the abode of evil is not so much of perceiving that it is a limitation but in actually dismantling it after arriving at such perception. The difficulty concerning the abode of the good is not so much in dismantling it as of perceiving that it is, in fact, a limitation." Eventually, good and bad impressions balance exactly and vanish from the mind, leaving it clear to reflect only the truth in the state of Illumination. At the end even the mind disappears in the state of Realization (Baba 1967, vol. I, 92–99; 1958, 55–62).

The law of karma offers yet another perspective on how divine purpose can include and resolve the problem of evil, pain, and suffering. A fundamental aspect of Eastern reincarnational systems, *karma* is a law as inevitable as gravity but operating in the realm of morals and values. Without this *"systematic connection between cause and effect in the world of values,"* there would be no *"moral order in the universe"* or rational scheme to human behavior. Baba writes, *"Karmic determination is the condition of true responsibility. It means that a man will reap as he sows."* The urgency to seek the opposite experience is not something imposed from without, but from within the person's own inner being. If one engages in an action that is not embraced fully by one's own inner self, it automatically invites its opposite. For example, an action such as killing sick animals or carelessly destroying plants may not resonate with what one would wish for oneself, and so one is drawn to balance such actions in future lives by caring for animals or plants. An evil done to another rebounds of necessity upon the perpetrator, because all beings are truly one. Baba writes, "*What [one] does for another he has also done for himself, although it may take time for him to realise that this is exactly so. The law of Karma might be said to be an expression of justice or a reflection of the unity of life in the world of duality*" (1967, vol. III, 90–91; 1958, 64–67).

Of special note is the suffering that comes to the Christ. According to Baba, it is not related to karma nor does it result from the vanquishing of good by evil. He writes, "It happens by divine will and is a form of divine compassion. He voluntarily takes upon himself the suffering of others in order to redeem those who are engulfed in gnawing cravings, unrelieved hatred and unabated jealousies" (Baba 1958, 58). The Christ undergoes "infinite agony eternally through [man's] ignorance" (Baba 1963, 64), a statement which offers yet another viewpoint on the underlying cause of evil and suffering: ignorance of one's own true nature.

COMPARISON TO SCIENCE AND OTHER RELIGIOUS PERSPECTIVES

Comparison to Science

The comprehensive system of thought presented by Meher Baba resonates in a number of interesting ways with ideas from both science and theology discussed earlier, and at the same time differs significantly from both. In some instances, where there is polarization into two scientific or theological camps, such as with the anthropic principle, Meher Baba's ideas seem to assert the truth of both sides.

The Nothing, the Whim, and Stages of Development   There are notable parallels with science in Meher Baba's description of the very early universe. Concepts that appear fruitful to explore are the role of "nothing" in producing a universe, concepts of emergence and expansion from a point, an initial interweaving of space and energy to form particles, and the existence of many distinct stages of rarefied matter before the atom developed.

Science envisions a primordial "vacuum" in which quantum fluctuations of energy arise and disappear randomly and by chance. It is possible for one to arise that had zero energy—its motion energy exactly offset by the work required to expand it against gravity. Such a fluctuation could conceivably become a whole universe and last an infinite time. Thus one could describe the universe as arising from "nothing" and expanding like a bubble from a point. If string theories are correct, individual particles obtained their properties from the shape of many tiny curled up dimensions of space in which their inner strings vibrated—thus energy was interwoven with space. In Meher Baba's system, there was also a chance-like event which gave rise to creation—God's whim—which had no prior cause or premeditation. The whim arose within that aspect of God's original transcendent being that Baba calls the "Ocean of Nothing," or the infinite unconsciousness, and "creation" was thus the emergence of innumerable forms out of nothing. The whim was like a burst of God's energy, or fire, which appeared at a single point and needed space to manifest and expand. Energy and space clashed and began to synchronize in wave-like motion, energy interweaving with space to form waves and bubbles of individuated drops.

In both systems there are very distinctive stages by which recognizable particles of matter developed from earlier more rarefied forms. For Meher Baba, the original fire descended through several levels of existence, each of which gave the descending drop a bubble, or body, of its own energy. It passed through over two hundred states of rarefied gases in the subtle realms and then through seven transitional forms at the physical level until the final seventh state of hydrogen manifested. Descent through each level represented a "stepping down" of the original energy. Science likewise has a progression of matter starting from an original superenergetic state—a "superforce" or "superparticle"—and descending down to less and less energetic levels at which particles could acquire new characteristics and interact in new ways. After one second the basic atomic particles had formed, and at age 300,000 years, they joined to form atoms. There is structural similarity in the processes, although there are certainly major differences—in the time scales, the role of God, the language, etc. It may well be inappropriate to compare Baba's transformation from divine energy down to a material particle on the one hand with science's purely material process on the other. Perhaps a more relevant comparison is between the seven physical forms of gases before hydrogen in Baba's system and the seven major eras of particle development during the first second of the Big Bang (see Chapter 4 and Figure 4.1).

Direction and Purpose to Evolution   Although nothing in Baba's system contradicts the basic scientific picture of cosmic evolution, there is clearly much more to his evolutionary scheme than what science can discern: a direction and final goal to evolution, the evolution of consciousness rather than form, the activation of consciousness at the very beginning, and the inner force within each being to move on to a higher form. It goes without saying that the emergence of the whole creation from within the being of God himself and the final goal of absorption into God are well beyond science's ability to evaluate.

Scientists would no doubt display a wide spectrum of reactions to Baba's ideas of cosmic direction and purpose, depending on their own philosophical persuasion. The ideas would without question be challenged by scientific materialists, who argue strongly that there is neither direction nor progress to evolution, and that the forces for change are so random and nonrepeatable that the development of intelligence and consciousness was purely a chance phenomenon. There is no discernable force or drive within living forms—and certainly not within stones! —that impels them to seek greater knowledge through the vehicle of more complex forms. Most scientists view consciousness as an "epiphenomenon," something added on that emerged at a very late stage in the cosmic process. On the other hand, a few scientists, such as de Duve and Conway Morris, do interpret evolutionary development as a "ladder of progress" and see a suggestion of purposive development toward intelligence in such phenomenon as convergence. Other scientists, such as Kauffman and Davies, focus on the possibility of laws of self-organization, complexification, and information or other new laws which would supplement known laws of physics and provide some kind of force or impetus for life's origin and development. One imagines that such scientists could find Baba's concepts provocative and potentially consonant with their own speculations.

Nonmaterial Levels and Laws   Scientists who deny the existence of anything beyond the physical would certainly reject schemes which integrate material and nonmaterial laws and levels of being, and, indeed, there seems to be no place within current science for understanding such ideas. Others might be neutral, considering them untestable and simply beyond the scope of science to address. Still others might observe a potential consonance between known physical mechanisms and the workings of the deeper levels that Baba describes, those layers of interior selves that surround the core of each being. These layers are the subtle and mental bodies that provide the momentum for onward progress and serve as the repository of all past experience of thought and feeling. These bodies develop as more and more experience is gathered, as energy is felt and used, and as impressions are made on the mind and stored in the form of sanskaric threads, woven together to form the invisible fabric of consciousness. These threads seem to operate as a kind of higher-level DNA forming the "tissue" of subtler layers of being. Just as DNA encodes for physical structure and process, so these threads carry all past experience of thought, feeling and action in some compressed and encoded form and contribute to forming ("are the consolidated mould of") the physical body of the next species or human incarnation.

Integrating this picture with ideas from the *Buddha's Explanation of the Universe*, presented in Chapter 3, leads to additional fruitful comparisons with science. Each level of creation is composed of a certain kind of matter and energy which vibrates at a particular rate and requires a particular kind of sensory apparatus to be detected. Thus, there are units of physical matter beating at a certain very fast frequency and detectable by the gross body and units of mind or thought beating at a much faster rate and detectable by the mental body. From this perspective advancing spiritually involves the increasing refinement of sensory apparatus so that it can bear the intensity of more and more powerful energy. The entire universe is thus all the same thing. It is a cosmos filled with nothing but the vibrant flowing force of divinity vibrating at different rates—light congealed into tissues of different density, stepped down and "frozen" into form. The divine fire, or *Tej*, descended into material form and became encased in, or congealed into, the "bubbles" of the mental, subtle and then physical bodies. As noted above, this notion suggests a potential comparison to the stepping down of the powerful energy of the Big Bang—as the universe expanded and cooled, as the forces of nature split into their separate realms of operation, and as energy congealed into particles. It is also consonant with ideas in string theory proposing that all particles of matter and energy are in truth the same thing, each a tiny loop or string whose different modes of vibration each produce a particular kind of particle.

All the processes of higher levels of matter and perception work according to very exact laws, which govern the mechanics of the subtle and mental realms much as gravity and electromagnetism govern the physical world. An example is the laws of attraction and repulsion that operate in the realm of karma and in the balancing of impressions at the end of each incarnation. These "higher laws" subsume the physical, perhaps in a manner analogous to the way Einstein's law of gravity encompasses Newton's at the physical level. Baba describes the workings of higher and lower laws as follows:

The mystery of the universe is hierarchic in structure. There are graded orders, one supervening upon the other. The spiritual panorama of the universe reveals itself as a gradient with laws upon laws. Superimposition of one type of law over the other implies elasticity and resilience of lower laws for the working out of higher superseding laws. Instead of lawlessness, it means a regime of graded laws adjusted with each other in such a manner that they all subserve the supreme purpose of God, the Creator. (Baba 1958, 33)

A single principle may unify the laws or forces at different levels, expressing itself one way at the physical level and another way at higher levels. A prime example for Baba is the force of love, which is the unifying principle in all of life and is present throughout the cosmos. It has different degrees of intensity throughout the levels of being in the universe—from inert matter up through the God-Realized soul. Even the law of gravitational attraction which all material objects exert upon each other is an elementary form of the force of love. As expressed in *The Discourses*:

Even the most rudimentary consciousness is always trying to burst out of its limitations and experience some kind of unity with other forms. Though each form is separate from other forms, in *reality* they are all forms of the same unity of life. The latent sense for this hidden inner reality indirectly makes itself felt even in the world of illusion through the attraction which one form has for another form.

The law of *gravitation*, to which all the planets and the stars are subject, is in its own way a dim reflection of the love which pervades every part of the universe. Even the forces of repulsion are in truth expressions of love, since things are repelled from each other because they are more powerfully attracted to some other things. Repulsion is a negative consequence of positive attraction. The forces of *cohesion* and *affinity* which prevail in the very constitution of matter are positive expressions of love. A striking example of love at this level is found in the attraction which the magnet exercises for iron. All these forms of love are of the lowest type, since they are necessarily conditioned by the rudimentary consciousness in which they appear. (Baba 1967, vol. I, 156–57)

Baba goes on to describe the impulsive and instinctual love that animals express and the enormously varied forms of human love from the lowest expressions of lust, greed, and anger to the highest and most self-sacrificing. The pinnacle is pure divine love, where no thought of individual self and no individual mind exist and the soul experiences absolute unity. Thus the same force of love pervades all levels of being, expressing itself in each sphere through laws suited to that level—physical gravitation and magnetism, animal instinct and predation, human attraction and repulsion, *karma* and impressional balancing at life's end, and the very fullest possible expression in Self-Realization.

The Anthropic Principle and Life Elsewhere in the Universe   An additional area where Baba's cosmology has an interesting and complex relationship to scientific thought is with the anthropic principle and life beyond Earth. His thought clearly implies an anthropic focus in the universe and supports the strong anthropic principle that the universe must be such as to produce life and intelligence and, Baba adds, higher consciousness. That is its whole purpose. It is imbued with a driving force to produce the form and structure of consciousness in which it can fully realize its own identity as God. This drive is latent from the beginning, deriving as it does from God's desire for self-knowledge. Thus the human form itself was latent in the very beginning in seed form in the most rudimentary inanimate forms as the prototype toward which all forms were

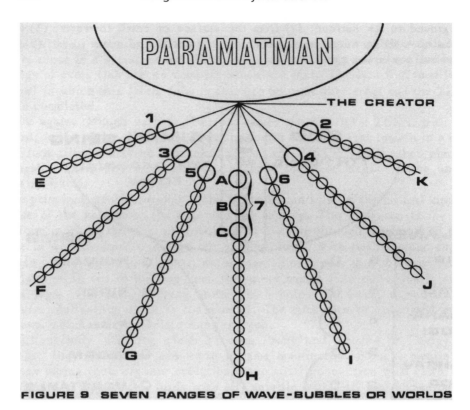

FIGURE 9   SEVEN RANGES OF WAVE-BUBBLES OR WORLDS

Figure 7.4 Diagram of Other Worlds in Creation. This chart depicts Meher Baba's explanation of the different levels of worlds in creation. Seven are closest to the Creator but differ in the forms inhabiting them—World 1 has only stones, but World 5 has stones, metals and vegetation, for instance. The seventh three-part world contains Earth as planet A, and two others, where humans have greater intellect than heart. Only on Earth can humans balance head and heart and advance spiritually to life's goal of union with God, and only on Earth does the Christ incarnate. (Reprinted by permission of the Avatar Meher Baba Perpetual Public Charitable Trust)

evolving. No wonder that in the first seconds of the universe, conditions were fine-tuned for human development!

Not only does the human being occupy a very special spiritual niche in creation, but so does planet Earth. According to Baba, Earth is the only location in the universe where advancement to higher planes of consciousness and God-realization are possible and the only location where the Christ incarnates in a physical form. As described in earlier chapters, in Baba's cosmology countless universes exist, each with billions of galaxies and millions of worlds. There are seven "ranges of worlds," each with a different level of advancement of matter and energy, the most elemental containing only stones (see Figure 7.4). In each successive range more stages are added: first wind, then metals, water, vegetation, animals, and finally in the seventh range are 18,000 worlds inhabited by human beings. Of these, one three-part world has prime importance for it contains Earth plus two others. The three differ by the balance of intellect and love their human inhabitants possess: on one, humans have 100 percent intellect and 0 percent love, on the second 75 percent intellect and 25 percent love, and on Earth a balance of the two. Thus, while Earth humans are inferior in intelligence, they are superior in terms of their emotional nature and capacity for love. Most important of all, only on Earth can humans advance in consciousness beyond the gross plane, traverse the higher subtle and mental planes, and fulfill creation's goal of overcoming the mind altogether and experiencing 100 percent love. Earth is nearest in some nonmaterial way to the Creator-Point, and it is only on Earth that the Christ incarnates (Abdulla 1954, 21–23; Baba 1973, 220).

It is hard to imagine a more extravagantly anthropocentric and Earth-centric viewpoint that this! Except for scientific creationists, ID proponents and other highly religious scientists such as Wallace, few in the scientific domain would presumably find this scheme credible. In some ways it extends anthropocentrism to new unimagined heights by suggesting that Earth may be special not just in our Galaxy, or even in our universe, but in a vast system of creation with multiple universes. Curiously, however, this view resonates with that of multiple universe scientists, even though the underlying motivations differ greatly. For Baba, an infinitude of multiple universes exists because they are but a reflection or shadow of the infinite God, and Earth just happens to be special spiritually as the last stepping stone of souls on their way back to God. For the scientists, the possibility of multiple universes exists because they are possibly implied by unified theories of physical reality, such as string theories, and because they also avoid the unpleasant nonscientific design implications of the anthropic principle by rendering life in one universe or world a statistical fluke. In an infinite megaverse of universes, chance alone dictates that somewhere conditions and laws had to be perfectly fine-tuned to permit life and intelligence.

Several aspects of Baba's thought qualify his own anthropocentrism and Earth-centrism. One is his assertion that the human form is not constant over time, having already been evolving for millions of years. With the basic form unchanged, it will continue to evolve into the future, becoming both brainier and smaller in stature (Kalchuri 1979, 1872). A second qualification is that despite the greater wholeness or capability of human consciousness on Earth, the notions of superiority and inferiority to other creatures should not apply. All creatures have an indispensable role in the ongoing movement of evolution, even ones like weeds that seem to hinder the growth of other organisms, for they help other plants to become more robust. Even so, such utilitarian considerations do not really determine spiritual status of creatures. That comes from their very being and their deeper relationship to both humans and the divine goal of all. The human structure of consciousness contains impressions of all subhuman forms, for it has been all of them; their impressions make up the human being. Conversely, the human form was latent in all other living and nonliving forms. In the face of such unity of being, the notions of superiority or inferiority and special or nonspecial status become meaningless. Baba indicates that to regard animal species as "lower" is to "take a purely anthropocentric view of creation... To view things in their right perspective, we have to see all forms, including human forms, as evolved for the fulfillment of the one eternal divine life" (Baba 1958, 22). Finally, Baba also tempers the Earth-centrism implied in his cosmology. The Earth too evolves and will, some millions or billions of years in the future, cool down and no longer be suitable as the site for the spiritual advancement of humans. Just as there have been many Earths before this one, so in time, others will take its place as the planet of spiritual awakening (Abdulla 1954, 23; Duce 1975, 395).

Time Scales and Summary Although he rarely discussed exact time scales, it is clear that in Baba's cosmology cosmic evolution occurs over a much vaster period of time than science suggests, at least until recently. One need only glance quickly at the number of individual subhuman and human forms the soul adopts on its evolutionary path to discern that the entire process is far lengthier than the whole span of the physical universe currently understood by science. With 8,400,000 forms adopted in each of the six subhuman kingdoms and another 8,400,000 in human lives, the total of forms alone is almost 60 billion. Assuming for simplicity that each "life" takes only a year in each form, the total time span would be 60 billion years, four to five times longer than the currently understood age of the universe. Since life spans for most forms would no doubt be longer than this, it is more realistic to estimate a figure that is an order of magnitude larger, or 600 billion years. Such a time scale is almost fifty times longer than the scientifically determined age of the present universe. For such a time period to be conceivable, given the Earth's age of 5 billion years, a single soul must go through evolution on many different planets, perhaps even in different universes, on its total journey through creation. Baba has suggested such a migratory path in discussing how more intellectual humans from other planets must eventually incarnate on this Earth to advance spiritually. It is also implied in his assertion that the human form has been evolving for millions of years, which is only possible if human evolution occurred elsewhere and/or humans emerged much earlier on Earth than science now estimates.

In sum, the total time span for the evolution of life suggested in Baba's cosmology is without question vastly different from science's 3.8 billion years on planet Earth. It is, however, conceivably consistent with science, if Earth were simply one of the latest sites where the evolutionary process developed—a stance exobiologists also take. Baba's time scales, as far as we can discern them, are also consistent with the scientific Earth age if souls migrate among many planets and even from much older universes to complete the growth of their consciousness. Such an idea is clearly beyond science's scope to determine, but science today does seriously entertain a number of untestable hypotheses, such as the existence of multiple universes themselves. What is, of course, extremely different is the idea that there may be a continuity of an individual soul's consciousness taking form on a multitude of worlds.

With respect to life in the universe Baba's cosmology agrees with exobiologists on the existence of advanced intelligence elsewhere but offers some interesting "in-between" answers to the questions about human uniqueness and special status. With 18,000 other worlds inhabited by humans, clearly we are not alone. With other planets inhabited by more intellectually advanced humans, neither are we unique nor special in terms of ordinary intelligence. However, spiritually, we are indeed unique and special in the extreme as the only beings in the present universe able to advance spiritually and realize our real nature as God. We inhabit the only planet in the whole system of creation where this passage is possible and where the Christ incarnates. However, neither Earth's special spiritual status nor our own exact form is permanent; both will change over time.

In sum, Baba's cosmology is at once potentially consonant with and yet strikingly different from the currently understood scientific scheme. It is possible to incorporate within his system almost all of what science presents, but the converse is obviously not true. Many things he asserts are well beyond science to assess: the much lengthier time scales, the entire spiritual dimension of existence, the many levels of being lying beyond but encompassing the physical, a system of higher laws of the mental and moral realm that integrate seamlessly with the physical laws, the strong forward flow of life toward its destined divine goal, and the identity of the whole system with aspects of God himself. Curiously, however, suggestive connections to the physical realm exist. His rich concept of "nothingness" as the source of created forms is intriguingly consonant with science. His concept of the descent of divine energy into material form has a structural parallel with science's stages of material development in the first few hundred thousand years of the Big Bang. Higher level mechanisms mirror patterns in the physical realm, such as the DNA-like sanskaric fibers of consciousness encoding all past impressions of action, thought, and feeling. A universe filled with light and matter vibrating at different rates and sensed by different levels of our interior selves seems consonant with string theory ideas of particles as strings vibrating at certain frequencies, some of which are not detectable with current scientific apparatus. Similarly, laws of attraction, laws balancing opposites in *karma* and laws of impressional processing resonate with physical laws of attraction and repulsion. It may be that just as with multiple universes, a now scientifically popular concept which Baba discussed many years ago, some of these ideas might prove fruitful for future scientific research, as Baba suggested on numerous occasions (Kalchuri 1979, 6005; Duce 1975, 395). Finally Baba's cosmology seems to both challenge and confirm science on some issues. He agrees with current ideas of multiple universes and an abundance of life but goes strongly against the grain of most scientific thinking in arguing for the special significance of human beings on planet Earth—thus agreeing in some way with both atheistic scientists and ID proponents.

Comparison to Other Metaphysical Systems and Theologies

Even though comparison of different metaphysical systems is beyond the scope of this book, the special interest taken by Baba in establishing unity among different religions warrants a few summary comments along these lines. Certain features of his cosmology are consonant with various philosophical and Western theological concepts discussed earlier. To name but a few, they include the presence of a hierarchy of laws and levels of being, the suggestion of a common core to all religions, the idea of the self-withdrawal of God, the existence of an interiority to all beings, the process notion of multiple influences on all events, and the notion of an inward pull and promise that draws all beings toward an ultimate fulfillment.

In *Science and Religion in Search of Cosmic Purpose* and other sources, Islamic scholar Nasr and Catholic theologian Haught both discuss the modern loss of the traditional religious belief in a hierarchy of material and spiritual levels of being and their replacement with a horizontal span of materialistic evolution and atomistic notion of reality. Haught calls on theologians and religious-minded scientists to heal this loss by finding ways to "marry" the two—to infuse the onward flow of evolution with the traditional hierarchy. It seems that Baba's cosmology does that with its integration of the eternal and unchanging God with the evolving and growing stream of life. His planes of consciousness seem very hierarchical, but the soul journeys through them in an evolutionary process, its various physical, subtle and mental bodies growing and changing to become ever better vehicles for the maturation of consciousness to eventual self-knowledge. The entire hierarchy, however, is not really changing, for the evolutionary process occurs totally within that aspect of God that has only illusory separate reality from his eternal and unchanging Self. Baba's hierarchy nonetheless differs considerably from a traditional one. Because the different levels are all encompassed within God, his hierarchy is better represented by an image of nested spheres than a ladder of being. And, of course, the identification of the process with God himself is a major difference from the monotheistic Western religions.

Baba's views also resonate with Nasr's expressed hope for a return to the "'perennial philosophy' . . . a belief in an ultimate source of being and meaning, manifesting itself in different ways in each tradition." Nasr writes, "The traditional interpretation of the *philosophia perennis* sees a single Divine Reality as the origin of all the millennial religions that have governed human life over the ages and have created the traditional civilizations with their sacred laws, social institutions, arts and sciences." He draws a beautiful analogy in likening this Divine Reality, "the origin of all that is sacred and the source of the teachings of each authentic faith," to a "mighty spring gushing forth atop a mountain," cascading down into the different streams of religious faith. Such thoughts harmonize well with many statements of Baba which emphasized a single divine truth manifesting in different ways over the ages according to the needs of humanity at a particular time. Nasr's views thus seem to accord well with Baba's mission to "bringing all the religions together like beads on a string" (Haught 2000b, 110–11).

There are also echoes in Baba's work of the concept of God's self-limitation—the idea of kenosis discussed in Christian theology and related to concepts in the Kabbalah—although the ideas have significant differences. In Christian thought God's self-limitation allows for something "other than" God to exist and have free will, which for one thing offers an explanation for the presence of evil and suffering. God limits his own omnipotence to give created beings power to find their way to him by free choice. In Baba's cosmology God's original whim to know himself caused one aspect of his being to "separate from" the rest and establish itself in duality as an infinitely tiny form with a most minuscule degree of consciousness and with illusory perception that it is "something different." This process resembles somewhat the idea of God withdrawing and limiting himself. The purpose seems similar as well—so that the long process of seeking union can take place. A major difference, of course, is that for Baba this process all happens within God, whereas in kenotic theologies God withdraws so that something truly other than him can exist. While this is a crucial distinction, there is suggestive similarity here, and the goal of both schemes is union with the ultimate reality.

Baba's thought is consonant also with aspects of process philosophy and the metaphysical system of Teilhard de Chardin in that all three posit an inner and outer reality to all beings and events and multiple influences in every event. Barbour describes process thought as follows:

[It sees] continuity as well as distinctiveness among levels of reality; the characteristics of each level have rudimentary forerunners at earlier and lower levels. Against a dualism of matter and mind, or a materialism that has no place for mind, process thought envisages two aspects of all events as seen from within and from without. (Barbour 1997, 104)

Similarly, Baba speaks of the inner and outer journeys and adds that the Mind is the intersection of the two (Duce 1975, 394). Teilhard de Chardin likewise asserts that "everything has an inside" and is activated by the two kinds of energy, an external one and an inner drive toward greater complexity, consciousness, and eventual consummation. In Baba's scheme, the inner subtle or "energy" body and the innermost level of divinity provide the momentum and drive toward the growth of consciousness, which is the central goal of the whole evolutionary process in both systems. This same urge toward a greater level of being also appears in the thought of Catholic theologian Rahner, who suggested that all beings have within themselves a drive for "self-transcendence," deriving from God's own transcendent nature immanent within them.

God's immanence within every event is an idea incorporated into many theological systems, but the process view of the triune nature of influences within each event bears striking similarity to Baba's view. "Process metaphysics," as Barbour expresses it, "understands every new event to be jointly the product of the entity's past, it own action, and the action of God" (Barbour 1997, 104). God acts not coercively but persuasively as the ground and source not only of order, but also of novelty, drawing the entity forward to new possibilities of growth. Similarly, with Baba, God's urge to know himself, immanent within all, provides the impetus for the soul to move on to new more complex potentialities of form and awareness. At any one time, then, any event experienced by the individual soul—any action, thought or feeling—is influenced by the sum total of all past experiences encoded in its sanskaric makeup, by its experiences in the present, and by the divine pull within.

Finally, Baba's theology joins almost all traditional religions in positing a future ultimate state of fulfillment and consummation. Called "God-Realization" in Baba's language, the Kingdom of God or the new creation by Christians, and Enlightenment or Nirvana in Eastern religion, it is in all cases the final goal in life. Many Christian theologians propose the idea that ultimate meaning of the present and of all of life and cosmic evolution is to be found in the future state towards which God is calling us. Although similar on the surface, obvious differences in particulars exist between the Christian and Eastern views of this future state. For Christians it happens at one temporal moment as a divine act of redemption for human beings who have lived but one life on Earth, and in some systems it does not happen for all. Typically, in Eastern religion and in Baba's theology the focus is on the individual soul's realization of the "God" state with the help of other divine beings and after countless lives on Earth and elsewhere, and the divine state is the eventual destiny of all souls everywhere. In essence, however, although the language differs, it seems possible that the different conceptions all refer the same state of being.

Comparison to Cosmic Evolution in Traditional Buddhism and Hinduism

Traditional Buddhism and Hinduism stress the ever-changing, transient nature of the outer world and incorporate ideas of evolutionary development to a varying degree, usually emphasizing cyclic change rather than progress. Buddhism conceives of a four-stage cycle of change that occurs within any one world or "universe system," even though in the vast universe of countless such systems there is no beginning or end. The stages are labeled variously as stasis, birth, growth, and death or, in the words of the fourteenth Dalai Lama, "emptiness . . . formation . . . abiding . . . and destruction" (The Dalai Lama 2005, 80–81). In *The Buddha's Explanation of the Universe* Ranasinghe describes how evolution is occurring at various stages on many worlds, and everywhere is driven only by the activity of "units of mind." Units of matter by themselves have no ability to assemble into more complex form and can do so only when processed into greater complexity by forces of mind (Ranasinghe 1957, 245–47). The Dalai Lama accepts many aspects of scientific evolution, affirming,

On the whole, I think the Darwinian theory of evolution, at least with the additional insights of modern genetics, gives us a fairly coherent account of the evolution of human life on earth. At the same time, I believe that karma can have a central role in understanding the origination of what Buddhism calls 'sentience,' through the media of energy and consciousness.

Even while generally endorsing evolutionary theory, he also finds it lacking for (1) its circular reasoning about survival of the fittest, (2) the weakness of the idea life originates by random events, (3) inadequate differentiation of sentient from nonsentient forms of life, (4) failure to account for altruism as well as competitive drives and (5) most importantly, failure to investigate the origin of consciousness. He stresses, "Until there is credible understanding of the nature of the origin of consciousness, the scientific story of the origins of life and the cosmos will not be complete" (The Dalai Lama 2005, 111–15).

In Hindu thought there are also strains of evolutionary thinking. Philosopher Anindita Balslev describes one ancient Hindu school, Sankhya, whose principle concept is cosmic evolution, conceived as an interaction between "*prakriti*, the ever-changing, ever-active principle of matter, and *purusha*, the principle of consciousness, unchanging and unchangeable, uncaused and indestructible." Interlaced with this is the notion of the dissolution of the cosmos—a return to an original, homogenous state from which a new cycle of creation begins (Balslev 2000, 61–62). Thus the sense of progressive change is similar to scientific cosmic evolution but the role of consciousness and the cyclical nature of temporal progression are different. A number of scholars also refer to the hint of evolution in Hindu myths of the avataric descents of divinity into a succession of animal forms—a Fish, Unicorn, Turtle, and Boar, then Man-lion, and Dwarf, and finally in full human form. While some scholars describe these avatars and the mythology surrounding them as having "uncanny parallel" with evolutionary biology, there is need for caution in not equating degrees of complexity and sophistication of creatures as evidence of different steps in a progressive evolution. It is the same kind of caution needed in interpreting Genesis in the light of science (Raman 2002, 189).

There are clearly strong parallels in all these ideas with Baba's cosmology, although he places much stronger emphasis on the idea of evolution, both of consciousness and of form. With great detail he delineates how and why consciousness drives the progressive development of all inanimate and animate forms. He is in clear consonance with the Buddhist affirmation of the primacy of consciousness and the impossibility of understanding evolution without it. With respect to Hinduism, Baba's thought seems to be a blend of AdvaitaVedantism, which emphasizes the identity of all with God, and the Sankhya philosophy of evolutionary change. He stresses evolution more than any other figure in Hinduism, with the possible exception of Sri Aurobindo Ghose, another eminent spiritual leader in modern India.

Aurobindo, like Baba, reformulated aspects of ancient Hinduism in the light of modern knowledge, also by incorporating concepts of evolution into Vedantic philosophy. In his major expository work, *The Life Divine*, he speaks of a spiritual evolution, which he describes as "an evolution of consciousness in Matter . . . till the form can reveal the indwelling Spirit . . . [this] is then the keynote, the central significant motive of the terrestrial existence" (Ghose 1990, 858–59). Evolution is thus a process in which conscious being slowly emerges in very gradual stages from "Inconscience," or an unconscious state of being, through long, slow stages of existing as matter, living forms and finally mentally aware beings. Evolution proceeds beyond the usual mental faculties of reason and intellect into levels of the higher mind until finally one transcends the mind by achieving supramental consciousness (See Primary Source No. 3). While his terminology differs from Baba's, it seems to be expressing very similar concepts. Aurobindo inspired many with a very exalted picture of how humanity would progress in the future, which will be discussed in the last chapter.

Chapter 8

The Future According to Science and Religion

The present epoch is often viewed as a pivotal moment in cosmic history for the Earth and humanity. The acceleration of evolution in the last half billion years and the recent appearance of humans, culminating in the explosion of knowledge and technology in the last two centuries seem to mark this epoch as a period of rapid and portentous change but also one of promising new potentialities. For the first time, humans have within their grasp the power not only to predict possible future events for the Earth but also the power to affect the whole course and even survival of humanity and all life on the planet. In the more distant future, scientists predict the destruction of Earth as the sun brightens and dies, the destruction of our Galaxy as we know it, the thinning out of the universe through expansion, the death of all stars, and eventually the disintegration of all cosmic bodies and matter itself. Numerous religious systems chime in that the present time is an era of enormous destruction and/or spiritual transformation. Some predict the imminent end of the world, while others foresee an enormous shift in human consciousness and modes of interaction heralding a new golden age. Religious ideas about eschatology—the final events and ultimate destiny of the world and humanity—vary from beliefs in an actual physical event fulfilling God's purposes on Earth or in the universe to the more mystical notion of the cessation of ordinary physical reality by reunion with the divine being.

THE FUTURE ACCORDING TO SCIENCE

Scientific predictions of the future fall into several categories: the immediate future of Earth and humanity, the more distant future of the sun and solar system, and the ultimate future of the cosmos predicted by cosmology. Because religious views relate to all of these time periods, each will be considered in turn.

The Immediate Future of Earth and Humanity

In this highly speculative subject, it is of course impossible to predict the future course of Earth in the next few centuries. Threatening trends loom, such as globally destructive nuclear warfare, resource depletion, and global warming, thought by some to be at a point of no return, unless corrective intervention occurs soon. Life's survival could also be threatened by unexpected astronomical events, such as asteroid and comet impacts, a supernova explosion or invisible massive star collapse within tens of light years, or the merger of neutron stars into a black hole within a thousand light years. These events have long statistical frequency periods of several hundred thousand to millions of years, but that means that scores of such events are predicted statistically over the next few billion years, and one could, of course, happen at any time (Stoeger 2000, 19–28). At the same time, promising future possibilities for Earth and humanity continue to multiply, as knowledge and technology advance in all areas, as global economic and communication networks grow, and as social change expands possibilities for individual human lives. It is far beyond the scope of this book to assess the present condition and immediate future possibilities for humanity and Earth, except to note that strong opposite trends exist: destructive and creative forces both appear to be at work (Musser et al. 2005). This backdrop will form the barest picture for comparison with religious views.

The Distant Future of Earth, Sun, and Solar System

Astronomers predict dire circumstances for Earth in the far future. One or 2 billion years down the road, the gradual increase of the sun's energy output will push Earth's temperatures to the point of no return. For the last 5 billion years, the sun has been shining by thermonuclear fusion in its core, every second converting several hundred million tons of hydrogen nuclei or protons into helium nuclei and releasing energy with every reaction. As hydrogen gradually depletes, the pressure weakens, contracting the core and raising its temperature and rate of energy production. The gradual brightening of the sun happens slowly, but in roughly 1 to 2 billion years, it will warm the Earth to an unbearable temperature, vaporizing water and signaling the end of life on the surface. At this point, the sun still appears normal on the outside, while the Earth is unrecognizable. Several factors influence exactly how long this process will take, most especially the future strength of the greenhouse effect from atmospheric carbon dioxide. Many scientists predict that in the far future, levels of carbon dioxide may fall off, reducing its greenhouse effect and the rate of heating by increased solar radiation. At this epoch, it is conceivable that Earth's inhabitants could comfortably emigrate to Mars.

Five billion years into the future, bigger changes are in store for the sun, as it ages into a red giant star. With its hydrogen depleted altogether, the core will contract and heat, igniting much fresh hydrogen in its outlying layers and causing the outermost regions of the star to expand enormously, cool off, and turn red. Estimates suggest that its energy output will increase a 1,000-fold and its diameter a hundred-fold, enlarging it to engulf the orbit of Mercury. Eventually, its core will fuse and exhaust helium and contract and heat again, creating a new wave of energy production and swelling its outer layers to even greater dimensions—this time expanding 500-fold to the current orbit of Mars. Throughout this process, the sun loses mass, weakening its gravity and ability to hold the planets, which slowly migrate outwards. At its more distant position, Earth may just escape the clutches of the swelling sun. Now Mars will also be too hot for human habitation, and the most likely site for survival in the solar system might be an ice and water laden Jovian moon, such as Europa (Talcott 2007, 28–33).

The same epoch marks the predicted demise of the Milky Way Galaxy, at least in its current form. Motion studies have revealed that the Milky Way is on a collision course with its nearest large neighboring galaxy, the Great Galaxy in Andromeda (see Figure 2.2), and in 7 billion years they will merge. Distortions in stellar orbits will begin much sooner, eventually propelling stars outward into a large spherical region. The central bulges of the two galaxies will merge, as long tidal streamers of stars will drift far behind. As the central supermassive black holes of the two galaxies combine, each weighing several million suns, a burst of gravity waves will ripple out into the cosmos. When all is done, our solar system will then be a part of a new giant elliptical galaxy, which may reveal little hint of its former self as separate spiral galaxies. Incredibly, because of the incredibly thin distribution of stars within galaxies, stars and their planetary systems will survive (Villard 2003).

The Far Distant Future of the Cosmos

One way or another, cosmological models and laws of physics predict that entire universe will ultimately reach a state of "death," where life as we know it will become impossible. Whether by an inevitable running down of energy, expansion to an ever colder and thinner state, or collapse into final fiery cataclysm, the universe and life within it seem destined ultimately to freeze or fry. Recent measurements indicate that an inexplicable accelerating expansion is underway and that we are headed for icy emptiness much sooner that previously thought. If multiple universes exist, however, a final end may never happen, if a multiplicity of new universes continually begins from scratch.

The concept of the heat death of the universe was first introduced in the mid-nineteenth century and sounded a note of pessimism against the belief in progress that had characterized Enlightenment thought and underlay Darwinian evolution and philosophies based on it. The phenomenon was first suggested by German physicist Hermann von Helmholtz. He noted that because of the Second Law of Thermodynamics, more and more energy is being converted into unusable random form—that is, entropy is increasing—and will continue to do so until all available energy becomes "heat," producing a final irreversible state of maximum entropy. This "Heat Death" became a cause for gloom, for those who cared, in the late nineteenth and twentieth centuries, as cosmologists such as Eddington and Jeans popularized the idea and philosophers and commentators grappled with its implications for human and cosmic purpose. Darwin himself was much troubled by it, writing that "it is an intolerable thought that [man] and all the other sentient beings are doomed to complete annihilation after such long-continued slow progress." In 1903 mathematician-philosopher Bertrand Russell expressed resignation:

... [The] world which science presents for our belief is even more purposeless, more void of meaning, [than a world in which God is malevolent] ... the whole temple of Man's achievement must inevitably be buried beneath the debris of a universe in ruins [and] ... only on the firm foundation of unyielding despair, can the soul's habitation henceforth be safely built.

But he remained unconcerned, arguing that most people would happily direct their energies to other activities. Those favoring a science-based philosophy, such as Teilhard de Chardin, faced the dilemma: "Either nature is closed to our demands for futurity, in which case thought, the fruit of millions of years of effort, is stifled or else an opening exists—that of the super-soul above our souls." British theologian William Inge reacted differently, seeing the heat death as "the new Gotterdammerung," the final destruction of all in Norse mythology. He welcomed it as a blow to modernist science-based philosophies with their immanent, evolving God and as confirmation of religious notions of everlasting life and an unchanging creator and sustainer God. British physicist Whittaker expressed similar views in his 1942 lectures, *The Beginning and End of the World,* where he argued that the world's creation and destruction only confirm a transcendent God beyond nature and man's destiny to love and serve God in a new creation (Barrow and Tipler 1988, 166–68).

Few disagreed with the heat death prognosis, but one contrary opinion came mid-century from British physicist Milne, who wrote, "I believe this conclusion to be mistaken; it is not an inevitable consequence of the second law of thermodynamics as applied to the universe as a whole, and I do not believe it to be true of the universes as it is." He questioned the extrapolation from available data to the entire universe and believed there were only local regions where entropy was increasing. The universe as a whole, in his theory of cosmology, was infinite and unending, a place of "infinite variety," where God took delight in countless different living forms (Worthing 1996, 180–81; Milne 1952, 146–50).

The early twentieth century saw the discovery of the expanding universe, and its last few years revealed a totally unexpected finding about its rate of growth. Early cosmological models based on Einstein's general theory of relativity predicted three different rates of expansion that corresponded to three different possible geometries to the universe—open, flat, and closed (see Figure 2.7). In the open and flat cases, the universe would expand forever and be subject to the heat death, whereas in the closed model, universal expansion would halt and reverse, drawing the universe back into a big crunch and a much more fiery heat death. In each case, the universe would slow down over time. These models were subject to various observational tests to determine which might be correct—for example, measuring the amount of matter and energy in the universe and ascertaining the expansion rate at early versus later epochs. At century's end, astrophysicists had one of the biggest surprises of the century. Using bright supernovae in distant galaxies, two independent teams led by astrophysicists Saul Perlmutter and Brian Schmidt simultaneously obtained the startling result that the expansion rate has actually been accelerating for the past 4 to 5 billion years, a finding confirmed by subsequent observations. Some form of mysterious energy spread evenly throughout space must be propelling it outwards, overcoming the expected tendency of the universe to slow down.

The existence of some form of antigravity force had once been postulated by Einstein, who incorporated it as a "cosmological constant" in his equations of general relativity to prevent the gravitational collapse of a supposedly static universe. The term was no longer needed when cosmic expansion was discovered, and Einstein regarded his introduction of this concept one of his greatest blunders, but now, ironically, it has regained great cosmic significance. Dubbed the "dark energy," theorists are endeavoring feverishly to understand what it is—perhaps a form of vacuum energy predicted by quantum mechanics, or the effect of the extradimensions in string or M-theory, or a result of variations in supposed constants of gravity or light speed, or simply some new form of energy unrelated to any of the above. A remarkable feature of this energy is that it comprises around 70 percent of all the stuff of the universe, with ordinary observable matter making up 5 percent or less, and the remainder being invisible "dark matter," whose exact nature is also still unknown. It is indeed remarkable that, according to this picture, somewhere around 95 percent of the universe's contents are completely unknown to science at the present time. Future implications are likewise uncertain. Considering what is known so far about dark energy, it could produce quite a range of possible futures: a continued accelerating expansion, an eventual collapse, or, if the energy grows with time, an accelerated acceleration that will rip the universe apart, cosmic bodies and particles alike, within 20 billion years or so (Krauss and Turner, 2004).

Scientific Eschatology

While most scientists are pessimistic about life's prospects in final states of the universe, a few scientists have ventured to address in some detail physical conditions and possible realities for living beings in the extreme far future, challenging the traditional gloomy view. Beginning in the late 1970s different studies speculated how life in some form might be able to continue existing indefinitely in the extreme conditions of a highly expanded and cooled off universe and also into the fiery Big Crunch. A more recent study delineates all the stages the universe might pass through from its Big Bang inception until its end and possible transformation 10^{100} years into the future!

The serious investigation of the prospects for intelligent life in the far distant future was initiated by physicist Freeman Dyson in his seminal 1979 paper "Time without End: Physics and Biology in an Open Universe" and later popularized in his book *Infinite in All Directions* (2004). He believes that an open universe allows "enormously much greater scope for the activities of life and intelligence" than a closed universe, although the latter is not ruled out if future technology could begin to guide cosmic forces. Thus a major assumption he makes is that one must account for the effect of intelligent life itself, which could direct the development of the universe technologically to suit its needs, or redesign itself through genetic engineering to meet the demands of a changing cosmos. He also assumes that matter's main constituent, the proton, is stable, although theoretical predictions currently challenge this. Even if protons do decay, he argues, "after the protons are gone, we shall still have the electrons and positrons (antimatter electrons with positive charge) and photons, and immaterial plasma may do as well as flesh and blood as a vehicle for the patterns of our thought" (Dyson 2004, 111–12). He concludes that even though life as we know it cannot continue indefinitely, some form of intelligent life could, one based a different structure that can hold the computing function of a brain, replicate itself and exist at low temperatures. In this way, "life and intelligence are potentially immortal...there could be a "universe growing without limit in richness and complexity, a universe of life surviving forever and making itself known to its neighbors across unimaginable gulfs of space and time" (Dyson 2004, 114; Worthing 1996, 168).

Another eschatological vision of life's continuity was proposed a decade later by physicist Tipler in his Vatican conference essay "The Omega Point Theory: A Model of an Evolving God" (1988) and in a later book entitled *The Physics of Immortality* (1995). His ideas built on his Final Anthropic Principle, which claims that "intelligent information-processing must come into existence in the Universe, and, once it comes into existence, it will never die out" (Barrow and Tipler 1986, 23). The universe is still quite young, having existed only 10–20 percent of its expected 100-billion-year lifespan in a closed universe, which eventually collapses to zero volume, and it will exist forever in an open universe. He considers it a "moral postulate" that life will live as long as the universe; were life to die out, the universe would have no significance. He goes on to define life in terms of physics as a "form of information processing" and the "human mind—and the human soul—[as] a very complex computer program." This is not all that life is, but it is accurate when life is examined at the elemental physics level alone. Life can be said to live forever if information processing can continue to the end, and an infinite amount of information can be processed and stored. He finds that these conditions are met only in a closed universe where all that has ever existed and been thought converges onto a single all-knowing point outside of time—the Omega Point. It represents the "collection of all experiences of all life that has, does, and will exist in the whole of universal history, together with all non-living instants." Life subsumed into this point thus becomes eternal and omniscient and thus provides a "model of an evolving God" (Tipler 1988, 313–28).

In a more recent book, *The Five Ages of the Universe: Inside the Physics of Eternity* (1999), astronomers Fred Adams and Greg Laughlin approach the same subject by presenting in novel detail far future states, processes of matter and prospects for life in every epoch. They mark the intervals of time in "cosmological decades," where each successive interval is ten times longer than the previous one. The first epoch is the Primordial Era, which lasts until the formation of atoms. Following that is the present Stelliferous ("full of stars") Era, which encompasses the birth, evolution and death of all stars and lasts a hundred trillion years into the future (see Figure 4.2). Gradually as more and more gas is used up in forming stars, and more and more stars die, both the stellar population and the chemical composition of the universe will undergo a fundamental transformation. Remnants of low-mass stars, white dwarfs and their final state, black dwarfs, will dominate—ultradense Earth-sized stars of helium, or carbon and oxygen, where a teaspoon of matter weighs a ton. The population of massive star remnants also grows—neutron stars the size of cities where a teaspoon outweighs a mountain and spin rates reach hundreds of times per second—and the heaviest of all, black holes.

As stars die out and no new ones form, there follows the Degenerate Era, when "dead stellar remnants capture dark matter, collide with each other, scatter into space and finally decay into nothingness," a process lasting until 10^{39} years into the future (Adams and Laughlin 1999, 73). The development of complex molecules, "building blocks for a new type of biology," they speculate, could conceivably develop in the atmospheres of white dwarf remnants, powered by dark matter annihilation (Adams and Laughlin 1999, 74). In its final phase of the degenerate era, protons themselves decay. With ordinary matter gone, black holes "inherit the universe" in the Black Hole Era, when they "warp space and time, evaporate their mass energy, and make an explosive exit" (Adams and Laughlin 1999, 107). Life in this era can be envisioned only in the most abstract way as having the "architecture" of life—information processing and self-replication-but not the matter of current life forms. They follow the ideas of Dyson in noting how rates of thinking might be scaled down to match the energy available at the ever colder temperatures. This era lasts until 10^{100} years when black holes the size of galaxies have finally evaporated. The universe comes to its bitter end in the Dark Era, when nothing remains but the waste products of proton and black hole decay: mostly very weak photons, neutrinos and a smattering of electrons and positrons—a situation oddly reminiscent of the primordial era, which likewise lacked large bodies. Thus in the final era, lasting until 10^{101} years, "the nearly moribund universe struggles with cosmological heat death and faces the possibility of universally transforming phase transitions" (Adams and Laughlin 1999, 152). Whether larger structures or chemistry could ever form from "positronium atoms," which combine electrons with positrons, is completely unknown, but they would be the only possible way to conceive of life developing in this era. It is complicated by the fact that although electron-positron combinations begin at enormous distances, they will eventually annihilate each other.

Perhaps the most speculative material in the book explains how the universe could take surprising turns in its final stages through a quantum tunneling process that would produce cosmological sized phase transitions, much like what happened during the inflationary epoch of the first second. A new bubble universe, with a different vacuum energy, could emerge and take over, producing any number of results depending on how the new vacuum energy relates to the old. New evolutionary processes could begin and eventually give rise to complexity and life, or the new bubble could collapse violently into a miniature Big Crunch. Stranger still is the possibility of other quantum tunneling processes causing the universe to give birth to a "child" universe. A new bubble universe with a different vacuum energy "nucleates", or expands rapidly and disconnects from the "mother" universe. (Such a process can also occur at the singularity of an existing black hole.) Strangest of all, both of these processes can conceivably be set off by external influences, a possibility which suggests that someday humans or other intelligent beings might be able to affect the evolutionary course of our universe and even other universes. A number of these possibilities are also discussed in Davies (1994) book *The Last Three Minutes.*

THE FUTURE ACCORDING TO RELIGION

Virtually all religions believe in some form of eternal life and transformation of the present earthly life into a future state of ultimate justice and fulfillment. In theistic systems this state represents a salvation, a redemption and reunion with the deity and the final accomplishment of God's purposes for all. The striking contrast between the religious eschatologies of faith in an everlasting life and the inevitable death scenarios of science make the search for consonance one of the most difficult challenges in the entire science and religion dialogue. Many issues must be examined, among them: whether the religious "life to come" is a continuation of the current physical reality or a supernatural new order of being, how the context and substance of this final state relates to the space-time and matter-energy of this world, and over what time period a comparison is truly relevant or meaningful. Also to be considered are the extent to which individuality of the human person is preserved and in what form, the role of the Messiah or Christ (or *Avatar*) in bringing about the ultimate state and whether the transformation happens even now for individuals or is a worldwide or universe-wide future event, or all three. How these issues are examined and related to science depends most assuredly on the particular religious tradition involved but also on the beliefs of individual theologians within the same tradition.

Biblical Eschatology—Judaism and Christianity

There are two major themes within the core Judeo-Christian belief in the establishment of God's Kingdom and the fulfillment of divine purpose. One is the hope for transcendence of death and eternal life in God's Realm, and the other is the promise of a future state of justice and righteousness on Earth. Willem Drees describes these two strands as concerns for "the *finiteness, creatureliness,* of existence" on the one hand and for "*injustice,* one might say the *brokenness* of existence" on the other (Drees 1990, 118). Barbour notes a similar dual concern in biblical writings with one strand dominating or both coexisting in different historical epochs. Old Testament early prophets foretold of a messiah to come from among their own people to guide them back to the covenant and restore peace and justice. Later, when Israelites were in exile, it appeared that only supernatural help would save them, and a more apocalyptic vision prevailed in which the messiah would wage a cosmic battle to defeat evil and oppression and bring about his Kingdom. In early Christianity Jesus' pronouncement that the "Kingdom of God" was "at hand" led to belief that he would return soon to establish it. When that did not happen, Christian views diverged into those who continued the apocalyptic tradition, as presented in the Book of Revelation, and those who believed that the Kingdom was being established right here on Earth in the community of faithful Christians experiencing the "living Christ," or, as later church fathers saw it, in the church itself. In medieval times there was much emphasis on the end times and the last judgment, and "Consummation" was added as a final stage to cosmic history (Barbour 1997, 217–18).

Christians today have similarly divergent interpretations of the Kingdom of God and the new heaven and Earth prophesied in New Testament scripture. Some conservative evangelical Christians hold to a literal interpretation of Revelation and other apocalyptic writings. For them, the current world situation indicates that prophecies of end times are now being fulfilled, and a whole sequence of apocalyptic events is about to unfold—from the rapture (simultaneous ascension of all Christians to join Christ), to a period of Great Tribulation on Earth, a final Battle of Armageddon in Israel, Christ's glorious return and thousand-year reign on Earth, Satan's final defeat, the Last Judgment, and ultimately the establishment of the New Heaven and Earth (Leyrer 1981). While beliefs differ between and within groups about the order and even occurrence of all these stages, all share a basic core belief about supernatural events—the second coming of Christ, a final apocalyptic battle between good and evil, a last judgment of all souls, and the creation of the everlasting home of the saved. In contrast, theologians in the neo-orthodox and existentialist traditions view the imminent Kingdom as an aspect of the present that symbolizes an urgent call to commit to God in the here and now. Many in the liberal Protestant and liberation theology traditions also focus on the present but more in the sense of lending one's own effort to achieve God's purposes of peace and social justice now—creating God's Kingdom here on Earth. Most moderate to liberal Christians probably adhere to a belief in both perspectives, recognizing the importance of laboring for truth and justice in the world today but also remembering that such work and its future outcomes really belong to God, both within the historical world and beyond it (Barbour 1997, 218). Today, Protestant and Catholic theologians address issues in religious eschatology over the whole range of possible futures from the decades and centuries immediately ahead to the far flung outer boundaries of cosmology.

Eschatology and the Earth   Drees is among those who find greater meaning in an eschatology focused on the present and on issues of justice, believing that "it is not realistic to relate the future on a cosmological scale . . . to perspectives for humankind" (1990, 117). The study of the future that has value is to evaluate the present and assess future possibilities in order to guide us toward the best course of ethical action. An example he describes is the "theology of mutual relation" of Isabel Carter Heyward, who rejects views that perfection can only come in the future through a long evolutionary process or that it can only be caused by God and found in a supernatural realm. These are forms of "escapism to heaven or to an indefinite future." Both Heyward and Drees argue for an "axiological eschatology," one fostering "reflection on values guiding decisions in the present" (Drees 1990, 150–51).

Three key features characterize Drees' axiological eschatology: evaluation of the present, a call to responsive action and comfort to victims of evil and suffering. The theology must have a metaphysical framework that is consonant with science and that incorporates the following: understanding how the possible and the actual differ, believing in the future as human-made, and conceiving of God as the source and ground of "values and possibilities." These conceptions, Drees notes, parallel closely the ideas of process philosophy. Finally, Drees suggests a present-oriented notion of perfection as "an affirmation of the goodness of finite being, an affirmation despite the anxieties of finite life." Such goodness derives from the existence of a "good God," just as individual moments of time and the present ultimately derive from "God's eternity." God is both the "ground of reality" joining existence in the present to the transcendent realm of eternity, and he is also the "source of values and possibilities, [which calls us to] conversion for the sake of a more just future." Thus two aspects of God's own being give rise to and merge the two levels of eschatological concern, the "dimensions of depth and future, of mysticism and history" (Drees 1990, 152–54).

The ecological spirituality of Rosemary Radford Ruether offers another view of eschatology focused on present concerns. In *Gaia and God* she first explores the historical religious and scientific roots of today's present ecological crisis, then reviews three Christian theologies that relate to ecological spirituality and finally offers her vision for healing and transformation. Her own views are consonant with those of Matthew Fox, Teilhard de Chardin, and process theology, although the first two have features she questions. Fox, an Episcopal priest and educator once dismissed from the Dominican order for his controversial views, affirms that the true nature of all creation is "original blessing" and goodness, and not original sin and fall. The focus of true Christian spirituality should be to celebrate the life-affirming interrelatedness of nature and all beings within it. The embodiment of this principle and source of the original blessing is the cosmic Christ, the divinity within all, and the goal toward which creation is moving. The Christ figure has appeared in many forms in many cultures, a truth which connects Christians to all other religions, including small-scale native spiritualities, in a "deep ecumenism." Interfaith and secular-religious dialogues are both part of his program for transformation, as is fostering a mystical response of awe towards creation and developing the capacity for intuitive, holistic experience in new expressions of worship. Ruether finds that Fox is "basically on target in these affirmations and values," but needs more rigor and depth in comparing his work with other thought.

Ruether also reviews the eschatological vision of Teilhard de Chardin, whose orientation toward present transformation and notions of a spirit-infused material world resonate with her views. For Teilhard the modern age that connects conscious humans in a global network is giving rise to a world mind or "noosphere," which is beginning to link all human minds into one "unitary Mind." The progression toward unitary Mind is the development of the divine spirit within creation—the cosmic Christ. The process will finally transcend organic, physical life and culminate in a state of ultimate oneness of Mind and Spirit, or God, who will subsume all that has ever lived into his eternal being. Ruether finds that Teilhard's belief expresses well the Gaia hypothesis that Earth is a living organism, but she rejects other aspects of his thought: the superiority of Western religion and culture, the lack of concern for species extinction, and finally the contradiction between his discarding of physical life in the eschatological consummation and his central affirmation of "consciousness as the interiority of complexified matter." The latter idea, suggesting the deep unity of mind and body and the presence of consciousness even at the simplest level, supports her vision of the need to cherish and nurture all of creation.

Process theology, as formulated by Christian theologians such as Cobb and Marjorie Suchocki from Whitehead's process philosophy, also has features that accord well with an ecological theology. As in Teilhard's system, all entities have an interiority or "mentality"—even particles jostling around in random movement. Process theology also affirms a compassionate God who works in a dual mode to offer, encourage and cherish right action, but then suffers with whatever choices are made and subsumes all in divine remembrance and redemption. As the primordial ground of all possibilities, God persuasively offers a best possibility or "initial aim," which the entity, in its subjectivity and freedom, accepts or not in its decision to act. In his other mode God then subsumes the action chosen into his own "consequent nature," holding its memory, immortalizing it, reconciling it in a total vision and thus redeeming it (Ruether 1994, 240–46).

In developing her own "ecofeminist theocosmology," Ruether affirms that the most important work ahead is to transform human consciousness into a mode of deeper relationship with the earth and its inhabitants, a mode in which the role of sustainer replaces the role of master and destroyer. She follows both Teilhard and process thought in affirming that consciousness is what links us to all living beings and the rest of nature, yet this same awareness yields understanding of the mortality of the "self." The quest for a mental self that survives death, a motivating force behind religion in the past, must be "recognized, not only as untenable, but as the source of much destructive behavior toward the earth and other humans." There is no eternal life for the individual in her scheme. The ecological theology she espouses should be based on three important precepts: "the transience of the selves, the living interdependency of all things, and the value of the personal in communion." As one fully integrates these perspectives and lives them, even in the face of one's own extinction, oneness with the personal centers of all beings grows, and one comes to identify with and celebrate the "great Self" in all (Ruether 1994, 250–53).

Another theologian who affirms the crucial role of humanity in the future is Lutheran theologian Philip Hefner (1989), who describes the human being as "God's created co-creator." He portrays a sweeping epic of evolution that leads to human freedom, which brings with it creative power and responsibility for fulfilling God's purposes. For Hefner, evolution includes cultural development and its current most important achievement, technology. Human beings bear great moral responsibility to use technology in ways to improve the well-being of all life and the Earth itself. But human activity in the present is only part of the grand "eschatological destiny of the creation." The entirety of cosmic evolution is God's "creative process" and "grand program" for creation. The status of humans as created beings is dynamic and growing continually by the pull of destiny. Our role as co-creator derives from God's own "creative thrust" energizing the human to participate "as a free, self-conscious creature in shaping the passage forward toward God's own *telos* or purpose which will appear in its fullness at the consummation, at the final perfection of the whole cosmic history of the creation." Hefner's concern for the present is thus embedded in a much wider evolutionary framework spanning the deep past and the indefinite future consummation, and his thought therefore encompasses both realms of eschatological concern.

Christian Eschatology and Scientific Cosmology   Theologians face a difficult challenge in relating Christian eschatology to the far distant future envisioned by scientific cosmology. Can religious hope for the coming of God and a transformation into eternal life be in any way consonant with the scientific picture of the destruction of Earth's life and the multi-billion year decay of the universe into ultimate freezing or frying? Most mainstream and liberal theologians might agree with Pannenberg, who calls it "one of the most obvious conflicts between a world view based on modern science and the Christian faith," or with Polkinghorne, who asks, "What does religion make of this? Don't these bleak predictions deny that there is a purpose at work in the world?" (1996b, 90). For Pannenberg the problem is not just eventual destruction but also the prospect of an inordinate temporal span:

Is the Christian affirmation of an imminent end of this world that in some way invades the present even now, reconcilable with scientific extrapolations of the continuing existence of the universe for billions of years ahead? To this question there are no easy answers. Scientific predictions that in some comfortably distant future the conditions for life will no longer continue on our planet, are hardly comparable to Biblical eschatology. (Pannenberg 1981, 14–15)

Many theologians acknowledge that the two futures may simply not be connected and therefore eschatological hope should not be situated in the world process itself. Polkinghorne expresses it well: "The scientific prediction of cosmic futility simply reminds us that a kind of evolutionary optimism—a belief that the unfolding of history must bring progress to fulfillment—is inadequate as a ground of hope. If there really is a true and lasting hope, it can only rest in the eternal being of God himself" (1996b, 91). These positions suggest that the two eschatologies should simply remain independent.

Those committed to developing a scholarly science-theology research program argue that theologians must take the future predictions of science seriously and continue striving to understand how the two eschatologies relate. It is the agreed upon methodology of their field to do so. They seek resolution of the impasse in a variety of ways, a number of which are reviewed by Drees (1990, chapter 4), Worthing (1996, chapter 5), and Russell (2000), whose summary guides the sketch to follow. One approach is to question predictions of life's inevitable end and to consider theologically the theories of Dyson and Tipler about physical immortality. Process philosophy also suggests a more open and positive view of the future. Other promising approaches focus on themes of transformation and time and eternity. Finally, some suggest that accommodation must await advances in science, which joint research could stimulate.

Reactions of theologians to the eschatological ideas of eternal intelligent life presented by Dyson, Tipler, and Barrow have been varied but mostly critical. Pannenberg epitomizes the most positive but still cautious reaction: "there are still a number of points that are difficult to reconcile with a Christian doctrine of creation and eschatology... [but] in the general thrust of Tipler's project there is a remarkable convergence with Christian theology" (Pannenberg 1989, 265–66).

Pannenberg notes agreement between Christian views of the final end state and some features of Tipler's Omega—God's full manifestation there, its transcendence and immanence, and its role in the original creation (Pannenberg, 1989). Drees agrees that it has elements of an "acceptable eschatology" in affirming resurrection and consummation in a realm transcending space and time (1990, 140–41). Both agree that it is a constructive step for both fields, and Pannenberg commends Tipler for "breaking a longstanding taboo by dealing as a physicist with the theological themes of God and immortality... [and insisting] that the world as described by physics is more open to interaction with biblical and theological perspectives than is often believed" (Pannenberg 1995).

Critiques of Tipler's eschatology note that his conception of both humans and God is not very Christian and restricts theology. Polkinghorne is especially forceful in his reaction: "it is a fantastic, and curiously chilling, programme," presenting many problems. Extrapolating to the extreme last moments of a collapsing universe when there is infinite computer power readily available and used only for good ends is wildly speculative at best, and the conceiving of humans as "computers made of meat... [is] unacceptable." Intuitively, Polkinghorne feels that the whole conception is just too mental and dry to satisfy real human hope. He admits "verbal parallel" with Pannenberg's "God as the power of the future" but remains "unconvinced by these fantastic speculations" (Polkinghorne 1996a, 165–66). For Barbour, the biblical conception of humans as a "unity of body, mind, and spirit," is destroyed, God is not active, and there is no transformation of individuals or societies (1997, 219). The whole scheme is more like Spinoza's "God in nature" pantheism than Christianity, contend Drees and Clayton, who also argues that it "puts a straightjacket on theological reflection" (Drees 1990, 140; Clayton 1997, 135–36). Peacocke also questions the disregard of values in preserving evil as well as the good unto eternity (1993, 345).

Process theologians present both universal and personal eschatological perspectives in discussing two kinds of immortality: objective and subjective. All agree that every experience achieves objective immortality through being absorbed and redeemed in God's consequent nature, which continues to grow and become enriched indefinitely through the ongoing experience of his creation. Some process thinkers also subscribe to subjective immortality, in which "the human self continues as a center of experience in a radically different environment but amid continuing change rather than a changeless eternity." One process theologian, Cobb, envisions a "new kind of community transcending individuality" (Barbour 1997, 304). Haught, who advances a future-oriented form of process theology, argues that theology has adjusted to all kinds of cosmologies in the past and will do so again. Furthermore, there are grounds for optimism emerging from science itself; it may not be predicting doom at all. The simplistic linear extrapolations and "abstract mathematical representations... inevitably overlook the elusive complexity and indeterminacy that open the cosmos to a genuinely novel future." The new sciences of chaos and complexity, suggesting unpredictable self-organization in nature, render "the prospect of precise scientific prediction of final cosmic catastrophe... shakier than ever." Hindsight and the anthropic principle reveal the seeds of promise from the very beginning of cosmic evolution; such seeds for future flowering and hope must exist today. Possibilities for novel outcomes flow from God, arising out of the future where he conceals himself as an "infinitely liberating source of new possibilities and new life" (Haught 1998, 237–41).

Most Christian theologians accept the science as it now stands but place their faith in God's final consummation of creation. God will somehow effect a transformation into a "new creation"—a new reality or different order of being that is "in God" or "in Christ." Scholars explore what this transformation might mean for both the individual and the universe, how time is conceived in each realm, and how it may or not relate to anything science can address.

In Christian thought hope for an ultimate reality involves continuity of the human person, whether it be after a single death or after billions of years of cosmic history. For Polkinghorne, "cosmic death and human death pose equivalent questions of what is God's intention for this creation." Theologians today are averse to a body-soul dualism, whereby the spiritual soul joins God after death or at the consummation, and consider different ideas of how the body-soul unity achieves a new embodiment in the transformed order. Polkinghorne gives his vision:

It seems to me that it is the essence of humanity to be embodied and that the soul is the immensely complex "information-bearing pattern" in which the ever-changing atoms of our bodies are arranged. It is surely a coherent hope that the pattern that is me will be remembered and re-embodied by God in his eschatological act of resurrection. The "matter" of that resurrected world will be the transformed matter of this dying universe, transmuted by God in his faithful act of cosmic resurrection. It will have new properties, consistent with the end of transience, death and suffering, because it will be part of a new creation... no longer standing apart from its Creator as the "other," . . . but fully integrated with the divine life through the universal reconciliation brought about by the Cosmic Christ. (Polkinghorne 1996c, 54–55)

He finds this idea of a "cosmic redemption. . . . immensely thrilling and deeply mysterious." It is an event beyond history that brings humanity and the cosmos together in a single transforming event.

It is important to realize that the "new heaven and earth" are in no way a re-run or "second attempt by God at what he had first tried to do in the old creation" but result from a wholly different kind of action by God and exist in a different order of time. The old creation has the nature appropriate to one undergoing evolution and allowed to have a certain independence from God and thus be subject to evil and suffering. The new creation has a very different character: it "represents the transformation of that universe when it enters freely into a new and closer relationship with its Creator, so that it becomes a totally sacramental world, suffused with the divine presence." Regarding the conception of time, Polkinghorne views the next world as "everlasting, but it will not be eternal in that special and mysterious timeless sense in which the word is applied to God himself." Heaven will not always be in the same state but will "involve the endless, dynamic exploration of the inexhaustible riches of the divine nature" (1996a, 163–70).

The concept of eternity, different "orders of time" and cosmic trans-formation are themes also in the writings of Pannenberg and Worthing. Pannenberg believes that the imminent biblical "end of the world" may be unapproachable by scientific inquiry and altogether different from a far distant cosmic end. Worthing agrees that it is not warranted to equate them: the biblical end could be a "renewal or transformation of the world rather than its absolute destruction and subsequent re-creation." This redirects the concern from the culminating event itself to the question of how both ordinary time and the cosmological future relate to and are subsumed into God's eternity—"an eternality of a decisively different order from that which the physical could potentially possess—which is precisely what Christian faith expects to happen" (Worthing 1996, 178, 198). Peters pursues similar ideas in presenting a holistic sense of time, in which he views the cosmos as a "unity of time and space" created from the future and awaiting its own consummation and transformation in the eschatological future, which transcends but includes creation (Russell 2000). These different ideas reveal the notion of time as an important area for joint research, especially since science itself has now challenged traditional ideas about time. The old notion of "flowing time" and a "universal present" that all observers experience has been questioned by special relativity, which showed that different observers can see the same event at different moments. Thus all space-time points exist together and not just a present moment—a notion called "block time." How this relates to God's time and how temporality relates to eternity are all rich questions for future dialogue.

The challenge remains of articulating Christian eschatology in intelligible detail in relation to scientific cosmology as it currently stands. Pannenberg concedes, "Perhaps one should rather accept a conflict in such an important issue, accept it as a challenge to the human mind to penetrate deeper still into the complexities of human experience and awareness." He envisions that an understanding of time and eternity may develop in ways that could accommodate both biblical eschatology and modern science (Pannenberg 1981, 15). Since that was written, major research projects have engaged in serious discourse seeking to bridge the gap between scientific and religious "futures." One three-year study held at Princeton and Heidelberg brought together an unusually diverse group of scientists, theologians, cultural historians, ethicists, and biblical scholars to explore common ground in their viewpoints about the future. Their essays, presented in *The End of the World and the Ends of God* (2000), reveal the nature of their work, a "multidisciplinary striving for a realistic eschatology which aims at coherence of insights while keeping the complexity of the topic in mind." All had to face the challenge of "continuity and discontinuity" in religious eschatology—what remains and what dies away in the "new creation"—and the need to wrestle with this notion emerged as the strongest theme of the study. They recognized that both science and theology must stretch themselves to expand their own approaches. Science must face areas and methods of inquiry beyond its usual competence, and theology must present and justify more fully its claims to truth, take science's findings seriously, and "reexamine its views of hope, joy, the divine future, the new creation, and eternal life . . . in a critical and self-critical dialogue with the social and natural sciences." In the shifting eschatological mood of the last half century where theologies of hope have given way to more apocalyptic modes of thinking, such efforts are both more challenging and more urgent (Polkinghorne and Welker 2000, 1–13).

The work of such research programs is exemplary of a genuine two-way dialogue that many see as a necessary path to progress. As an energetic proponent and leader of such discourse, Russell has developed his own methodological path for deepening the dialogue. As with Pannenberg, he suspects that the level of knowledge in science may be the heart of the difficulty—that presuppositions underlying physics produce "an insufficiently rich cosmology" for accommodation with theology. He envisions an iterative process where theologians develop a richer conception of time in the light of relativity theory and use those ideas to suggest new assumptions for science, which could lead to richer cosmological theories more amenable to theology, should scientists be willing to undertake them. In such an effort science and theology would engage in a joint quest to answer significant questions raised in both fields but not answerable fully in either alone.

The Far Future in Hinduism, Buddhism, and the Cosmology of Meher Baba

In Eastern religions the total cosmos continues forever, but within it all material things ultimately come to an end. Over the very grandest of time scales, even the entire universe undergoes dissolution but then is reborn again in a new cycle, and thus the cosmos continues. With such ideas, Eastern systems seem to synthesize the two rival cosmologies of Western philosophy and science—first, that the present universe had a starting point and will progress in linear fashion towards an end, and second, that the greater cosmos, although experiencing cycles of creation and dissolution, is unending. Even with the latest discoveries of modern cosmology, the same duality persists. Science's new "accelerating" model seems to suggest never ending expansion and a "one-shot" universe, but at the same time, there is serious scientific consideration of an unending series of multiple universes, an idea which also resonates with Eastern thought.

Vast periods of time are envisioned in Hindu and Buddhist thought. The grand sweep of Hindu chronology begins with *yugas,* four successive ages of human history in which adherence to spiritual ideals and harmony among men gradually decreases. The current *Kali Yuga*, or Iron Age, marks the nadir of the cycle and lasts for 432,000 years. Before that was the *Dvapara Yuga* or Bronze Age, the *Treta Yuga* or Silver Age and, at the zenith of the cycle, the *Krita Yuga* or Golden Age. Each previous stage lasted twice as long as the one which followed. The entire cycle of four ages, called a *Mahayuga*, thus lasts 4,320,000 years. One thousand *Mahayugas*, or 4.32 billion years, forms a *Kalpa*, also called a Day of *Brahma*. As the Creator God, *Brahma* is one aspect of the Hindu trinity together with *Vishnu*, the Sustainer, and *Shiva*, the Transformer or Dissolver of the created universe. At the end of one Brahmanic day, there is a *"laya"* or *"Mahapralaya,"* in which the universe is dissolved back into the mind of Brahma, where it remains for another whole *kalpa*. Thus is formed a full Brahmanic day and night of 8.64 billion years-twice the age of the Earth and an appreciable fraction of the estimated age of the entire universe. In an interesting coincidence, the acceleration of the universe has been occurring just during the last day of *Brahma*, or approximately since the Earth was formed. Hindu chronology continues well beyond this, however, envisioning first the year of *Brahma*, just over three trillion years, and finally a century or lifetime of *Brahma*, 311 trillion years. In this enormous time span, we are in the fifty-first year of *Brahma*, just beginning the second half of the immense time period of the present creation, which exceeds the scientifically estimated age by a factor of more than 10,000! (Morales 1997, 1–3; Raman 2002, 187–88).

Buddhists speak of similarly staggering time periods, describing how the "Earth and the planets and stars in our skies are ... in the habit of dissolving themselves once in about every 100 trillion years," of which two-thirds has already elapsed—figures that differ somewhat but are of the same order of magnitude as the Hindu (Ranasinghe 1957, 263). They also offer descriptions of the cosmic dissolution. The explanation in *The Buddha's Explanation of the Universe* begins with a reminder of the primacy of the mind in structuring and evolving the universe, which is composed both of units of mind and units of matter. Only units of mind can sustain the forms of the cosmos, and when their force weakens, the forms dissolve and units of matter spread out into a homogeneous state of blank emptiness and heat energy. As units of mind absorb and suffer in this heat, they become purified and strengthened and finally capable of once again gathering units of matter together to form planets, stars and a universe with material structure. Beings whose mind is already strong can escape into spiritual realms and avoid this suffering. This is, of course, one of the central teachings of Buddhism—that through detachment one can learn to avoid the inevitable suffering that desires bring and achieve the state of enlightenment.

The Buddhist description above strikes a very resonant chord with modern scientific descriptions of the heat death of the universe and the cyclic version of the Big Bang. Scientific scenarios of heat death and eventual decay, made more explicit by the more recent extrapolations of Adams and Laughlin, also end with particles distributed evenly in a "state of blank darkness." Of course, a fundamental and crucial difference is the role of mind, which is generally absent in the scientific picture, although one might wonder if some aspects of the speculations of Dyson, Tipler, Adams, and Laughlin, different as they are in purpose, are pointing in that direction. Further, the state of blank darkness as an invisible field of energy seems to have parallels with the concept of science's "vacuum energy"—a state of matter with no discernible form but teeming with the potential to become a universe.

Meher Baba also speaks of universal dissolution, or "*Mahapralaya*," in ways that resemble the Buddhist conception but suggest additional aspects. He describes two kinds of dissolution. One is a "*pralaya*," which occurs routinely to a person while unconscious in deep sleep, and the other, *Mahapralaya*, which occurs to the universe as a whole, as he explains:

These universes come out of the Sadguru [Perfect Master] and merge back into him after aeons. This is called the *mahapralaya*. When the universes disappear in

Mahapralaya, they are no longer in gross form, but they remain within the *universal mind*. Every individual gross mind rests in the universal mind. Though this is all an illusion, still the individual souls of all beings that inhabit the universes remain within the universal mind. And after aeons the evolution starts again and every *jeevatma* (incarnate soul) gradually takes form in accordance with the consciousness he had before the cosmic event of Mahapralaya occurred. (Kalchuri 1986, 989)

This suggests dissolution and reconstitution of the universe that brings forward a soul's higher subtle and mental bodies into the next cycle but may not even be perceptible to the individual. As Baba further explained:

[Mahapralaya is] the breathing in and out of all creation by God . . . When you are breathed in [by God], you are in effect stored in exactly your present situation until creation is breathed out again. Then, when some planetary system has reached the proper stage of development, you are reincarnated and proceed with your involution from the point where you had left off. (Kalchuri 1986, 6286–6287)

Other statements seem to imply a grander end of all ends, occurring

when the original cosmic Creation . . . is finally destroyed by the play of cosmic opposite impressions of God [and] the whole cosmic Creation as Nothingness is absorbed infinitely by the Everything . . . Like a manifested tree, [it] recedes into the unmanifested seed form of non-existent non-existence, only to be manifested anew once more in the very next moment of eternity. (Baba 1997, 111, 233)

Is it possible to relate all these ideas of immense time scales and universal dissolutions and recreations to the modern scientific and other spiritual frameworks? The problem already arose with Baba's cosmology in realizing that the time required for the soul's journey through creation greatly exceeded the 14-billion-year age of the present universe. Eastern religious cosmology resolves this problem by affirming a much longer "period of creation" of 300 trillion years, although it raises other issues. If our present universe really is only 14 billion years old but over half of the "lifetime of Brahma"—150 trillion years—has already passed, then most cosmic activity had to occur in other, earlier "universes," which may have come and gone. This notion resonates well with some current scientific hypotheses of multiple universes that are an outcome of string theory. One example was the new endlessly cycling universe of periodically colliding three-dimensional "branes" floating in a higher dimensional universe, discussed in Chapter 3. Both this and the more traditional standard Big Bang oscillating universe model are consonant with the Eastern notion of cycles of creation and dissolution. On the other hand, if *Mahapralaya* is something that God could initiate at any time, as a kind of breathing spell for creation, which is not detectable on the physical plane, it is hard to understand how science could have any grasp of it. However, the idea could conceivably be of interest to Western theologians, who grapple with the issue of how God acts in creation. The paradox of time stopping and then picking right up "in the next moment of eternity" and the notion of a person's identity being carried forward to be re-embodied later might also interest theologians, who are attempting to understand how God's eternity relates to human time and how he will resurrect humans in the eschatological future.

So far the discussion of the future in Eastern religious thought has centered only on cycles of time and the eventual fate of the cosmos rather than the more religious eschatological issues that concern Western theologians. In Eastern thought there is much more focus on the journey of the individual soul toward enlightenment as the spiritual goal rather than the salvation or redemption of whole communities or groups or all of humanity at once. And there is less focus on the idea that there might be a single supernatural event by which God consummates his creation and brings some or all souls to him. For Baba, Hinduism and Buddhism, the goal of all creation—enlightenment or God-realization—occurs on a very individual basis and could be fulfilled at any time or place, given the readiness of the spiritual aspirant. To one degree or another, all three emphasize the transience, impermanence, and illusoriness of the outer world of forms and stress the goal of unveiling the reality—the light of the divine—hidden at the deepest level of all beings. Thus the world forms the backdrop for the divine drama of the soul's journey.

The Near Future in the Thought of Aurobindo Ghose and Meher Baba

Both Aurobindo and Meher Baba describe an enormous spiritual upheaval and transition that is taking place in our time and is a prelude to a new more spiritual age. Aurobindo describes the present evolutionary crisis and its future resolution in *The Future Evolution of Man.* He argues that neither reason nor traditional religion can solve the crisis, which arises from a "disparity between the limited faculties of man—mental, ethical and spiritual—and the technical and economical means at his disposal . . . without an inner change man can no longer cope with the gigantic development of the outer life." Real progress can only arise from the development of true spirituality, which opens individual souls to greater reality. But the goal of evolution on Earth is not just to ascend into higher, more spiritual realms of existence beyond the earthly plane but to attain a "greater perfection in the earthly existence itself." This will require the descent of a "supramental Force" into the lower nature to initiate the "divine life upon the Earth." This is the real goal of evolution—to bring about a conscious "identity of the individual self with the transcendent Self, the supreme Reality." The being who has this awareness is a new type of humanity, as different from the present day man as man today differs from the animal (Ghose 1990, viii–xv). These words resonate with Polkinghorne's vision of God's future as a "totally sacramental world, suffused with divine presence."

The need to transcend the limited self is the essence of Baba's message about the future. In his discourse, "The New Humanity," he likewise defines the present spiritual crisis and its remedy:

As in all great critical periods of human history, humanity is now going through the agonising travail of spiritual rebirth. Great forces of destruction are afoot and seem to be dominant at the moment, but constructive and creative forces which will redeem humanity are also being released through several channels. (Baba 1967, vol. 1, 17)

Baba goes on to explain that the root cause of destructive forces is identification with the ego and separative existence focused on the limited self. Gradually humanity will come to see the limitations of this way of being, and a different kind of life will unfold in which people are more aligned with lasting spiritual values, deeper truth and even a different mode of knowing—intuition. In the coming transformation science will play a vital role as a tool for self-knowledge but only as long as its knowledge is "properly fitted into a larger spiritual understanding . . . All-sided progress of humanity can be assured only if science and religion proceed hand in hand." There will be a new golden age of the "New Humanity . . . which will learn the art of co-operative and harmonious life . . . enjoy peace and abiding happiness . . . [and] be initiated in the life of Eternity" (Baba 1967, vol. 1, 17–25).

Such transformations occur at the end of important cycles of time, and Baba has identified the present time as the end of a great "cycle of cycles." He often spoke of the 700–1,400 year cycles in which a great world religious teacher appears as the *Avatar* or Christ. These figures help humanity advance spiritually in whatever way is needed at the time. He indicated that the world was in such a period now, and that he was fulfilling that role in the present cycle, a claim which many followers believe. The present transformation is an especially powerful one, coming at the transition point between the darkest of ages, the *Kali Yuga*, and a new Golden Age, and also marking the end of a much longer "cycle of cycles." Hindu conceptions also affirm the significance of this time as a turning point—the halfway mark for the present system of creation in its total life of several hundred trillion years.

THE FUTURE OF SCIENCE AND RELIGION

An evaluation of the current and future status of science and religion can be enriched by re-examining the themes of scientific discovery and its relation to spirituality. They are the discrepancy of appearance and reality ("things are not what they seem"), the increasingly peripheral cosmic location of humans, the place of God in science's world, and the search for unity. The dialogue between science and religion has uncovered new dimensions of these themes.

Revelations of the illusory appearance of the material world have only increased, as scientists present to the world a more and more bizarre picture of cosmic reality—from empty atoms to wave-like electrons to tiny vibrating strings that constitute every particle and an accelerating universe composed almost completely of dark forms of energy and matter still unknown. Spiritual cosmologies reflect many of these same conceptions of matter. This is an unseen world, indeed, and, more than ever before, science shares with religion the challenge of understanding and communicating the mysteries of realms beyond direct comprehension.

As for humanity's cosmic location and significance, new voices from both science and religion have suggested two things: either that we are ever more peripheral in a "mega-cosmos" of multiple universes, or that a major shift in Copernican thinking is in order—or both. Science describes how the fine-tuning of physical laws and Earth's near-perfect cosmic location made life and intelligence possible, while some religious views have also affirmed the special spiritual significance of planet Earth, at least during this epoch.

The search for God in the world of science has given rise to rich new conceptions of the Supreme Being. The former all-powerful monarch who fashioned everything directly at the beginning has given way to a God deeply involved in nature's processes, one who may limit his own power to allow nature and humanity freedom of growth and choice. More organic and relational metaphors have arisen—God as the world's mind, creative artist or mother giving birth. In some systems, God is himself the whole process, even as he is also the wholly transcendent being beyond all.

Finally, the search for unity has intensified—within science, within religion and in the dialogue that joins them. Unity has expressed itself both in the discovery of hidden unities and also in the process of people working together to seek new understanding. In religion, the idea of an underlying unity or common core in all faiths has emerged in "perennial philosophy" and in certain integrative systems. Interfaith dialogue has greatly increased, often to discuss issues raised by science. On the science side, discoveries show that basic elements of the physical world are unified—forces, matter and energy, space and time—and the drive for a final unified theory continues. Underlying unity and even the integration of consciousness in the physical world have been suggested by certain paradoxes and puzzles in quantum physics. In science, too, the process of unified effort is crucial—only by merging discoveries of many separate scientific disciplines did the story of cosmic evolution arise. In the same way, in the science and religion dialogue, it has become increasingly clear that only joint inquiry by scientists and theologians can yield satisfying answers to the big questions of the cosmos and human significance. The quest for unity in science and religion has found common ground, uncovered consonances and parallels, led to integrated systems uniting physical and spiritual realities and forged a unified framework for studying science and religion together as a new academic discipline. The groundwork for such a discipline has been laid by a group of scholars who have deep training in both fields and a methodology for engaging in "creative mutual interaction."

Many additional factors have contributed to the tremendous growth of the current dialogue. Among them are the enthusiastic leadership of such prominent religious figures as Pope John Paul II and the Dalai Lama, the global proliferation of centers and programs, the emergence of fully integrated cosmologies—from process theologians, neo-Thomistic scholars, Teilhard de Chardin, Meher Baba, and Aurobindo Ghose—and the burgeoning interest and writings of thousands of scholars, clergy, and laymen worldwide. The dialogue has indeed blossomed and intensified into a rich, interreligious global exchange. The time seems ripe for significant advancement.

Primary Sources

The following sources have been selected so that the reader may experience firsthand the writings of major religious figures and a leading scientist on the subject of cosmic evolution and the relation between science and religion. Eastern spiritual perspectives on cosmic evolution are illustrated in selected passages from the writings of the modern figures, Meher Baba, and Aurobindo Ghose, and an ancient Buddhist text. Views of the relation between science and religion in the modern world are highlighted in the writings of Pope John Paul II and Albert Einstein.

—1—
Meher Baba, *God Speaks: The Theme of Creation and Its Purpose.* Second edition, revised and enlarged. Walnut Creek, CA: Sufism Reoriented, 1997, pages 220–24.

In the following selection, "The Divine Theme," Meher Baba gives a summary of the spiritual cosmology he describes comprehensively in his major work, *God Speaks.* In this book, he traces the journey of the soul through all fives stages in the evolutionary path—creation, evolution, reincarnation, involution, and God-Realization.

The Divine Theme

By Meher Baba
Evolution, Reincarnation and the Path to Realization

A soul becomes perfect* after passing through evolution, reincarnation and the process of realization. To gain full consciousness, it gets increasing *sanskaras* in the process of evolution, till in the human form, it gets full consciousness as well as all the gross *sanskaras.*

In the process of reincarnation, this soul retains its full consciousness and exchanges (*i.e.,* alternatingly experiences) the diverse *sanskaras* in itself; and in the process of realization, this soul retains its full consciousness, but its *sanskaras* become fainter and fainter till they all disappear and only consciousness remains. While becoming faint, gross *sanskaras* become subtle *sanskaras*, subtle *sanskaras* become mental *sanskaras;* and finally they all disappear.

Up to the human form, the winding process of *sanskaras* becomes stronger and stronger in the process of evolution. In the human form, in the process of reincarnation, the winding retains its full strength; but in the process of realization, the *sanskaras* gradually unwind themselves, till in the God-state, they are completely unwound.

God, the Over-Soul, alone is real. Nothing exists but God. The different souls are in the Over-Soul and one with it. The processes of evolution, reincarnation and realization are all necessary in order to enable the soul to gain self-consciousness. In the process of winding, *sanskaras* become instrumental for the evolution of consciousness though they also give *sanskaric* bindings; and in the process of unwinding, *sanskaric* attachments are annihilated, though the consciousness which has been gained is fully retained.

In the process of the winding of *sanskaras,* the soul goes through seven stages of **descent;** and in the process of unwinding, the soul goes through seven stages of **ascent.** But the phenomena of descent as well as ascent are both illusory. The soul is everywhere and indivisibly infinite; and it does not move or descend or ascend.

The souls of all men and women, of all nationalities, castes and creeds, are really one; and their experiences of good and evil, of fighting and helping, of waging wars and living in peace are all a part of illusion and delusion, because all these experiences are gained through bodies and minds, which in themselves are nothing.

Before the world of forms and duality came into existence, there was nothing but God, *i.e.,* an indivisible and boundless ocean of Power, Knowledge and Bliss. But this ocean was unconscious of itself. Picture to yourself this ocean as absolutely still and calm, unconscious of its Power,

Knowledge and Bliss and unconscious that it is the ocean. The billions of drops which are in the ocean do not have any consciousness; they do not know that they are drops nor that they are in the ocean nor that they are a part of the ocean. This represents the original state of Reality.

This original state of Reality comes to be disturbed by an urge to know itself. This urge was always latent in the ocean; and when it begins to express itself, it endows the drops with individuality. When this urge makes the still water move, there immediately spring up numerous bubbles or forms around the drops; and it is these bubbles which give individuality to the drops. The bubbles do not and cannot actually divide the indivisible ocean; they cannot separate the drop from the ocean; they merely give to these drops a feeling of separateness or limited individuality.

Now let us study the life of one drop-soul through its different stages. Owing to the arising of the bubble, the drop-soul which was completely unconscious is invested with individuality (or a feeling of separateness) as well as with very slight consciousness. This consciousness, which has sprung up in the drop-soul, is not of itself nor of the ocean; but it is of the bubble or the form, which in itself is nothing. This imperfect bubble at this stage is represented by the **form** of a stone. After some time, this bubble or form bursts and there springs up in its place another bubble or form. Now, when a bubble bursts, two things happen: (1) there is an increase in consciousness and (2) there is a twist or consolidation of impressions or *sanskaras* accumulated during the life of the previous bubble. The consciousness of the drop-soul has now slightly increased; but the drop-soul is still conscious only of this new bubble or form and not of itself nor of the ocean. This new bubble is represented by the form of the metal. This new bubble or form also bursts in due course of time; and simultaneously there is a further increase in consciousness and a fresh twist or consolidation of *sanskaras*, which gives rise to the emergence of another type of bubble or form.

This process continues right through the course of evolution, which covers the stages of stones, metals, vegetables, worms, fishes, birds and animals. Every time that the previous bubble or form bursts, it gains more consciousness and adds one twist to the already accumulated *sanskaras*, until it reaches the human bubble or form, in which the ever-increasing consciousness becomes full and complete. The process of the winding up of *sanskaras* consists of these regular twists; and it is these twists which keep the consciousness, gained by the drop-soul, directed and fixed towards the bubble or the form instead of towards its real Self, even when consciousness is fully developed in the human form.

On gaining the human form, the second process begins; this process is that of reincarnation. At this point, the process of the winding up of *sanskaras* comes to an end. The drop-soul takes numerous human forms one by one; and these forms are exactly eighty-four *lakhs* in number. These human forms are sometimes those of man and sometimes those of woman; and they change nationalities, appearance, colour and creed. The drop-soul through human incarnations experiences itself sometimes as a beggar and sometimes as a king, and thus gathers experiences of the opposites of happiness or misery according to its good or bad *sanskaras*. In reincarnation (*i.e.,* in its successive and several human forms) the drop-soul retains its full consciousness but continues to have alternating experiences of opposite *sanskaras,* till the process of realization begins. And during this process of realization the *sanskaras* get unwound. In reincarnations, there is a spending up of *sanskaras;* but this spending up is quite different from the unwinding of the *sanskaras,* which takes place during the process of realization. The spending up of *sanskaras* itself creates new *sanskaras,* which bind the soul; but the unwinding of *sanskaras* does not itself create fresh *sanskaras;* and it is intended to undo the very strong grip of *sanskaras,* in which the drop-soul is caught.

Up to the human form, the winding up of *sanskaras* becomes stronger and stronger during the process of evolution. In the human forms of reincarnation, the winding continues to operate as a limiting factor; but with every change of the human bubble or form, the tight twists, gained during the process of winding, get loosened through eighty-four *lakhs* of shakings,* before they are ready to unwind in the process of realization.

Now begins the third process of realization, which is a process of ascent. Here, the drop-soul undergoes the gradual unwinding of the *sanskaras.* During this process of unwinding, the *sanskaras* become fainter and fainter; and at the same time, the consciousness of the drop-soul gets directed more and more towards itself; and thus, the drop-soul passes through the subtle and mental planes till all the *sanskaras* disappear completely, enabling it to become conscious of itself as the ocean.

In the infinite ocean of the Over-Soul, you are the drop or the soul. You are the soul in the ordinary state; and you use your consciousness in seeing and experiencing the bubble or the form. Through the gross layer of the bubble, you experience that part of the huge gross bubble which is the earth. You are eternally lodged and indivisibly one with the Over-Soul; but you do not experience it. In the advanced stage, up through the third plane, you use your consciousness in seeing and experiencing the huge subtle bubble called the subtle world, through the subtle bubble or form called the subtle body; but you do not see and experience the Over-Soul which you are in, since your consciousness is not now directed towards the Over-Soul. In the advanced stage from the fourth through the sixth plane, you use your consciousness in seeing and experiencing the huge mental bubble, which is called the mental world, through the mental bubble or form which is called the mental body, but even now you do not experience the Over-Soul. But in the God-realized state, you continually use your consciousness for seeing and experiencing the Over-Soul; and then all the forms are known as being nothing but bubbles.

So, now, picture yourself as the soul-drop, lodged in the Over-Soul, behind five layers after the gross body. You, the soul-drop, are now looking at the gross body and through it at the gross world. When you look at the second layer and through it, the first layer will appear to you as nothing but a layer only, and thus, looking behind each layer, you will find all these layers as only your shadow covers; and finally, when you (*i.e.*, the soul-drop) look at and get merged in the Over-Soul, you realize that only you were real and all that you were seeing and experiencing till now was your own shadow and nothing else.

—2—
Meher Baba, "The Original Whim," in *Beams from Meher Baba on the Spiritual Panorama.* San Francisco, CA: Sufism Reoriented, 1958, pages 7–11.

In response to queries from readers about concepts in *God Speaks*, Meher Baba gave an additional set of discourses, *Beams*, clarifying and elaborating on some of its themes. In this passage he gives additional insight into God's "original whim" to know himself, which gave birth to creation.

The Whim from the Beyond

THROUGH the ages, the human mind has been profoundly restless in its search for final explanations about first things. The history of these endeavors to grasp first things through the intellect is a tale of recurrent failures. The redeeming feature of these great efforts is that instead of being disheartened by the confessed failures of past thinkers, others are inspired to make fresh attempts. All these philosophical explanations are creations of the mind that has never succeeded in passing beyond itself. Thus they are confessed though inspiring failures; nonetheless each such failure is a partial contribution to knowledge of the Beyond. Only those who have gone beyond the mind know the Truth in its reality. If they sometimes explain what they know, which they very rarely do, those explanations also being in words are limited but these words illumine the mind; they do not fill it with novel ideas.

The unitarian Beyond is an indivisible and indescribable infinity. It *seeks to know itself.* It is of no use to ask why it does so. To attempt to give a reason for this is to be involved in further questions and thus to start an unending chain of reasons for reasons, reasons for these reasons and so on *ad infinitum.* The plain truth about this initial urge to know itself is best called a whim *(Lahar).* A whim is not a whim if it can be explained or rationalized. And just as no one may usefully ask why it arises, so no one may ask when it arises. "When" implies a time series with past, present and future. All these are absent in the eternal Beyond. So let us call this initial urge to know a "whim." You may call this an explanation if you like or you may call it an affirmation of its inherent inexplicability.

The initial whim is completely independent of reason, intellect or imagination, all of which are by-products of this whim. Reason, intellect and imagination depend upon the initial whim and not *vice versa.* Because the whim is not dependent upon reason, intellect or imagination, it can neither be understood nor interpreted in terms of any of these faculties of the limited mind.

The first whim to know instantaneously implies a duality, an apparent differentiation (not amounting to a breaking up) into two separate aspects, both of which are infinite as aspects of the Infinite. The first aspect is that of infinite consciousness and the second aspect is that of infinite unconsciousness. The duality strives to overcome itself and to restore the apparently lost unity; the infinite unconsciousness tries to unite with infinite consciousness. Both aspects are precipitated by the whim. This whim of the Infinite is in a way comparable to an infinite question, calling forth an infinite answer.

With the infinite question, there arises also the infinite answer. The infinite question is infinite unconsciousness; the infinite answer is infinite consciousness. But the infinite question and the infinite answer do not simply annul each other and relapse into the original unity of the Beyond. The two aspects have now descended into the primal duality which can resolve itself only by fulfilling the entire game of duality and not by any shortcut. The infinite unconsciousness cannot overlap on infinite consciousness; such coalescence is impossible.

To reach out towards infinite consciousness the infinite unconsciousness first has to fathom its own depths. It must experience itself first as infinitely finite, and gradually evolve into limited and limiting consciousness. With the evolution of the limited and limiting consciousness, there is also the evolution of the *illusion* which limits this limiting consciousness. The two processes keep pace with each other.

When the infinite unconsciousness tries to reach out to the infinite consciousness, the process is not instantaneous because of the infinite disparity between the two. The process takes an infinitely long time and eternity gets seemingly broken into the unending past, the transient present and the uncertain future. Instead of embracing the infinite consciousness in one timeless act the infinite unconsciousness reaches out towards it through a long-drawn-out temporal process of evolution, with all of its innumerable steps. It first attempts to fathom its own depths, then by backward treads it seeks and ultimately finds the infinite consciousness through numberless steps, thus fulfilling the whim from the Beyond.

—3—
Aurobindo Ghose, "Man and the Evolution," in *The Life Divine.* Twin Lakes, WI: Lotus Press, 1990.

Like Meher Baba, Hindu spiritual teacher Aurobindo Ghose presents a spiritual evolutionary scheme in which the driving force is the evolution of consciousness—an "invisible process of soul evolution." The scientific and the spiritual are woven together in one overarching process of growth and evolution toward a final goal.

Chapter XXIII

Man and the Evolution

The one Godhead secret in all beings, all-pervading, the inner Self of all, presiding over all action, witness, conscious knower and absolute . . . the One in control over the many who are passive to Nature, fashions one seed in many ways. *Swetaswatara Upanishad.*

The Godhead moves in this Field modifying each web of things separately in many ways. . . . One, he presides over all wombs and natures; himself the womb of all, he is that which brings to ripeness the nature of the being and he gives to all who have to be matured their result of development and appoints all qualities to their workings. *Swetaswatara Upanishad.*

He fashions one form of things in many ways. *Katha Upanishad*

Who has perceived this truth occult, that the Child gives being to the Mothers by the workings of his nature? An offspring from the lap of many Waters, he comes forth from them a seer possessed of his whole law of nature. Manifested, he grows in the lap of their crookednesses and becomes high, beautiful and glorious. *Rig Veda.*

From the non-being to true being, from the darkness to the Light, from death to Immortality. *Brihadaranyaka Upanishad.*

A spiritual evolution, an evolution of consciousness in Matter in a constant developing self-formation till the form can reveal the indwelling Spirit, is then the keynote, the central significant motive of the terrestrial existence. This significance is concealed at the outset by the involution of the Spirit, the Divine Reality, in a dense material Inconscience; a veil of Inconscience, a veil of insensibility of Matter hides the universal Consciousness-Force which works within it, so that the Energy, which is the first form the Force of creation assumes in the physical universe, appears to be itself inconscient and yet does the works of a vast occult Intelligence. The obscure mysterious creatrix ends indeed by delivering the secret consciousness out of its thick and tenebrous prison; but she delivers it slowly, little by little, in minute infinitesimal drops, in thin jets, in small vibrant concretions of energy and substance, of life, of mind, as if that were all she could get out through the crass obstacle, the dull reluctant medium of an inconscient stuff of existence. At first she houses herself in forms of Matter which appear to be altogether unconscious, then struggles towards mentality in the guise of living Matter and attains to it imperfectly in the conscious animal. This consciousness is at first rudimentary, mostly a half subconscious or just conscious instinct; it develops slowly till in more organised forms of living Matter it reaches its climax of intelligence and exceeds itself in Man, the thinking animal who develops into the reasoning mental being but carries along with him even at his highest elevation the mould of original animality, the dead weight of sub conscience of body, the downward pull of gravitation towards the original Inertia and Nescience, the control of an inconscient material Nature over his conscious evolution, its power for limitation, its law of difficult development, its immense force for retardation and frustration. This control by the original Inconscience over the consciousness emerging from it takes the general shape of a mentality struggling towards knowledge but itself, in what seems to be its fundamental nature, an Ignorance. Thus hampered and burdened, mental man has still to evolve out of himself the fully conscious being, a divine manhood or a spiritual and supramental supermanhood which shall be the next product of the evolution. That transition will mark the passage from the evolution in the Ignorance to a greater evolution in the Knowledge; founded and proceeding in the light of the Superconscient and no longer in the darkness of the Ignorance and Inconscience.

This terrestrial evolutionary working of Nature from Matter to Mind and beyond it has a double process: there is an outward visible process of physical evolution with birth as its machinery, – for each evolved form of body housing its own evolved power of consciousness is maintained and kept in continuity by heredity; there is, at the same time, an invisible process of soul evolution with rebirth into ascending grades of form and consciousness as its machinery. The first by itself would mean only a cosmic evolution; for the individual would be a quickly perishing instrument, and the race, a more abiding collective formulation, would be the real step in the progressive manifestation of the cosmic Inhabitant, the universal Spirit: rebirth is an indispensable condition for any long duration and evolution of the individual being in the earth-existence. Each grade of cosmic manifestation, each type of form that can house the indwelling Spirit, is turned by rebirth into a means for the individual soul, the psychic entity, to manifest more and more of its concealed consciousness; each life becomes a step in a victory over Matter by a greater progression of consciousness in it which shall make eventually Matter itself a means for the full manifestation of the Spirit.

—4—
C. P. Ranasinghe, *The Buddha's Explanation of the Universe.* Colombo, Ceylon (Sri Lanka): Lanka Bauddha Mandalaya Fund, 1957, pages 254–63.

This work contains material from the Abhidhamma section of the Buddha's teachings, which was translated from Pali texts in the last century by the Sri Lankan Buddhist scholar C. P. Ranasinghe. The passage below presents the eschatology of the teachings and the process by which universes dissolve and are reborn. It emphasizes a traditional Eastern view of the primacy of consciousness and mind as the directing force in the cosmos.

Mind Builds and Maintains the Universe

Each of the units of mind, in its building of its material structure, grasps more of one kind of units of elements of matter, and less of another. This causes concentrations at some places and vacuums at other places. The neighbouring units of matter move to establish a homogeneous condition where vacuums are created, and the established units move out to provide room for the concentrations to exist. This process repeats itself an infinite number of times in every part of the infinite universe. Thus every part of the infinite universe is moving, developing substances at some places, and deteriorating substances at others; making stars at some places, and dispersing stars at others. As units of mind circulate in the universe, the various abstract elements circulate, and abstract heat, particularly, becomes very swift in activity. When the abstract elements circulate, the various planets and stars in space also circulate.

Each unit of mind, according to the strength of its evolution current, develops a gravitational force so as to keep in concentration a measure of units of abstract elements. Every being in the material universe which exists above levels of elementary life, contributes a measure of gravitational force to the universe. The higher beings, such as the human beings and the animals, contribute a larger measure, and the lesser beings, contribute a proportionately smaller measure. All the beings living on this Earth are contributing towards the sustenance of the gravitational force of this Earth. And if all the beings existing on this Earth, disappear from it, there would be no gravitational force on this Earth. And without gravitational force, the Earth cannot exist. Water would evaporate but rain would not fall. Heat would concentrate and, ultimately, the Earth would blow up and dissolve in space.

When Present Universe is at Its End

The Buddha's description of the last stages of the present universe amply illustrates the manner in which the material universe depends on the strength of evolution of the units of mind for its existence. A period will come, in the distant future, when the combined strength of evolution of . the minds of the material beings in the universe becomes insufficient to hold the planets and stars in due position and the conditions of life in the universe becomes very difficult.

Those beings who understand the impending disaster at that time would make haste to increase the force of evolution in their minds and thus escape into a spiritual sphere of existence, but the majority would sink lower and lower in strength of evolution and, ultimately, establish themselves in elementary life.

Gravitational Deterioration and Harsh Conditions

As the combined force of evolution of the units of mind decreases, the total quantity of units of abstract matter held in their control also decreases. In other words, the units of mind lose their gravitational force to control a sufficiency of units of abstract elements, and, consequently, the planets and stars lose their strength to keep together the various substances on them. More of the substances from the planets and stars will radiate into space and disappear, and the conditions on the planets will become increasingly difficult.

Harsh climate, followed by scarcity of food makes the beings in the universe increasingly violent and cruel, and the increase of cruelty amongst the beings in the universe reduces further their strength of evolution of mind. The lower strength of evolution causes conditions on the planets and stars to deteriorate further. And when planets and stars deteriorate, the beings deteriorate; and when the beings deteriorate, the planets and stars deteriorate and this process continues until the point when the universe loses its equilibrium and shatters itself in splinters and crashes.

Last Scenes on This Earth

The Buddha's description of the last scenes of the present universe as they would appear to a being then living on this Earth is extremely interesting. The Earth would then have its harsh climate, and the beings would be with smaller physiques and a very low morality. The standards of climate, physique, and morality descend lower and lower, and many of the beings on this Earth die and disappear. Fresh beings arriving to settle on this Earth keep dwindling in numbers and further deterioration takes place. After further lapse of time beings on this Earth are reduced to very small numbers and a period of long drought with intensive heat occurs. Owing to this drought, rivers dry up, and crops fail; and, in consequence, most of the remaining beings die and disappear.

Probably by encountering an inter-stellar cloud of water vapour, a spell of rain then begins to occur. This rain which pours down continuously for seven consecutive days, begins with slow drizzling and, at its worst, rain drops several feet in diameter fall. This rain causes floods all over, the only points escaping inundation being a few mountain tops. The floods kill a further number of the remaining beings on this Earth.

Then follows a second drought and within a few weeks all the flood waters evaporate and shortly afterwards all the oceans too dry up. By this time, the universe is in chaos and the stars are crumbling all over, and our solar system is also crumbling in a mad dash. Our solar system meets other solar systems and at its final stage, the Earth comes within the orbit of seven suns.

Fire that Burns out the Universe

The Earth at this stage is without a drop of water and her envelope of air is in tatters. It is heat all over, and the surface is burning at many places. And finally, the seven suns radiate on to this Earth so much of heat, that the whole Earth turns ablaze. After burning for a short while in this manner, the Earth blows up and ceases to exist.

In the meantime, all the other planets and stars in the universe too blow up and disappear, some in the same manner as this Earth and the others in different other ways. Ultimately, the entire universe of space becomes completely void of all planets and stars. The condition of the universe remains for some time in this confused state of blank darkness.

During the Darkness of Time

Although during this period the whole universe turns into a condition of blank darkness, no part of it gets destroyed. All the units of mind and all the abstract units of matter remain in it perfectly intact. All that gets destroyed in this catastrophe, are the various substances and the various material formations of beings. The units of abstract matter, during this period, establish the homogeneous dispersal over space towards which they always keep pulling themselves, and as in this state the units of abstract matter are too much stretched out in space to remain in atomic combinations, they exist only in the form of energy (paramattha). Since matter in the form of energy is not discernible to environment sense organs, the only way we could visualize the condition of the universe then is as space in a state of blank darkness.

Excepting those beings who escape into planes of spiritual existence (Brahma), all the other beings have, by this time, reduced in the strength of their evolution to the level of elementary existence. They, therefore, continue their existence by taking conception in units of free abstract heat. Free abstract heat does not produce a mass and, therefore, the beings existing in them are material beings without mass. Their presence, therefore, is also not discernible to the environment sense organs.

Worst Suffering

Existence in the universe of free abstract heat is the most acute form of suffering in the universe. All the beings that enter into life in units of free abstract heat on the dissolution of a material universe, therefore, suffer intensely. As we have noted, suffering reduces the forces of defilements, and the beings remaining in such conditions liquidate most of their defilements.

After a considerable lapse of time, these beings liquidate their defilements so much, and also, in the meanwhile, whatever remnants of impressions of purity remaining in the evolution current of their mind becomes so increased in growth, that many units of mind acquire a sufficiency of evolution strength to cover their suffering physical systems of free abstract heat with a few units of the other abstract elements. When such strength is gained by these beings, they form themselves into physical structures of tiny atoms which remain floating about in space.

Rebirth of the Universe

As the strength of the forces of evolution in the minds of these beings increases further and further, the physical systems they evolve become larger, and, in course of time, clouds of such material beings appear all over space. At this stage the activity of these tiny beings disturbs the homogeneous equilibrium of the material universe, and everywhere the abstract elements begin to move about. As the beings evolve further, planets and stars get formed, and once again the material universe gets established.

Beginning in this manner, the new universe grows, matures, decays, dies, and ends again in a catastrophic dissolution, and the process goes on without end.

Betrayal

This universe is thus the result of the co-operative effort of all the units of mind existing in the universe. Every being in the universe has performed its part in the establishment and the bringing into being of this universe. Every being in the universe according to its strength, wields an oar to move this universe and, as we are riding on it, we are also riding on the efforts of all the other beings of the universe.

But the misfortune is that all the beings that make the material universe, find themselves betrayed by the same material universe in the end. This material universe is a dangerous pet that we beings rear. All beings contribute to making this universe, and they do so with great enthusiasm and industry. But, ultimately, the very thing they made and nurtured bites them and kills them. This is the folly of rearing this venomous pet.

Escape

This pet does not give any being a measurable satisfaction, and " so the wise people are cautioned against this useless pursuit." Escape from matter, the ever clinging devil that pulls all beings towards suffering all the time, and enter that peaceful and pleasant plane of existence where the inferior and low matter does not gather and existence is perfect! There will not be the cruel interferences from matter there. There will be no formations of substances, no pulls towards defilements, no risks of births in lower forms of life, no wants, no greed, no hatred, and no ignorance.

It is all pure and untainted pleasure, the maximum happiness, fullest contentment, and everlasting joy. It is that life of nonalliance with matter, the fullest and most supreme form of life, the life with the mind charged with the maximum forces of purity to the total exclusion of all defilements, and with propulsion solely by purity. This ideal and permanent state of existence is only reached by those who purify their minds beyond the point of achievement of the state of Arahat. This existence is only for those who have reached the illumination of Nibbana, the supreme state of perpetual bliss.

GUIDANCE TO WISDOM

WE have seen that this universe is one vast and infinite, self-winding, machine. There are wheels in it both large and small and, in each of these wheels, there run an infinity of smaller and smaller wheels.

The universe is an infinite phenomenon: it shrinks with the infinity of contraction and expands to the infinity of extension. Any section or portion of the universe is only a part of the infinite expanse of the universe, and every portion of it consists of an infinity of infinitely contracted portions of the universe. As such, we could say that even the most minute part of the universe is the whole universe in a miniature form, and the whole universe is a combination of an infinity of miniature universes. This is the fundamental truth about the formation and the structure of the universe.

What is true of the universe as a whole, is also true of all beings and man. Each material fibre of a being is a miniature form of the whole being, and, as such, each material fibre constituting a human being, is a miniature material form of the whole human being. Thus is the material construction of all beings and substances.

Flux of Mind and Matter

The smallest conceivable, and by far the most powerful, of all the wheels of the universe, are the wheels of the units of mind. Each beat of the mind is a rotation of this wheel, and we have noted that its speed is about 3 billionth part of the duration of a flash of lightning. Even whilst turning at this extreme speed, it systematically and accurately passes through the 17 different stages, and does not miss a single point of time. The units of matter, too, revolve in the same cycle, but their speed is 17 times slower than the speed of rotation of the units of mind.

There is one constant feature in all rotations of the wheels of every unit of mind and matter: they begin, develop, exist, decay, and finally die. Although the rotations of the units of mind have a beginning, the units of mind or matter themselves never began their manifestations and, therefore, no unit of mind or matter will ever cease to exist, nor does any unit of mind or matter in the material universe ever remain unchanged. Time turns every wheel of the units of mind and matter; and, as time never remains constant, so no wheel of the units of mind or matter in the universe ever remains constant. The wheels of the units of mind and matter in the universe, therefore, keep on repeating the turns of their cycles from each beginning to its ending; and from each ending to a new beginning, closely skipping from the tip of each ending to the tip of the new beginning.

The universe, thus, is in a state of flux all over, and there is no continuance of a uniform state in any unit of mind or matter in the material universe. Although the manifestations of the units of mind and matter are in themselves without any beginning or cessation, the cycles which turn these units have beginnings and cessations—each birth is a beginning and, therefore, leads to a death which is the cessation and, conversely, each cessation or death leads to a new beginning or birth. The law is that the phenomena that begin in the universe invariably move to their end, all phenomena that end invariably cause the beginning of fresh phenomena.

This law which is common to all units of mind and matter in the material universe is also common to all substances of matter and material beings in the universe. The occurrence of a state of continuous flux is a universal law in the material universe, and no combination of matter, or matter and mind, can ever escape from the continuous operation of this universal natural law.

Consequent on Birth is Death

In the same way as the beat of mind begins and ends and the beats of elements of matter begin and end, every substance or being that begins in this universe also ends. This is true of rocks, mountains, and all planets and stars, and this is also true of all beings, including animals and man. Birth of every form in this universe invariably leads ultimately to death; there is no way out; and the fact being that we were born in the material universe, we have no way of escape, having to face death sooner or later.

No phenomenon in this universe is permanent. Those things which we normally consider to be permanent are far from possessing any permanency. This Earth and the planets and stars in our skies are not permanent— they are in the habit of dissolving themselves in space once in about 100,000,000,000,000 years, a great universe period (maha kappa). About 2/3 of this period has already lapsed, and the balance 1/3 still remains to run, which, when completed, will culminate in the universe invariably dissolving itself into energy in the state of blank darkness.

Some planets and stars, comparatively, are very small and their spans of duration are in most cases very short. There exist in the universe other planets and stars of much bigger size and of more solidified substances, which exist through almost incalculable periods of time. But this Earth and all the other planets and stars of whatever substance or of whatever duration they may be, are subject to the same law and they will end in dissolution in due course.

—5—

Albert Einstein, *Ideas and Opinions*. Ed. Cal Seelig and others. New trans. and rev. by Sonja Bargmann. New York: The Modern Library, 1994. "Religion and Science," pages 39–43 and "Science and Religion," pages 47–53. Copyright 1954 and renewed 1982 by Crown Publishers, Inc. Used by permission of Crown Publishers, a division of Random House, Inc.

In these passages the most renowned scientist of the last century gives his view of how religion develops from earlier stages to its highest expression a "cosmic religious feeling," an inner experience known especially to inspired men of science. In the second passage, he describes how science and religion work together in an interdependent fashion and how science's drive for and revelation of unity helps in the spiritual task of lifting humanity above the prison of individual, limited existence.

RELIGION AND SCIENCE

Everything that the human race has done and thought is concerned with the satisfaction of deeply felt needs and the assuagement of pain. One has to keep this constantly in mind if one wishes to understand spiritual movements and their development. Feeling and longing are the motive force behind all human endeavor and human creation, in however exalted a guise the latter may present themselves to us. Now what are the feelings and needs that have led men to religious thought and belief in the widest sense of the words? A little consideration will suffice to show us that the most varying emotions preside over the birth of religious thought and experience. With primitive man it is above all fear that evokes religious notions—fear of hunger, wild beasts, sickness, death. Since at this stage of existence understanding of causal connections is usually poorly developed, the human mind creates illusory beings more or less analogous to itself on whose wills and actions these fearful happenings depend. Thus one tries to secure the favor of these beings by carrying out actions and offering sacrifices which, according to the tradition handed down from generation to generation, propitiate them or make them well disposed toward a mortal. In this sense I am speaking of a religion of fear. This, though not created, is in an important degree stabilized by the formation of a special priestly caste which sets itself up as a mediator between the people and the beings they fear, and erects a hegemony on this basis. In many cases a leader or ruler or a privileged class whose position rests on other factors combines priestly functions with its secular authority in order to make the latter more secure; or the political rulers and the priestly caste make common cause in their own interests.

The social impulses are another source of the crystallization of religion. Fathers and mothers and the leaders of larger human communities are mortal and fallible. The desire for guidance, love, and support prompts men to form the social or moral conception of God. This is the God of Providence, who protects, disposes, rewards, and punishes; the God who, according to the limits of the believer's outlook, loves and cherishes the life of the tribe or of the human race, or even life itself; the comforter in sorrow and unsatisfied longing; he who preserves the souls of the dead. This is the social or moral conception of God.

The Jewish scriptures admirably illustrate the development from the religion of fear to moral religion, a development continued in the New Testament. The religions of all civilized peoples, especially the peoples of the Orient, are primarily moral religions. The development from a religion of fear to moral religion is a great step in peoples' lives. And yet, that primitive religions are based entirely on fear and the religions of civilized peoples purely on morality is a prejudice against which we must be on our guard. The truth is that all religions are a varying blend of both types, with this differentiation: that on the higher levels of social life the religion of morality predominates.

Common to all these types is the anthropomorphic character of their conception of God. In general, only individuals of exceptional endowments, and exceptionally high-minded communities, rise to any considerable extent above this level. But there is a third stage of religious experience which belongs to all of them, even though it is rarely found in a pure form: I shall call it cosmic religious feeling. It is very difficult to elucidate this feeling to anyone who is entirely without it, especially as there is no anthropomorphic conception of God corresponding to it.

The individual feels the futility of human desires and aims and the sublimity and marvelous order which reveal themselves both in nature and in the world of thought. Individual existence impresses him as a sort of prison and he wants to experience the universe as a single significant whole. The beginnings of cosmic religious feeling already appear at an early stage of development, e.g., in many of the Psalms of David and in some of the Prophets. Buddhism, as we have learned especially from the wonderful writings of Schopenhauer, contains a much stronger element of this.

The religious geniuses of all ages have been distinguished by this kind of religious feeling, which knows no dogma and no God conceived in man's image; so that there can be no church whose central teachings are based on it. Hence it is precisely among the heretics of every age that we find men who were filled with this highest kind of religious feeling and were in many cases regarded by their contemporaries as atheists, sometimes also as saints. Looked at in this light, men like Democritus, Francis of Assisi, and Spinoza are closely akin to one another.

How can cosmic religious feeling be communicated from one person to another, if it can give rise to no definite notion of a God and no theology? In my view, it is the most important function of art and science to awaken this feeling and keep it alive in those who are receptive to it.

We thus arrive at a conception of the relation of science to religion very different from the usual one. When one views the matter historically, one is inclined to look upon science and religion as irreconcilable antagonists, and for a very obvious reason. The man who is thoroughly convinced of the universal operation of the law of causation cannot for a moment entertain the idea of a being who interferes in the course of events—provided, of course, that he takes the hypothesis of causality really seriously. He has no use for the religion of fear and equally little for social or moral religion. A God who rewards and punishes is inconceivable to him for the simple reason that a man's actions are determined by necessity, external and internal, so that in God's eyes he cannot be responsible, any more than an inanimate object is responsible for the motions it undergoes. Science has therefore been charged with undermining morality, but the charge is unjust. A man's ethical behavior should be based effectually on sympathy, education, and social ties and needs; no religious basis is necessary. Man would indeed be in a poor way if he had to be restrained by fear of punishment and hope of reward after death.

It is therefore easy to see why the churches have always fought science and persecuted its devotees. On the other hand, I maintain that the cosmic religious feeling is the strongest and noblest motive for scientific research. Only those who realize the immense efforts and, above all, the devotion without which pioneer work in theoretical science cannot be achieved are able to grasp the strength of the emotion out of which alone such work, remote as it is from the immediate realities of life, can issue. What a deep conviction of the rationality of the universe and what a yearning to understand, were it but a feeble reflection of the mind revealed in this world, Kepler and Newton must have had to enable them to spend years of solitary labor in disentangling the principles of celestial mechanics! Those whose acquaintance with scientific research is derived chiefly from its practical results easily develop a completely false notion of the mentality of the men who, surrounded by a skeptical world, have shown the way to kindred spirits scattered wide through the world and the centuries. Only one who has devoted his life to similar ends can have a vivid realization of what has inspired these men and given them the strength to remain true to their purpose in spite of countless failures. It is cosmic religious feeling that gives a man such strength. A contemporary has said, not unjustly, that in this materialistic age of ours the serious scientific workers are the only profoundly religious people.

Science and Religion

II.

It would not be difficult to come to an agreement as to what we understand by science. Science is the century-old endeavor to bring together by means of systematic thought the perceptible phenomena of this world into as thoroughgoing an association as possible. To put it boldly, it is the attempt at the posterior reconstruction of existence by the process of conceptualization. But when asking myself what religion is I cannot think of the answer so easily. And even after finding an answer which may satisfy me at this particular moment, I still remain convinced that I can never under any circumstances bring together, even to a slight extent, the thoughts of all those who have given this question serious consideration.

At first, then, instead of asking what religion is I should prefer to ask what characterizes the aspirations of a person who gives me the impression of being religious: a person who is religiously enlightened appears to me to be one who has, to the best of his ability, liberated himself from the fetters of his selfish desires and is preoccupied with thoughts, feelings, and aspirations to which he clings because of their superpersonal value. It seems to me that what is important is the force of this superpersonal content and the depth of the conviction concerning its overpowering meaningfulness, regardless of whether any attempt is made to unite this content with a divine Being, for otherwise it would not be possible to count Buddha and Spinoza as religious personalities. Accordingly, a religious person is devout in the sense that he has no doubt of the significance and loftiness of those superpersonal objects and goals which neither require nor are capable of rational foundation. They exist with the same necessity and matter-of-factness as he himself. In this sense religion is the age-old endeavor of mankind to become clearly and completely conscious of these values and goals and constantly to strengthen and extend their effect. If one conceives of religion and science according to these definitions then a conflict between them appears impossible. For science can only ascertain what *is*, but not what *should be*, and outside of its domain value judgments of all kinds remain necessary. Religion, on the, other hand, deals only with evaluations of human thought and action: it cannot justifiably speak of facts and relationships between facts. According to this interpretation the well-known conflicts between religion and science in the past must all be ascribed to a misapprehension of the situation which has been described.

For example, a conflict arises when a religious community insists on the absolute truthfulness of all statements recorded in the Bible. This means an intervention on the part of religion into the sphere of science; this is where the struggle of the Church against the doctrines of Galileo and Darwin belongs. On the other hand, representatives of science have often made an attempt to arrive at fundamental judgments with respect to values and ends on the basis of scientific method, and in this way have set themselves in opposition to religion. These conflicts have all sprung from fatal errors.

Now, even though the realms of religion and science in themselves are clearly marked off from each other, nevertheless there exist between the two strong reciprocal relationships and dependencies. Though religion may be that which determines the goal, it has, nevertheless, learned from science, in the broadest sense, what means will contribute to the attainment of the goals it has set up. But science can only be created by those who are thoroughly imbued with the aspiration toward truth and understanding. This source of feeling, however, springs from the sphere of religion. To this there also belongs the faith in the possibility that the regulations valid for the world of existence are rational, that is, comprehensible to reason. I cannot conceive of a genuine scientist without that profound faith. The situation may be expressed by an image: science without religion is lame, religion without science is blind.

Though I have asserted above that in truth a legitimate conflict between religion and science cannot exist, I must nevertheless qualify this assertion once again on an essential point, with reference to the actual content of historical religions. This qualification has to do with the concept of God. During the youthful period of mankind's spiritual evolution human fantasy created gods in man's own image, who, by the operations of their will were supposed to determine, or at any rate to influence, the phenomenal world. Man sought to alter the disposition of these gods in his own favor by means of magic and prayer. The idea of God in the religions taught at present is a sublimation of that old concept of the gods. Its anthropomorphic character is shown, for instance, by the fact that men appeal to the Divine Being in prayers and plead for the fulfillment of their wishes.

Nobody, certainly, will deny that the idea of the existence of an omnipotent, just, and omnibeneficent personal God is able to accord man solace, help, and guidance; also, by virtue of its simplicity it is accessible to the most undeveloped mind. But, on the other hand, there are decisive weaknesses attached to this idea in itself, which have been painfully felt since the beginning of history. That is, if this being is omnipotent, then every occurrence, including every human action, every human thought, and every human feeling and aspiration is also His work; how is it possible to think of holding men responsible for their deeds and thoughts before such an almighty Being? In giving out punishment and rewards He would to a certain extent be passing judgment on Himself. How can this be combined with the goodness and righteousness ascribed to Him?

The main source of the present-day conflicts between the spheres of religion and of science lies in this concept of a personal God. It is the aim of science to establish general rules which determine the reciprocal connection of objects and events in time and space. For these rules, or laws of nature, absolutely general validity is required—not proven. It is mainly a program, and faith in the possibility of its accomplishment in principle is only founded on partial successes. But hardly anyone could be found who would deny these partial successes and ascribe them to human self-deception. The fact that on the basis of such laws we are able to predict the temporal behavior of phenomena in certain domains with great precision and certainty is deeply embedded in the consciousness of the modern man, even though he may have grasped very little of the contents of those laws. He need only consider that planetary courses within the solar system may be calculated in advance with great exactitude on the basis of a limited number of simple laws. In a similar way, though not with the same precision, it is possible to calculate in advance the mode of operation of an electric motor, a transmission system, or of a wireless apparatus, even when dealing with a novel development.

To be sure, when the number of factors coming into play in a phenomenological complex is too large, scientific method in most cases fails us. One need only think of the weather, in which case prediction even for a few days ahead is impossible. Nevertheless no one doubts that we are confronted with a causal connection whose causal components are in the main known to us. Occurrences in this domain are beyond the reach of exact prediction because of the variety of factors in operation, not because of any lack of order in nature.

We have penetrated far less deeply into the regularities obtaining within the realm of living things, but deeply enough nevertheless to sense at least the rule of fixed necessity. One need only think of the systematic order in heredity, and in the effect of poisons, as for instance alcohol, on the behavior of organic beings. What is still lacking here is a grasp of connections of profound generality, but not a knowledge of order in itself.

The more a man is imbued with the ordered regularity of all events the firmer becomes his conviction that there is no room left by the side of this ordered regularity for causes of a different nature. For him neither the rule of human nor the rule of divine will exists as an independent cause of natural events. To be sure, the doctrine of a personal God interfering with natural events could never be *refuted*, in the real sense, by science, for this doctrine can always take refuge in those domains in which scientific knowledge has not yet been able to set foot.

But I am persuaded that such behavior on the part of the representatives of religion would not only be unworthy but also fatal. For a doctrine which is able to maintain itself not in clear light but only in the dark, will of necessity lose its effect on mankind, with incalculable harm to human progress. In their struggle for the ethical good, teachers of religion must have the stature to give up the doctrine of a personal God, that is, give up that source of fear and hope which in the past placed such vast power in the hands of priests. In their labors they will have to avail themselves of those forces which are capable of cultivating the Good, the True, and the Beautiful in humanity itself. This is, to be sure, a more difficult but an incomparably more worthy task. After religious teachers accomplish the refining process indicated they will surely recognize with joy that true religion has been ennobled and made more profound by scientific knowledge.

If it is one of the goals of religion to liberate mankind as far as possible from the bondage of egocentric cravings, desires, and fears, scientific reasoning can aid religion in yet another sense. Although it is true that it is the goal of science to discover rules which permit the association and foretelling of facts, this is not its only aim. It also seeks to reduce the connections discovered to the smallest possible number of mutually independent conceptual elements. It is in this striving after the rational unification of the manifold that it encounters its greatest successes, even though it is precisely this attempt which causes it to run the greatest risk of falling a prey to illusions. But whoever has undergone the intense experience of successful advances made in this domain is moved by profound reverence for the rationality made manifest in existence. By way of the understanding he achieves a far-reaching emancipation from the shackles of personal hopes and desires, and thereby attains that humble attitude of mind toward the grandeur of reason incarnate in existence, and which, in its profoundest depths, is inaccessible to man. This attitude, however, appears to me to be religious, in the highest sense of the word. And so it seems to me that science not only purifies the religious impulse of the dross of its anthropomorphism but also contributes to a religious spiritualization of our understanding of life.

The further the spiritual evolution of mankind advances, the more certain it seems to me that the path to genuine religiosity does not lie through the fear of life, and the fear of death, and blind faith, but through striving after rational knowledge. In this sense I believe that the priest must become a teacher if he wishes to do justice to his lofty educational mission

—6—
"Message His Holiness John Paul II," June 1, 1988. *John Paul II on Science and Religion.* Ed. Robert J. Russell, William R. Stoeger, S. J., and George V. Coyne. Vatican City State: Vatican Observatory Publications, 1990, pages M1–14.

On the three hundredth anniversary of the publication of Newton's *Principia* Pope John Paul II sought to organize a study group that would further his own efforts to foster dialogue between the cultures of science and religious belief. This wish came to fruition in a research Study Week on physics, philosophy and theology convened at the Papal summer residence in September, 1987. After reflecting on the discussion and research which resulted from this and later meetings, the Pope delivered the following message about how science and religion should work together.

To the Reverend George V. Coyne, S.J.

Director of the Vatican Observatory

"Grace to you and peace from God our Father and the Lord Jesus Christ" (Eph 1:2).

As you prepare to publish the papers presented at the Study Week held at Castelgandolfo on 21–26 September 1987, I take the occasion to express my gratitude to you and through you to all who contributed to that important initiative. I am confident that the publication of these papers will ensure that the fruits of that endeavour will be further enriched.

The three hundredth anniversary of the publication of Newton's *Philosophiae Naturalis Principia Mathematica* provided an appropriate occasion for the Holy See to sponsor a Study Week that investigated the multiple relationships among theology, philosophy and the natural sciences. The man so honoured, Sir Isaac Newton, had himself devoted much of his life to these same issues, and his reflections upon them can be found throughout his major works, his unfinished manuscripts and his vast correspondence. The publication of your own papers from this Study Week, taking up again some of the same questions which this great genius explored, affords me the opportunity to thank you for the efforts you devoted to a subject of such paramount importance. The theme of your conference, "Our Knowledge of God and Nature: Physics, Philosophy and Theology", is assuredly a crucial one for the contemporary world. Because of its importance, I should like to address some issues which the interactions among natural science, philosophy, and theology present to the Church and to human society in general.

The Church and the Academy engage one another as two very different but major institutions within human civilization and world culture. We bear before God enormous responsibilities for the human condition because historically we have had and continue to have a major influence on the development of ideas and values and on the course of human action. We both have histories stretching back over thousands of years: the learned, academic community dating back to the origins of culture, to the city and the library and the school, and the Church with her historical roots in ancient Israel. We have come into contact often during these centuries, sometimes in mutual support, at other times in those needless conflicts which have marred both our histories. In your conference we met again, and it was altogether fitting that as we approach the close of this millennium we initiated a series of reflections together upon the world as we touch it and as it shapes and challenges our actions.

So much of our world seems to be in fragments, in disjointed pieces. So much of human life is passed in isolation or in hostility. The division between rich nations and poor nations continues to grow; the contrast between northern and southern regions of our planet becomes ever more marked and intolerable. The antagonism between races and religions splits countries into warring camps; historical animosities show no signs of abating. Even within the academic community, the separation between truth and values persists, and the isolation of their several cultures - scientific, humanistic and religious—makes common discourse difficult if not at times impossible.

But at the same time we see in large sectors of the human community a growing critical openness towards people of different cultures and backgrounds, different competencies and viewpoints. More and more frequently, people are seeking intellectual coherence and collaboration, and are discovering values and experiences they have in common even within their diversities. This openness, this dynamic interchange, is a notable feature of the international scientific communities themselves, and is based on common interests, common goals and a common enterprise, along with a deep awareness that the insights and attainments of one are often important for the progress of the other. In a similar but more subtle way this has occurred and is continuing to occur among more diverse groups—among the communities that make up the Church, and even between the scientific community and the Church herself. This drive is essentially a movement towards the kind of unity which resists homogenization and relishes diversity. Such community is determined by a common meaning and by a shared understanding that evokes a sense of mutual involvement. Two groups which may seem initially to have nothing in common can begin to enter into community with one another by discovering a common goal, and this in turn can lead to broader areas of shared understanding and concern.

As never before in her history, the Church has entered into the movement for the union of all Christians, fostering common study, prayer, and discussions that "all may be one" (Jn 17:20). She has attempted to rid herself of every vestige of anti-semitism and to emphasize her origins in and her religious debt to Judaism. In reflection and prayer, she has reached out to the great world religions, recognizing the values we all hold in common and our universal and utter dependence upon God.

Within the Church herself, there is a growing sense of "world-church", so much in evidence at the last Ecumenical Council in which bishops native to every continent—no longer predominantly of European or even Western origin—assumed for the first time their common responsibility for the entire Church. The documents from that Council and of the magisterium have reflected this new world-consciousness both in their content and in their attempt to address all people of good will. During this century, we have witnessed a dynamic tendency to reconciliation and unity that has taken many forms within the Church.

Nor should such a development be surprising. The Christian community in moving so emphatically in this direction is realizing in greater intensity the activity of Christ within her: "For God was in Christ, reconciling the world to himself" (2 Cor 5:19). We ourselves are called to be a continuation of this reconciliation of human beings, one with another and all with God. Our very nature as Church entails this commitment to unity.

Turning to the relationship between religion and science, there has been a definite, though still fragile and provisional, movement towards a new and more nuanced interchange. We have begun to talk to one another on deeper levels than before, and with greater openness towards one another's perspectives. We have begun to search together for a more thorough understanding of one another's disciplines, with their competencies and their limitations, and especially for areas of common ground. In doing so we have uncovered important questions which concern both of us, and which are vital to the larger human community we both serve. It is crucial that this common search based on critical openness and interchange should not only continue but also grow and deepen in its quality and scope.

For the impact each has, and will continue to have, on the course of civilization and on the world itself, cannot be overestimated, and there is so much that each can offer the other. There is, of course, the vision of the unity of all things and all peoples in Christ, who is active and present with us in our daily lives—in our struggles, our sufferings, our joys and in our searchings—and who is the focus of the Church's life and witness. This vision carries with it into the larger community a deep reverence for all that is, a hope and assurance that the fragile goodness, beauty and life we see in the universe is moving towards a completion and fulfilment which will not be overwhelmed by the forces of dissolution and death. This vision also provides a strong support for the values which are emerging both from our knowledge and appreciation of creation and of ourselves as the products, knowers and stewards of creation.

The scientific disciplines too, as is obvious, are endowing us with an understanding and appreciation of our universe as a whole and of the incredibly rich variety of intricately related processes and structures which constitute its animate and inanimate components. This knowledge has given us a more thorough understanding of ourselves and of our humble yet unique role within creation. Through technology it also has given us the capacity to travel, to communicate, to build, to cure, and to probe in ways which would have been almost unimaginable to our ancestors. Such knowledge and power, as we have discovered, can be used greatly to enhance and improve our lives or they can be exploited to diminish and destroy human life and the environment even on a global scale.

The unity we perceive in creation on the basis of our faith in Jesus Christ as Lord of the universe, and the correlative unity for which we strive in our human communities, seems to be reflected and even reinforced in what contemporary science is revealing to us. As we behold the incredible development of scientific research we detect an underlying movement towards the discovery of levels of law and process which unify created reality and which at the same time have given rise to the vast diversity of structures and organisms which constitute the physical and biological, and even the psychological and sociological, worlds.

Contemporary physics furnishes a striking example. The quest for the unification of all four fundamental physical forces—gravitation, electromagnetism, the strong and weak nuclear interactions—has met with increasing success. This unification may well combine discoveries from the sub-atomic and the cosmological domains and shed light both on the origin of the universe and, eventually, on the origin of the laws and constants which govern its evolution. Physicists possess a detailed though incomplete and provisional knowledge of elementary particles and of the fundamental forces through which they interact at low and intermediate energies. They now have an acceptable theory unifying the electro-magnetic and weak nuclear force, along with much less adequate but still promising grand unified field theories which attempt to incorporate the strong nuclear interaction as well. Further in the line of this same development there are already several detailed suggestions for the final stage, superunification, that is, the unification of all four fundamental forces, including gravity. Is it not important for us to note that in a world of such detailed specialization as contemporary physics there exists this drive towards convergence?

In the life sciences, too, something similar has happened. Molecular biologists have probed the structure of living material, its functions and its processes of replication. They have discovered that the same underlying constituents serve in the make-up of all living organisms on earth and constitute both the genes and the proteins which these genes code. This is another impressive manifestation of the unity of nature.

By encouraging openness between the Church and the scientific communities, we are not envisioning a disciplinary unity between theology and science like that which exists within a given scientific field or within theology proper. As dialogue and common searching continue, there will be growth towards mutual understanding and a gradual uncovering of common concerns which will provide the basis for further research and discussion. Exactly what form that will take must be left to the future. What is important, as we have already stressed, is that the dialogue should continue and grow in depth and scope. In the process we must overcome every regressive tendency to a unilateral reductionism, to fear, and to self-imposed isolation. What is critically important is that each discipline should continue to enrich, nourish and challenge the other to be more fully what it can be and to contribute to our vision of who we are and who we are becoming.

We might ask whether or not we are ready for this crucial endeavour. Is the community of world religions, including the Church, ready to enter into a more thorough-going dialogue with the scientific community, a dialogue in which the integrity of both religion and science is supported and the advance of each is fostered? Is the scientific community now prepared to open itself to Christianity, and indeed to all the great world religions, working with us all to build a culture that is more humane and in that way more divine? Do we dare to risk the honesty and the courage that this task demands? We must ask ourselves whether both science and religion will contribute to the integration of human culture or to its fragmentation. It is a single choice and it confronts us all.

For a simple neutrality is no longer acceptable. If they are to grow and mature, peoples cannot continue to live in separate compartments, pursing totally divergent interests from which they evaluate and judge their world. A divided community fosters a fragmented vision of the world; a community of interchange encourages its members to expand their partial perspectives and form a new unified vision.

Yet the unity that we seek, as we have already stressed, is not identity. The Church does not propose that science should become religion or religion science. On the contrary, unity always presupposes the diversity and the integrity of its elements. Each of these members should become not less itself but more itself in a dynamic interchange, for a unity in which one of the elements is reduced to the other is destructive, false in its promises of harmony, and ruinous of the integrity of its components. We are asked to become one. We are not asked to become each other.

To be more specific, both religion and science must preserve their autonomy and their distinctiveness. Religion is not founded on science nor is science an extension of religion. Each should possess its own principles, its pattern of procedures, its diversities of interpretation and its own conclusions. Christianity possesses the source of its justification within itself and does not expect science to constitute its primary apologetic. Science must bear witness to its own worth. While each can and should support the other as distinct dimensions of a common human culture, neither ought to assume that it forms a necessary premise for the other. The unprecedented opportunity we have today is for a common interactive relationship in which each discipline retains its integrity and yet is radically open to the discoveries and insights of the other.

But why is critical openness and mutual interchange a value for both of us? Unity involves the drive of the human mind towards understanding and the desire of the human spirit for love. When human beings seek to understand the multiplicities that surround them, when they seek to make sense of experience, they do so by bringing many factors into a common vision. Understanding is achieved when many data are unified by a common structure. The one illuminates the many; it makes sense of the whole. Simple multiplicity is chaos; an insight, a single model, can give that chaos structure and draw it into intelligibility. We move towards unity as we move towards meaning in our lives. Unity is also the consequence of love. If love is genuine, it moves not towards the assimilation of the other but towards union with the other. Human community begins in desire when that union has not been achieved, and it is completed in joy when those who have been apart are now united.

In the Church's earliest documents, the realization of community, in the radical sense of that word, was seen as the promise and goal of the Gospel: "That which we have seen and heard we proclaim also to you, so that you may have fellowship with us; and our fellowship is with the Father and with his Son Jesus Christ. And we are writing this that our joy may be complete" (1 Jn 1:3–3). Later the Church reached out to the sciences and to the arts, founding great universities and building monuments of surpassing beauty so that all things might be recapitulated in Christ (cf. Eph 1:10).

What, then, does the Church encourage in this relational unity between science and religion? First and foremost that they should come to understand one another. For too long a time they have been at arm's length. Theology has been defined as an effort of faith to achieve understanding, as *fides quaerens intellectum*. As such, it must be in vital interchange today with science just as it always has been with philosophy and other forms of learning. Theology will have to call on the findings of science to one degree or another as it pursues its primary concern for the human person, the reaches of freedom, the possibilities of Christian community, the nature of belief and the intelligibility of nature and history. The vitality and significance of theology for humanity will in a profound way be reflected in its ability to incorporate these findings.

Now this is a point of delicate importance, and it has to be carefully qualified. Theology is not to incorporate indifferently each new philosophical or scientific theory. As these findings become part of the intellectual culture of the time, however, theologians must understand them and test their value in bringing out from Christian belief some of the possibilities which have not yet been realized. The hylomorphism of Aristotelian natural philosophy, for example, was adopted by the medieval theologians to help them explore the nature of the sacraments and the hypostatic union. This did not mean that the Church adjudicated the truth or falsity of the Aristotelian insight, since that is not her concern. It did mean that this was one of the rich insights offered by Greek culture, that it needed to be understood and taken seriously and tested for its value in illuminating various areas of theology. Theologians might well ask, with respect to contemporary science, philosophy and the other areas of human knowing, if they have accomplished this extraordinarily difficult process as well as did these medieval masters.

If the cosmologies of the ancient Near Eastern world could be purified and assimilated into the first chapters of Genesis, might contemporary cosmology have something to offer to our reflections upon creation? Does an evolutionary perspective bring any light to bear upon theological anthropology, the meaning of the human person as the *imago Dei,* the problem of Christology—and even upon the development of doctrine itself? What, if any, are the eschatological implications of contemporary cosmology, especially in light of the vast future of our universe? Can theological method fruitfully appropriate insights from scientific methodology and the philosophy of science?

Questions of this kind can be suggested in abundance. Pursuing them further would require the sort of intense dialogue with contemporary science that has, on the whole, been lacking among those engaged in theological research and teaching. It would entail that some theologians, at least, should be sufficiently well-versed in the sciences to make authentic and creative use of the resources that the best-established theories may offer them. Such an expertise would prevent them from making uncritical and overhasty use for apologetic purposes of such recent theories as that of the "Big Bang" in cosmology. Yet it would equally keep them from discounting altogether the potential relevance of such theories to the deepening of understanding in traditional areas of theological inquiry.

In this process of mutual learning, those members of the Church who are themselves either active scientists or, in some special cases, both scientists and theologians could serve as a key resource. They can also provide a much-needed ministry to others struggling to integrate the worlds of science and religion in their own intellectual and spiritual lives, as well as to those who face difficult moral decisions in matters of technological research and application. Such bridging ministries must be nurtured and encouraged. The Church long ago recognized the importance of such links by establishing the Pontifical Academy of Sciences, in which some of the world's leading scientists meet together regularly to discuss their researches and to convey to the larger community where the directions of discovery are tending. But much more is needed.

The matter is urgent. Contemporary developments in science challenge theology far more deeply than did the introduction of Aristotle into

Western Europe in the thirteenth century. Yet these developments also offer to theology a potentially important resource. Just as Aristotelian philosophy, through the ministry of such great scholars as St Thomas Aquinas, ultimately came to shape some of the most profound expressions of theological doctrine, so can we not hope that the sciences of today, along with all forms of human knowing, may invigorate and inform those parts of the theological enterprise that bear on the relation of nature, humanity and God?

Can science also benefit from this interchange? It would seem that it should. For science develops best when its concepts and conclusions are integrated into the broader human culture and its concerns for ultimate meaning and value. Scientists cannot, therefore, hold themselves entirely aloof from the sorts of issues dealt with by philosophers and theologians. By devoting to these issues something of the energy and care they give to their research in science, they can help others realize more fully the human potentialities of their discoveries. They can also come to appreciate for themselves that these discoveries cannot be a genuine substitute for knowledge of the truly ultimate. Science can purify religion from error and superstition; religion can purify science from idolatry and false absolutes. Each can draw the other into a wider world, a world in which both can flourish.

For the truth of the matter is that the Church and the scientific community will inevitably interact; their options do not include isolation. Christians will inevitably assimilate the prevailing ideas about the world, and today these are deeply shaped by science. The only question is whether they will do this critically or unreflectively, with depth and nuance or with a shallowness that debases the Gospel and leaves us ashamed before history. Scientists, like all human beings, will make decisions upon what ultimately gives meaning and value to their lives and to their work. This they will do well or poorly, with the reflective depth that theological wisdom can help them attain, or with an unconsidered absolutizing of their results beyond their reasonable and proper limits.

Both the Church and the scientific community are faced with such inescapable alternatives. We shall make our choices much better if we live in a collaborative interaction in which we are called continually to be more. Only a dynamic relationship between theology and science can reveal those limits which support the integrity of either discipline, so that theology does not profess a pseudo-science and science does not become an unconscious theology. Our knowledge of each other can lead us to be more authentically ourselves. No one can read the history of the past century and not realize that crisis is upon us both. The uses of science have on more than one occasion proven massively destructive, and the reflections on religion have too often been sterile. We need each other to be what we must be, what we are called to be.

And so on this occasion of the Newton Tricentennial, the Church speaking through my ministry calls upon herself and the scientific community to intensify their constructive relations of interchange through unity. You are called to learn from one another, to renew the context in which science is done and to nourish the inculturation which vital theology demands. Each of you has everything to gain from such an interaction, and the human community which we both serve has a right to demand it from us.

Upon all who participated in the Study Week sponsored by the Holy See and upon all who will read and study the papers herein published I invoke wisdom and peace in our Lord Jesus Christ and cordially impart my Apostolic Blessing.

From the Vatican, 1 June, 1988

—7—
Meher Baba, "The New Humanity," *The Discourses.* Kingsport, TN: Kingsport Press, 1967, vol. 1, pages 16–25.

Meher Baba's *Discourses,* originally given to his close followers in the period 1938–1943, give much detailed information and guidance for the aspirant about how to incorporate daily life into following the spiritual path. In this first and probably best known discourse, Meher Baba's offers an explanation of the spiritual significance of the present suffering of humanity and Earth's current transformation into a new age. He also outlines how science can contribute to this ongoing spiritual development.

The New Humanity

As in all great critical periods of human history, humanity is now going through the agonising travail of spiritual rebirth. Great forces of destruction are afoot and seem to be dominant at the moment, but constructive and creative forces which will redeem humanity are also being released through several channels. Although the working of these forces of light is chiefly silent, they are eventually bound to bring about those transformations which will make the further spiritual advance of humanity safe and steady. *It is all a part of the divine plan, which is to give to the hungry and weary world a fresh dispensation of the eternal and only Truth.*

At present the urgent problem facing humanity is to devise ways and means of eliminating competition, conflict and rivalry in all the subtle and gross forms which they assume in the various spheres of life. Military wars are, of course, the most obvious sources of chaos and destruction. However, *wars in themselves do not constitute the central problem for humanity, but are rather the external symptoms of something graver at their root.* Wars and the suffering they bring cannot be completely avoided by mere propaganda against war; if they are to disappear from human history it will be necessary to tackle their root-cause. Even when military wars are not being waged, individuals or groups of individuals are constantly engaged in *economic or some other subtle form of warfare.* Military wars, with all the cruelty which they involve, arise only when these underground causes are aggravated.

The root-cause of the chaos which precipitates itself in wars is that most persons are in the grip of egoism and selfish considerations, and they express their *egoism and self-interest* individually as well as collectively. *This is the life of illusory values in which men are caught.* To face the Truth is to realise that life is one, in and through its manifold manifestations. To have this understanding is to forget the limiting self in the realisation of the unity of life.

With the dawn of true understanding the problem of wars would immediately disappear. *Wars have to be so clearly seen as both unnecessary and unreasonable that the immediate problem would not be how to stop wars but to wage them spiritually against the attitude of mind responsible for such a cruel and painful state of things.* In the light of the Truth of the unity of all life, co-operative and harmonious action becomes natural and inevitable. Hence, the chief task before those who are deeply concerned with the rebuilding of humanity, is to do their utmost to dispel the spiritual ignorance which envelops humanity.

Wars do not arise merely to secure material adjustment; they are often the product of uncritical identification with narrow interests which through association come to be included in that part of the world which is regarded as "mine." *Material adjustment is only part of the wider problem of establishing spiritual adjustment, but spiritual adjustment requires the elimination of self not only from the material aspects of life but also from those spheres which affect the intellectual, emotional and cultural life of man.*

To understand the problem of humanity as merely a problem of bread is to reduce humanity to the level of animality. But even when man sets himself to the limited task of securing purely material adjustment, he can only succeed in this attempt if he has spiritual understanding.

Economic adjustment is impossible unless people realise that there can be no planned and co-operative action in economic matters until self-interest gives place to selfgiving love. Otherwise, with the best of equipment and efficiency in the material spheres, humanity cannot avoid conflict and insufficiency.

The NEW HUMANITY, which emerges from the travail of present struggle and suffering, will not ignore science or its practical attainments; it is a mistake to look upon science as anti-spiritual. *Science is a help or hindrance to spirituality according to the use to which it is put.* Just as true art expresses spirituality, so science, when properly handled, can be the expression and fulfillment of the spirit. Scientific truths concerning the physical body and its life in the gross world can become a medium for the soul to know itself; but to serve this purpose they must be properly fitted into the larger spiritual understanding. This includes a steady perception of true and lasting values. In the absence of such spiritual understanding, scientific truths and attainments are liable to be used for mutual destruction and for a life which will tend to strengthen the chains which bind the spirit. All-sided progress of humanity can be assured only if science and religion proceed hand in hand.

The coming civilisation of the New Humanity shall be ensouled not by dry intellectual doctrines, but by living spiritual experience. Spiritual experience has a hold on the deeper truths which are inaccessible to mere intellect; it cannot be born of unaided intellect. Spiritual truth can often be stated and expressed through the intellect, and the intellect surely is of some help for the communication of spiritual experience. But by itself, the intellect is insufficient to enable man to have spiritual experience or to communicate it to others. If two persons have had headaches they can co-operatively examine their experience of headache and make it explicit to themselves through the work of the intellect. If a person has never experienced a headache, no amount of intellectual explanation will suffice for making him understand what a headache is. Intellectual explanation can never be a substitute for spiritual experience; it can at best prepare the ground for it.

Spiritual experience involves more than can be grasped by mere intellect. This is often emphasised by calling it a mystical experience. Mysticism is often regarded as something anti-intellectual, obscure and confused, or impractical and unconnected with experience. In fact, true mysticism is none of these. *There is nothing irrational in true mysticism when it is, as it should be, a vision of Reality. It is a firm of perception which is absolutely unclouded, and so practical that it can be lived every moment of life and expressed in every-day duties. Its connection with experience is so deep that, in one sense, it is the final understanding if all experience.* When spiritual experience is described as mystical one should not assume that it is something supernatural or entirely beyond the grasp of human consciousness. All that is meant is that it is not accessible to limited human intellect until it transcends its limits and is illumined by direct realisation of the Infinite. Christ pointed out the way to spiritual experience when he said, "Leave all and follow me." This means that man must leave limitations and establish himself in the infinite life of God. Real spiritual experience involves not only realisation of the soul on higher planes, but also a right attitude towards worldly duties. If it loses its connection with the different phases of life, what we have is a neurotic reaction that is far from being a spiritual experience.

The spiritual experience that is to enliven and energise the New Humanity cannot be a reaction to the stern and uncompromising demands made by the realities of life. Those without the capacity for adjustment to the flow of life have a tendency to recoil from the realities of life and to seek shelter and protection in a self-created fortress of illusions. Such reaction is an attempt to perpetuate one's separate existence by protecting it from the demands made by life. It can only give a pseudo-solution to the problems of life by providing a false sense of security and selfcompleteness. It is not even an advance towards the real and lasting solution; on the contrary, it is a sidetracking from the true Path. *Man will be dislodged again and again from his illusory shelters by fresh and irresistible waves if life, and will invite upon himself fresh forms if suffering by seeking to protect his separative existence through escape.*

Just as a person may seek to hold onto his separative experience through escape, he may also seek to hold it through uncritical identification with forms, ceremonies and rituals or with traditions and conventions. Forms, ceremonies and rituals, traditions and conventions are in most cases fetters to the release of infinite life. If they were a pliant medium for the expression of unlimited life, they would be an asset rather than a handicap for securing the fulfillment of divine life on earth; but they mostly have a tendency to gather prestige and claims in their own right, independently of the life which they might express. When this happens, any attachment to them must eventually lead to a drastic curtailment and restriction of life. *The New Humanity will be freed from a life of limitations, allowing unhampered scope for the creative life of the spirit; and it will break the attachment to external forms and learn to subordinate them to the claims of the spirit.* The limited life of illusions and false values will then be replaced by unlimited life in the Truth, and the limitations, through which the separative self lives, will wither away at the touch of true understanding.

Just as a person may seek to hold onto his separative existence through escape or identification with external forms, he may seek to hold it through identification with some narrow class, creed, sect or religion, or with the divisions based upon sex. Here the individual may seem to have lost his separative existence through identification with a larger whole. But, in fact, he is often *expressing* his separative existence through such an identification, which enables him to delight in his feeling of being separate from others who belong to another class, nationality, creed, sect, religion or sex.

Separative existence derives its being and strength by identifying itself with one opposite and contrasting itself with the other. A man may seek to protect his separate existence through identification with one ideology rather than another or with his conception of good as contrasted with his idea of evil. *What results from identification with narrow groups or limited ideals is not a real merging of the separative self, but only a semblance of it. A real merging of the limited self in the ocean of universal life involves complete surrender of separative existence in all its forms.*

The large mass of humanity is caught up in the clutches of separative and assertive tendencies. For one who is overpowered by the spectacle of these fetters of humanity, there is bound to be nothing but unrelieved despair about its future. One must look deeper into the realities of the day if one is to get a correct perspective on the present distress of humanity. The real possibilities of the New Humanity are hidden to those who look only at the surface of the world-situation, but they exist and only need the spark of spiritual understanding to come into full play and effect. The forces of lust, hate and greed produce incalculable suffering and chaos, but *the one redeeming feature about human nature is that even in the midst of disruptive forces there invariably exists some form of love.*

Even wars require co-operative functioning, but the scope of this co-operative functioning is artificially restricted by identification with a limited group or ideal. *Wars often are carried on by a form of love, but it is a love which has not been understood properly. In order that love should come into its own, it must be untrammeled and unlimited.* Love does exist in all phases of human life, but it is latent or is limited and poisoned by personal ambition, racial pride, narrow loyalties and rivalries, and attachment to sex, nationality, sect, caste or religion. If there is to be a resurrection of humanity, the heart of man will have to be unlocked so that a new love is born into it—*a love which knows no corruption and is entirely free from individual or collective greed.*

The New Humanity will come into existence through a release of love in measureless abundance, and this release of love can come through spiritual awakening brought about by the Masters. *Love cannot be born of mere determination, through the exercise of will one can at best be dutiful.* Through struggle and effort, one may succeed in assuring that one's external action is in conformity with one's concept of what is right; but such action is spiritually barren because it lacks the inward beauty of spontaneous love. Love has to spring spontaneously from within; it is in no way amenable to any form of inner or outer force. Love and coercion can never go together, but while love cannot be forced upon anyone, it can be awakened through love itself. *Love is essentially self-communicative; those who do not have it catch it from those who have it.* Those who receive love from others cannot be its recipients without giving a response which, in itself, is the nature of love. True *love* is unconquerable and irresistible. It goes on gathering power and spreading itself until eventually it transforms everyone it touches. *Humanity will attain to a new mode of being and life through the free and unhampered interplay of pure love from heart to heart.*

When it is recognised that there are no claims greater than the claims of the universal divine life which, without exception, includes everyone and everything, love will not only establish peace, harmony and happiness in social, national and international spheres, but it will shine in its own purity and beauty. Divine love is unassailable to the onslaughts of duality and is an expression of divinity itself. It is through divine love that the New Humanity will tune in with the divine plan. Divine love will not only introduce imperishable sweetness and infinite bliss into personal life, but it will also make possible an era of New Humanity. *Through divine love the New Humanity will learn the art of co-operative and harmonious life; it will free itself from the tyranny of dead forms and release the creative life of spiritual wisdom; it will shed all illusions and get established in the Truth; it will enjoy peace and abiding happiness; it will be initiated in the life of Eternity.*

Bibliography

Abdulla, Ramjoo. 1954."How It All Happened." *The Silent Teachings of Meher Baba.* Reprinted as a special edition of *The Awakener Magazine* XV (1 and 2), 1954: 1–23.

Adams, Fred, and Greg Laughlin. 1999. *The Five Ages of the Universe: Inside the Physics of Eternity.* New York: The Free Press.

Adams, Fred. 2004. *Our Living Multiverse: A Book of Genesis in 0+7 Chapters.* New York: Pi Press.

Alpher, Ralph A., and Robert Herman. 2001. *Genesis of the Big Bang.* Oxford: Oxford University Press.

Baba, Meher. 1958. *Beams from Meher Baba on the Spiritual Panorama.* San Francisco: Sufism Reoriented, Inc.

———. 1963. *The Everything and the Nothing.* Berkeley, CA: The Beguine Library.

———. 1967. *The Discourses,* 3 vols. Kingsport, TN: Kingsport Press.

———. 1973, 1997 (last printing). *God Speaks: The Theme of Creation and Its Purpose,* 2nd ed. rev. and enlarged. Walnut Creek, CA: Sufism Reoriented.

Balslev, Anindita. 2000. "Cosmos and Consciousness: Indian Perspectives. Pp. 58–68 in John Haught, ed., *Science and Religion: In Search of Cosmic Purpose.* Washington, DC: Georgetown University Press.

Barbour, Ian G. 1971. *Issues in Science and Religion.* New York: Harper and Row. (Originally published in 1966 by Prentice Hall.)

———. 1974. *Myths, Models and Paradigms: A Comparative Study in Science and Religion.* New York: Harper & Row.

———. 1990. *Religion in an Age of Science: The Gifford Lectures: 1989–1990,* Vol. 1. San Francisco: Harper & Row.

———. 1997. *Religion and Science: Historical and Contemporary Issues* (revised and expanded edition of *Religion an Age of Science*). San Francisco: HarperSanFrancisco.

———. 2000. *When Science Meets Religion.* San Francisco: HarperSanFrancisco.

Barnett, Lincoln. 1957. *The Universe and Dr. Einstein.* 2nd rev. ed. Mineola, NY: Dover Publications.

Barr, Stephen N. 2003. *Modern Physics and Ancient Faith.* Notre Dame, IN: University of Notre Dame Press.

Barrow, John, and Frank Tipler. 1988. *The Anthropic Cosmological Principle.* Oxford: Oxford University Press (paperback).

Basalla, George. 2006. *Civilized Life in the Universe: Scientists on Intelligent Extraterrestrials.* New York: Oxford University Press.

Bohm, David. 1983. *Wholeness and the Implicate Order.* London: Ark Paperbacks.

Brooke, John Hedley. 1991. *Science and Religion: Some Historical Perspectives.* New York: Cambridge University Press.

Bucaille, Dr. Maurice. 2003. *The Bible, the Qur'an and Science: The Holy Scriptures Examined in the Light of Modern Knowledge.* Seventh edition, revised and expanded. Translated by Alastair D. Pannel and the Author. Elmhurst, NY: Tahrike Tarsile Quar'an, Inc.

Cabezon, Jose Ignacio. 2003. "Buddhism and Science: On the Nature of the Dialogue." Pp. 35–68 in Alan B. Wallace, ed., *Buddhism and Science: Breaking New Ground.* New York: Columbia University Press.

Cantor, Geoffrey, and Marc Swetlitz, eds. 2006. *Jewish Tradition and the Challenge of Darwinism.* Chicago: University of Chicago Press.

Capra, Fritjof. 1991. *The Tao of Physics.* Boston: Shambhala.

Clarke, J. J. 1997. *Oriental Enlightenment: The Encounter between Asian and Western Thought.* London: Routledge.

Clarke, W. N., S. J. 1988. "Is a Natural Theology Still Possible Today?" Pp. 103–123 in R. J. Russell, W. R. Stoeger, and G. V. Coyne, eds., *Physics, Philosophy, and Theology: A Common Quest for Understanding.* Vatican City State: Vatican Observatory.

Clayton, Philip. 1997. *God and Contemporary Physics.* Grand Rapids, MI: William. B. Eerdmans Publishing Company.

Cobb, John B., Jr. and David Ray Griffin. 1976. *Process Theology: An Introductory Exposition.* Philadelphia: The Westminster Press.

Collins, Francis S. 2006. *The Language of God: A Scientist Presents Evidence for Belief.* New York: Free Press.

Coulson, C. A. 1955. *Science and Christian Belief.* Chapel Hill, NC: University of North Carolina Press.

The Dalai Lama, H. H. 2005. *The Universe in a Single Atom: The Convergence of Science and Spirituality.* New York: Morgan Road Books.

Davies, P. C. W. 1982. *The Accidental Universe.* Cambridge, UK: Cambridge University Press.

———. 1983. *God and the New Physics.* New York: Touchstone Book.

———. 1988. *The Cosmic Blueprint: New Discoveries in Nature's Creative Ability to Order the Universe.* New York: Simon and Schuster.

———. 1992. *The Mind of God: The Scientific Basis for a Rational World.* New York: Simon and Schuster.

———. 1994. *The Last Three Minutes: Conjectures about the Ultimate Fate of the Universe.* New York: Basic Books.

———. 1995. *Are We Alone? Philosophical Implications of the Discovery of Extraterrestrial Life*. New York: Basic Books.

———. 2000. "Biological Determinism, Information Theory, and the Origin of Life." Pp. 15–28 in Steven J. Dick, ed., *Many Worlds: The New Universe, Extraterrestrial Life and the Theological Implications*. Philadelphia: Templeton Foundation Press.

———. 2003. "E. T. and God: Could Earthly Religions Survive the Discovery of Life Elsewhere in the Universe." *The Atlantic Monthly* 292.2 (Sept. 2003): 112(6). *Expanded Academic ASAP*. Thompson Gale. Los Medanos College. 6 Sep. 2006.

Davies, P. C. W., and J. Brown, eds. 1988. *Superstrings: A Theory of Everything*. Cambridge, UK: Cambridge University Press.

de Duve, Christian. 2000. "Lessons of Life." Pp. 3–13 in Steven J. Dick, ed., *Many Worlds: The New Universe, Extraterrestrial Life and the Theological Implications*. Philadelphia: Templeton Foundation Press.

Dembski, William A. 1999. *Intelligent Design: The Bridge between Science and Theology*. Downers Grove, IL: InterVarsity Press.

Dick, Steven J. 1996. *The Biological Universe: The Twentieth-Century Extraterrestrial Life Debate and the Limits of Science*. New York: Cambridge University Press.

———. 1998. *Life on Other Worlds: The 20th-Century Extraterrestrial Life Debate*. New York: Cambridge University Press.

———. 2000. "Cosmotheology: Theological Implications of the New Universe." Pp. 191–210 in Steven J. Dick, ed., *Many Worlds: The New Universe, Extraterrestrial Life and the Theological Implications*. Philadelphia: Templeton Foundation Press.

Drees, Willem. 1990. *Beyond the Big Bang: Quantum Cosmologies and God*. LaSalle, IL: Open Court.

Duce, Ivy Oneita. 1975. *How A Master Works*. Walnut Creek, CA: Sufism Reoriented.

Dyson, Freeman. 2004. *Infinite in All Directions*. New York: Perennial.

Eddington, Arthur S. 1928. *The Nature of the Physical World*. New York: The Macmillan Company.

———. 1929. *Science and the Unseen World*. New York: The Macmillan Company.

———. 1930. *Why I Believe in God: Science and Religion, as a Scientist Sees It*. Girard, KS: Haldeman-Julius Publications.

———. 1933. *The Expansion of the Universe*. New York: Cambridge University Press.

Eliade, Mircea. 1965. *The Two and the One*. Translated by J. M. Cohen. Chicago: The University of Chicago Press.

Ellis, George. 1993a. *Before the Beginning: Cosmology Explained*. London: Boyars/Bowerdean.

———. 1993b. "The Theology of the Anthropic Principle." Pp. 367–406 in R. Russell, N. Murphy, and C. Isham, eds., *Quantum Cosmology and the Laws of Nature: Scientific Perspectives on Divine Action*. Vatican City State: Vatican Observatory Publications and Berkeley, CA: Center for Theology and Natural Sciences.

Fabel, Arthur, and Donald St. John, eds. 2003. *Teilhard in the 21st Century: The Emerging Spirit of the Earth*. Maryknoll, NY: Orbis Books.

Farrington, Benjamin. 1953. *Greek Science*. Baltimore, MD: Penguin Books.

Faulkner, Danny R. 1998. "The Current State of Creation Astronomy." Institute for Creation Research. Available at http://www.icr.org/research/index/researchp_df_r01/.

Ferris, Timothy. 1997. *The Whole She-Bang: A State of the Universe(s) Report*. New York: Simon and Schuster, 1997.

———.1985. *The Creation of the Universe*. Produced by Northstar Productions. Videocassetts. PBS Home Video.

Galilei, Galileo. 1953. *Dialogue Concerning the Two Chief World Systems: Ptolemaic and Copernican*. Berkeley: University of California Press.

Gamow, George. 1985. *The Thirty Years that Shook Physics: The Story of Quantum Theory*. New York: Dover Publications, Inc.

———. 2004. *The Creation of the Universe*. Mineola, NY: Dover Publications.

Ghose, Sri Aurobindo. 1990. *The Future Evolution of Man: The Divine Life upon Earth*. Pondicherry. India: Sri Aurobindo Ashram.

———. 1990. *The Life Divine*. Twin Lakes, WI: Lotus Press.

Gilkey, Langdon. 1959. *Maker of Heaven and Earth: A Study of the Christian Doctrine of Creation*. New York: Doubleday 7 Company, Inc.

———. 1993. *Nature, Reality and the Sacred: The Nexus of Science and Religion*. Minneapolis: Fortress Press.

Gingerich, Owen. 2006. *God's Universe*. Cambridge, MA: The Belknap Press of Harvard University Press, 2006.

Gleiser, Marcelo. 1997. *The Dancing Universe: From Creation Myths to the Big Bang*. New York: A Dutton Book.

Gonzalez, Guillermo, and Jay W. Richards. 2004. *The Privileged Planet: How Our Place in the Cosmos Is Designed for Discovery*. Washington, DC: Regnery Publishing, Inc.

González, Roberto. 2001. *Zapotec Science: Farming and Food in the Northern Sierra of Oaxaca*. Austin: University of Texas Press.

Goodenough, Ursula. 1998. *The Sacred Depths of Nature*. Oxford: Oxford University Press.

Gosling, David L. 1976. *Science and Religion in India*. Madras: Christian Literature Society(published for Christian Institute for the Study of Religion and Society, Bangalore).

Goswami, Amit. 1997. *Science and Spirituality: A Quantum Integration*. New Delhi: Project of History of Indian Science, Philosophy and Culture.

Gould, Stephen J. 1999. *Rocks of Ages: Science and Religion in the Fullness of Life*. New York: The Ballantine Publishing Group.

Greene, Brian. 1999. *The Elegant Universe: Superstrings, Hidden Dimensions, and the Quest for the Ultimate Theory*. New York: W. W. Norton & Company.

———. 2004. *The Fabric of the Cosmos: Space, Time, and the Texture of Reality*. New York: Alfred A. Knopf.

Greenstein, George. 1988. *The Symbiotic Universe: Life and Mind in the Cosmos*. New York: William Morrow and Company, Inc.

Gribbin, John. 1986. *In Search of the Big Bang: Quantum Physics and Cosmology*. New York: Bantam Books.

————. 1998. *The Search for Superstrings, Symmetry, and the Theory of Everything.* New York: Little, Brown and Company.

Grim John, and Mary Evelyn Tucker. 2003."Introduction." *Teilhard in the 21st Century: The Emerging Spirit of the Earth.* Maryknoll, NY: Orbis Books.

Guiderdoni, Bruno. 2003. "Islam, Contemporary Issues in Science and Religion." Pp. 465–469 in J. Wentzel van Huyssteen, ed., *Encyclopedia of Science and Religion*, vol. 1. New York: Macmillan Reference.

Guth, Alan H. 1997. *The Inflationary Universe: The Quest for a New Theory of Cosmic Origins.* New York: Helix Books.

Halpern, Paul. 2004. *The Great Beyond: Higher Dimensions, Parallel universes, and the Extraordinary Search for a Theory of Everything.* Hoboken, NJ: John Wiley & Sons, Inc.

Harris, Steven J. 2002. "Roman Catholicism since Trent." Pp. 247–260 in Gary B. Ferngren, ed., *Science and Religion: A Historical Introduction.* Baltimore: Johns Hopkins University Press.

Haught, John F. 1984. *The Cosmic Adventure: Science, Religion and the Quest for Purpose.* New York: Paulist Press.

————. 1995. *Science and Religion: From Conflict to Conversation.* Mahwah, NJ: Paulist Press.

————. 1998. "Evolution, Tragedy, and Hope." Pp. 228–243 in Ted Peters, ed., *Science and Theology: The New Consonance.* Boulder, CO: Westview Press.

————. 2000a. *God after Darwin: A Theology of Evolution.* Boulder, CO: Westview Press.

————, ed. 2000b. *Science and Religion in Search of Cosmic Purpose.* Washington, DC: Georgetown University Press.

Hawking, Stephen. 1993. *Black Holes and Baby Universes and Other Essays.* New York: Bantam.

————. 1988. *A Brief History of Time: From the Big Bang to Black Holes.* New York: Bantam Books.

————. 1998. *A Brief History of Time: The Updated and Expanded Tenth Anniversary Edition.* New York: Bantam Books.

Hearn, Walter. 1986. *Teaching Science in a Climate of Controversy: A View from the American Scientific Affiliation.* Ipswich, MA: American Scientific Affiliation.

Hefner, Philip. 1989. "Evolution of the Created Co-Creator." Pp. 212–233 in Ted Peters, ed., *Cosmos as Creation: Theology and Science in Consonance.* Nashville, TN: Abingdon Press.

Heisenberg, Werner. 2007. *Physics and Philosophy: The Revolution in Modern Science.* New York: Harper Perennial Modern Classics.

Herbert, Nick. 1986. *Quantum Reality: Beyond the New Physics.* New York: Doubleday.

Hogan, Craig. 1998. *The Little Book of the Big Bang: A Cosmic Primer.* New York: Copernicus, an Imprint of Springer-Verlag.

Hoodboy, P. 1991. *Islam and Science: Religious Orthodoxy and the Battle for Rationality.* London: Zed Books Ltd.

Hoyle, Fred. 1950. *The Nature of the Universe.* New York: Harper.

Hoyle, Fred. 1981. "The Universe: Past and Present Reflections." *Engineering and Science* 44: 8–12.

Iqbal, Muzzafar. 2002. *Islam and Science*. Hampshire, England: Ashgate.

Jaki, Stanley. 1989. *God and the Cosmologists*. Washington, DC: Regnery Gateway.

Jammer, Max. 1999. *Einstein and Religion: Physics and Theology*. Princeton, NJ: Princeton University Press.

Jastrow, Robert. 1967. *Red Giants and White Dwarfs: The Evolution of Stars, Planets and Life*. New York: Harper and Row.

———. 1992. *God and the Astronomers*. New and expanded edition. NY: W. W. Norton & Company, Inc. (1st ed., 1978).

Jayatilleke, K. N. 1971. *Facets of Buddhist Thought: Six Essays*. Kandy, Sri Lanka: Buddhist Publication Society.

Jeans, James. 1978. *The Mysterious Universe*. Cambridge, UK: Cambridge University Press, 1948.

Jitatmananda, Swami. 2006. *Modern Physics and Vedanta*. Mumbai: Bharatiya Vidya Bhavan.

Johnson, Philip E. 1991. *Darwin on Trial*. Washington, DC: Regnery Gateway.

———. 2000. *The Wedge of Truth: Splitting the Foundations of Naturalism*. Downers Grove, IL: InterVarsity Press.

Jones, Richard H. 1986. *Science and Mysticism: A Comparative Study of Western Natural Science, Theravada Buddhism, and Advaita Vedanta*. Lewisburg: Bucknell University Press.

Kalchuri, Bhau. 1981. *The Nothing and the Everything*. North Myrtle Beach, CA: Manifestation, Inc.

———. 1979 (ed. 1986–2001). *Lord Meher: The Biography of the Avatar of the Age, Meher Baba*. Myrtle Beach, SC: Manifestation, Inc., 20 vols.

Kragh, Helge. 1999. *Cosmology and Controversy: The Historical Development of Two Theories of the Universe*. Princeton, NJ: Princeton University Press.

Krauss, Lawrence M., and Michael S. Turner. September 2004. "A Cosmic Conundrum." *Scientific American* 291(3): 71–77.

Kuhn, Thomas. 1996. *The Structure of Scientific Revolution*, 3rd edition. Chicago: The University of Chicago Press.

Kuppers, Bernd-Olaf. 2000. "The World of Biological Complexity: The Origin and Evolution of Life." Pp. 31–43 in Steven J. 2004, ed., *Many Worlds: The New Universe, Extraterrestrial Life and the Theological Implications*. Philadelphia: Templeton Foundation Press.

Lemaitre, Georges. 1965. "The Primeval Atom." Pp. 339–353 in Milton K. Munitz, ed., *Theories of the Universe: From Babylonian Myth to Modern Science*. New York: The Free Press.

Lerner, Eric. 1991. *The Big Bang Never Happened*. New York: Times Books.

Leslie, John. 1989. *Universes*. London: Routledge.

Leyrer, Carl W. 1981. "Endtime Theology and the "Rapture." Paper presented at Arizona-California Pastoral Conference, Tucson, Arizona, October 1981. Available at http://www.wls.essays.net.

Lightman, Alan. 2005. *The Discoveries: Great Breakthroughs in 20th Century Science*. New York: Pantheon Books.

Loder, James E. and W. Jim Neidhardt. 1996. "Barth, Bohr, and Dialectic." Pp. 271–289 in Mark W. Richardson, and Wesley J. Wildman, eds., *Religion and Science: History, Method, Dialogue*. New York: Routledge.

MacQuarrie, John. 2002. *Twentieth Century Religious Thought*, new ed. Harrisburg, PA: Trinity Press International.

Matsumura, Molleen, ed. 1995. *Voices for Evolution*. Berkeley, CA: The National Center for Science Education, Inc.

Matt, Daniel C. 1996. *God and the Big Bang: Discovering Harmony between Science and Spirituality*. Woodstock, VT: Jewish Lights Publishing.

Matthews, Clifford N., and Roy Abraham Varghese, eds. 1995. *Cosmic Beginnings and Human Ends*. Chicago: Open Court.

Matthews, Clifford N., Mary Evelyn Tucker, and Philip Hefner, eds. 2002. *When Worlds Converge: What Science and Religion Tell Us about the Story of the Universe and Our Place in It*. Chicago: Open Court.

McMullin, Ernan. 1981. "How Should Cosmology Relate to Cosmology?" Pp. 17–57 in A. R. Peacocke, ed., *The Sciences and Theology in the 20th Century*. Notre Dame, IN: University of Notre Dame Press.

———. 1988. "Natural Science and Belief in a Creator: Historical Notes." Pp. 41–79 in R. J. Russell, W. R. Stoeger, S. J., and G. V. Coyne, S. J., eds., *Physics, Philosophy, and Theology: A Common Quest for Understanding*. Vatican City State: Vatican Observatory.

———. 2000. "Life and Intelligence Far from Earth: Formulating Theological Issues." Pp. 151–175 in Steven J. Dick, ed., *Many Worlds: The New Universe, Extraterrestrial Life and the Theological Implications*. Philadelphia: Templeton Foundation Press.

Milne, E. A. 1952. *Modern Cosmology and the Christian Idea of God*. Oxford: Oxford University Press.

Monod, Jacques. 1972. *Chance and Necessity: An Essay on the Natural Philosophy of Modern Biology*. Translated by A. Wainhouse. London: Collins.

Mooney, Christopher F., S. J. 1996. *Theology and Scientific Knowledge: Changing Models of God's Presence in the World*. Notre Dame, IN: University of Notre Dame Press.

Moore, James R. 1979. *The Post-Darwinian Controversies: A Study of the Protestant Struggle to Come to Terms with Darwin in Great Britain and America, 1870–1900*. Cambridge, UK: Cambridge University Press.

Moore, Walter. 1992. *Schrodinger: Life and Thought*. Cambridge, UK: Cambridge University Press.

Morales, Joseph F. 1997. "The Hindu Theory of World Cycles in the Light of Modern Science." Available at http://baharna.com/karma/yuga.htm.

Morris, Henry M. 1980. "The Tenets of Creationism." *Impact* #85. Available at www.icr.org.

———. 1985. *Scientific Creationism*, general ed. El Cajon, CA: Master Books.

Morris, Henry M., and Gary E. Parker. 1987. *What is Creation Science?* Revised and Expanded. El Cajon, CA: Master Books.

Murphy, Nancey. 1995. "Divine Action in the Natural Order: Buridan's Ass and Schrodinger's Cat." Pp. 325–358 in R. J. Russell, N. Murphy, and A. R. Peacocke, eds., *Chaos and Complexity: Scientific Perspectives on Divine Action*.

Vatican City State: Vatican Observatory Publications and Berkeley, CA: Center for Theology and Natural Sciences.

Musser, George. September 2005. "The Climax of Humanity." *Scientific American Special Issue*: *Crossroads for Planet Earth* 293.3: 44–47.

Nasr, Sayyed Hussein. 1993. *The Need for a Sacred Science*. Albany, NY: State University of New York Press.

Numbers, Ronald L. November 1982. "Creationism in 20th Century America." *Science* 218 (Reprint Series 5): 538–544.

———. 1992. *The Creationists*. New York: Alfred A. Knopf, Inc.

Olson, Richard. 2000. "Physics." Pp. 247–313 in Gary B. Ferngren, ed., *Science and Religion: A Historical Introduction*. Baltimore: Johns Hopkins University Press.

Pagels, Heinz. 1982. *The Cosmic Code*. New York: Bantam Books.

———. 1985. *Perfect Symmetry: The Search for the Beginning of Time*. New York: Bantam Books.

Pannenberg, Wolfhart. 1981. "Theological Questions to Scientists." Pp. 1–16 in A. R. Peacocke, ed., *The Sciences and Theology in the Twentieth Century*. Notre Dame, IN: University of Notre Dame Press.

———. 1989. "Theological appropriation of Scientific Understandings: Response to Hefner, Wicken, Eaves, and Tipler." *Zygon* 24.2, June 1989: 255–271.

———. June 1995. "Breaking a Taboo: Frank Tipler's *The Physics of Immortality*." *Zygon* 30(2): 309–314.

Peacocke, Arthur. 1993. *Theology for a Scientific Age: Being and Becoming–Natural, Divine, and Human*, enlarged ed. Minneapolis, MN: Fortress Press.

———. 1998. "Biological Evolution: A Positive Theological Appraisal." Pp. 357–376 in R. J. Russell, W. R. Stoeger, and F. J. Ayala, eds., *Evolution and Molecular Biology: Scientific Perspectives on Divine Action*. Vatican City State: Vatican Observatory Publications and Berkeley, CA: Center for Theology and Natural Sciences.

———. 2001. *Paths from Science towards God: The End of All Our Exploring*. Oxford: Oneworld.

———. 2004. *Creation and the World of Science The Reshaping of Belief*. New York: Oxford University Press.

Peters, Ted. 1989. "Cosmos as Creation." Pp. 45–113 in Ted Peters, ed., *Cosmos as Creation: Theology and Science in Consonance*. Nashville, TN: Abingdon Press.

———, ed. 1998. *Science and Theology: The New Consonance*. Boulder, CO: Westview Press.

———. 2003. *Science, Theology and Ethics*. Hants, England: Ashgate.

Peters, Ted, and Gaymon Bennett, eds. 2003. *Bridging Science and Religion*. Minneapolis, MN: Fortress Press.

Peters, Ted, and Martinez Hewlett. 2003. *Evolution from Creation to New Creation: Conflict, Conversation, and Convergence*. Nashville, TN: Abingdon Press.

———. 1998. "On Creating the Cosmos." Pp. 273–296 in R. J. Russell, W. R. Stoeger, and G. V. Coyne, eds., *Physics, Philosophy and Theology: A Common Quest for Understanding*. Vatican City State: Vatican Observatory.

Peters, Ted, Muzaffar Iqbal, and Homanul S. Haq (eds.). 2002. *God, Life and the Cosmos: Christian and Islamic Perspectives*. Aldershot, England: Ashgate.

Polkinghorne, John. 1986. *The Quantum World.* Harmondsworth: Penguin Books.
———. 1994. "A Potent Universe." Pp. 105–115 in Templeton, John Marks, ed., *Evidence of Purpose: Scientists Discover the Creator.* New York: Continuum.
———. 1995. "The Metaphysics of Divine Action." Pp. 147–156 in R. J. Russell, N. Murphy, and A. R. Peacocke, eds., *Chaos and Complexity: Scientific Perspectives on Divine Action* Vatican City State: Vatican Observatory Publications and Berkeley, CA: Center for Theology and Natural Sciences.
———. 1996a. *Faith of a Physicist: Reflections of a Bottom-Up Thinker.* Minneapolis, MN: Fortress Press.
———. 1996b. *Quarks, Chaos, and Christianity: Questions to Science and Religion.* New York: Crossroad.
———. 1996c. *Scientists as Theologians: A Comparison of the Writings of Ian Barbour, Arthur Peacock, and John Polkinghorne.* London: SPCK.
———. 2007. *Quantum Physics and Theology: An Unexpected Kinship.* New Haven, CT: Yale University Press.
Polkinghorne, John, and Michael Welker, eds. 2000. *The End of the World and the Ends of God: Science and Theology on Eschatology.* Harrisburg, PA: Trinity Press International.
Primack, Joel R. 1997. "Cosmology and Culture." *CTNS Bulletin* 17.3: 9–15.
Primack, Joel R., and Nancy Ellen Abrams. 2006. *The View from the Center: Discovering Our Extraoredinary Place in the Cosmos.* New York: Riverhead Books.
Raman, Varadaraja V. 2002. "Traditional Hinduism and Modern Science." Pp. 185–195 in Ted Peters and Gaymon Bennett, eds., *Bridging Science and Religion.* London, UK: SCM Press.
Ranasinghe, C. P. 1957. *The Buddha's Explanation of the Universe.* Colombo, Sri Lanka: Lanka Bauddha Mandalaya Fund.
Randall, Lisa. 2005. *Warped Passages: Unraveling the Mysteries of the Universe's Hidden Dimensions.* New York: Harper Collins Publishers.
Rees, Martin. 1997. *Before the Beginning: Our Universe and Others.* Reading, MA: Addison-Wesley Helix Books.
———. 2001. *Our Cosmic Habitat.* Princeton, NJ: Princeton University Press.
Reese, William L. 1996. *Dictionary of Philosophy and Religion: Eastern and Western Thought,* expanded ed. Amherst, New York: Humanity Books.
Reiter, Lawrence, ed. 2004. *Lord Buddha's Explanation of the Universe.* Adapted from Ranasinghe, C. P. *The Buddha's Explanation of the Universe,* Myrtle Beach, SC: MANifestation, Inc.
Ricard, Matthieu, and Trinh Xuan Thuan. 2001. *The Quantum and the Lotus: A Journey to the Frontiers Where Science and Buddhism Meet.* New York: Three Rivers Press.
Richardson, W. Mark, Robert John Russell, Philip Clayton, and Kirk Wegter-McNelly, eds. 2002. *Science and the Spiritual Quest: New Essays by Leading Scientists.* London: Routledge.
Ross, Hugh. 1995. *The Creator and the Cosmos: How the Greatest Scientific Discoveries of The Century Reveal God.* Colorado Springs, Col: NavPress.
———. 2000. *The Fingerprint of God,* new ed. New Kensington, PA: Whitaker House.
Ruether, Rosemary Radford. 1994. *Gaia and God: An Ecofeminist Theology of Earth Healing.* San Francisco: HarperSanFrancisco.

Russell, Robert John. 1989. "Cosmology, Creation, and Contingency." Pp. 177–209 in Ted Peters, ed., *Cosmos as Creation*. Nashville, TN: Abingdon Press.

———. 1995. "Introduction." Pp. 1–31 in R. J. Russell, N. Murphy, and A. R. Peacocke, eds. *Chaos and Complexity: Scientific Perspectives*. Vatican City State: Vatican Observatory Publications and Berkeley, CA: Center for Theology and Natural Sciences.

———. "T = 0: Is It Theologically Significant?" 1996. Pp. 201–224 in Mark W. Richardson and Wesley J. Wildman, eds., *Religion and Science: History, Method, Dialogue*. New York: Routledge.

———. 2000. "Theology and Science: Current Issues and Future Directions." *CTNS: Publications* 2000. Available at http://www.ctns.org/russell_article.html, June 15, 2007.

———. 2001. "Did God Create Our Universe? Theological Reflections on the Big Bang, Inflation, and Quantum Cosmologies." Pp. 108–127 in James B. Miller, ed., *Cosmic Questions. Annals of the New York Academy of Sciences*, Vol. 950. New York: The New York Academy of Sciences.

———. 2004. "Preface," "Introduction" and "Ian Barbour's Methodological Breakthrough." Pp. xii–xvi, 1–16, 45–59 in R. J. Russell, ed. *Fifty Years in Science and Religion*. Hants, England: Ashgate.

Russell, Robert John, Nancey Murphy, and C. J. Isham, eds. 1993. *Quantum Cosmology and the Laws of Nature: Scientific Perspectives on Divine Action*. Vatican City State: Vatican Observatory Publications and Berkeley, CA: Center for Theology and Natural Sciences.

Russell, Robert John, Philip Clayton, Kirk Wegter-McNelly, and John Polkinghorne, eds. 2001. *Quantum Mechanics: Scientific Perspectives on Divine Action*, Vol. 5. Vatican City State: Vatican Observatory Publications and Berkeley, CA: Center for Theology and Natural Sciences.

Sadakata, Akira. 1997. *Buddhist Cosmology: Philosophy and Origins*. Translated by Gaynor Sekimori. Tokyo: Kosei Publishing Co.

Sagan, Carl. 1980. *Cosmos*. New York: Random House.

———. *Cosmos: Collectors' Edition*. 2000. Originally written, directed and produced by Carl Sagan. For Collectors' edition produced by Ann Druyan and Kent Gibson. Videocassette. Cosmos Studios.

Sagan, Carl and Ann Druyan, eds. 2006. *The Varieties of Scientific Experience: A Personal View of the Search for God*. New York: Penguin Press.

Schroeder, Gerald L. 1992. *Genesis and the Big Bang: The Discovery of Harmony between Modern Science and the Bible*. New York: Bantam Books.

Schrödinger, Erwin. 1964. *My View of the World*. Cambridge at the University Press.

Scott, Eugenie C. 2004. *Evolution vs. Creationism: An Introduction*. Berkeley, CA: University of California Press.

Seventy-two Nobel Laureates, Seventeen State Academies of Science and Seven other Scientific Organizations. *Amicus Curiae* Brief in support of *Appelles Don Aguilard et al. v. Edwin Edwards* in his official capacity as Governor of Louisiana et al., 1986.

Shafer, Ingrid. 2002. "Being Human: A Personal and Mostly Catholic Perspective." Pp. 343–356 in C. N. Matthews, M. E. Tucker, and P. Hefner, eds.,

When Worlds Converge: What Science and Religion Tell Us about the Story of the Universe and Our Place in It. Chicago: Open Court.

Shapley, Harlow. 1964. *Of Stars and Men: The Human Response to an Expanding Universe.* Boston: Beacon Press.

Silk, Joseph. 1997. *A Short History of the Universe.* New York: Scientific American Library.

———. 2001. *The Big Bang.* 3rd ed. New York: W. H. Freeman and Company (1st ed., 1980; rev ed., 1989).

———. 2005. *On the Shores of the Unknown: A Short History of the Universe.* Cambridge, UK: Cambridge University Press.

———. 2006. *The Infinite Cosmos: Questions from the Frontiers of Cosmology.* Oxford: Oxford University Press.

Smith, Howard. 1996. *Let There be Light: Modern Cosmology and Kabbalah, a New Conversation between Science and Religion.* Novato, CA: New World Library.

Smolin, Lee. 1997. *The Life of the Cosmos.* New York: Oxford University Press.

———. 2001. *Three Roads to Quantum Gravity.* New York: Basic Books.

Southgate, Christopher, Celia Deane-Drummond, Paul D. Murray, Michael Robert Negus, Lawrence Osborn, Michael Poole, Jacqui Stewart, and Fraser Watts. 1999. *God, Humanity and the Cosmos: A Textbook in Science and Religion.* Edinburgh: T & T Clark.

Steinhardt, Paul J., and Neil Turok. 2007. *The Endless Universe: Beyond the Big Bang.* New York: Doubleday.

Stenmark, Mikael. 2004. *How to Relate Science and Religion: a Multidimensional Model.* Grand Rapids, MI: Williamm B. Eerdmans Publishing Company.

Stoeger, William R., S. J. 1988. "Contemporary Cosmology and Its Implications for the Science–Religion Dialogue." Pp. 219–247 in R. J. Russell, W. R. Stoeger, S. J., and G. V. Coyne, S. J., eds., *Physics, Philosophy, and Theology: A Common Quest for Understanding.* Vatican City State: Vatican Observatory.

———. 1993. "Contemporary Physics and the Ontological Status of the Laws of Nature." Pp. 209–234 in R. J. Russell, N. Murphy, and C. Isham, eds., *Quantum Cosmology and the Laws of Nature: Scientific Perspectives on Divine Action.* Vatican City State: Vatican Observatory Publications and Berkeley, CA: Center for Theology and Natural Sciences.

———. 1995. "Describing God's Action in the World in Light of Scientific Knowledge of Reality." Pp. 239–261 in R. J. Russell, N. Murphy, and Arthur Peacocke, eds., *Chaos and Complexity: Scientific Perspectives on Divine Action.* Vatican City State: Vatican Observatory Publications and Berkeley, CA: Center for Theology and Natural Sciences.

———. 1998. "The Immanent Directionality of the Evolutionary Process, and its Relationship to Teleology." Pp. 163–190 in R. J. Russell, W. R. Stoeger, and F. J. Ayala, eds., *Evolutionary and Molecular Biology: Scientific Perspectives on Divine Action.* Vatican City State: Vatican Observatory Publications and Berkeley, CA: Center for Theology and Natural Sciences.

———. 2000. "Scientific Accounts of Ultimate Catastrophes in Our Life-Bearing Universe." Pp. 19–28 in John Polkinghorne and Michael Welker, eds., *The*

End of the World and the Ends of God: Science and Theology on Eschatology.
    Harrisburg, PA: Trinity Press International.
Susskind, Leonard. 2006. *The Cosmic Landscape: String Theory and the Illusion of
    Intelligent Design.* New York: Little, Brown and Company.
Swimme, Brian. 1996. *The Hidden Heart of the Cosmos: Humanity and the New Story.*
    Maryknoll, NY: Orbis Books.
Swimme, Brian and Thomas Berry. 1992. *The Universe Story: From the Primordial
    Flaring Forth to the Ecozoic Era, A Celebration of the Unfolding of the Cosmos.*
    San Francisco, CA: Harper.
Talcott, Richard. 2007. "Earth's Deadly Future." *Astronomy* 35(7): 28–33.
Tarter, Jill Cornell. 2000. "SETI and the Religions of the Universe." Pp. 143–149 in
    Steven Dick, ed., *Many Worlds: The New Universe, Extraterrestrial Life and the
    Theological Implications.* Philadelphia: Templeton Foundation Press.
Teilhard de Chardin, Pierre. 1971. "How I Believe." In *Christianity and Evolution.*
    New York: Harcourt Brace Jovanovich.
———. 1999. *The Human Phenomenon.* New edition and translation by Sarah
    Appleton-Weber. Brighton, England: Sussex Academic Press.
Tertullian. 1896–1903. Ad nations. (Holmes, Peter, trans.). In Roberts, Alexander,
    and Donaldson, James (eds.), *The Anti-Nicene Fathers*, Vol. 3. New York:
    Charles Scribner's Sons.
The Gallup Poll. Evolution, Creationism and Intelligent Design, 2007. Available at
    http://www.galluppoll.com/content/default.aspx?ci=21814.
Thaxton, Charles. 1984. *The Mystery of Life's Origin: Reassessing Current Theories.*
    NY: Philosophical Library.
Thuan, Trinh Xuan. 1995. *The Secret Melody: And Man Created the Universe.* Trans-
    lated by Storm Dunlop. New York: Oxford University Press.
Tipler, Frank. 1988. "The Omega Point Theory: A Model of an Evolving God."
    Pp. 313–328 in R. J. Russell, W. R. Stoeger, S. J., and George V. Coyne,
    S. J., eds., *Physics, Philosophy and Theology: a Common Quest for Understanding.*
    Vatican City State: Vatican Observatory.
Tipler, Frank. 1995. *The Physics of Immortality: Modern Cosmology, God, and the Res-
    urrection of the Dead.* New York: Anchor Books.
Toumey, Christopher. 1994. *God's Own Scientists: Creationists in a Secular World.*
    New Brunswick, NJ: Rutgers University Press.
Townes, Charles H. 1998. "Logic and Uncertainties in Science and Religion."
    Pp. 43–55 in Ted Peters, ed., *Science and Theology: The New Consonance.* Boul-
    der, CO: Westview Press.
Trefil, James S. 1983. *The Moment of Creation: Big Bang Physics from before the First
    Millisecond to the Present Universe.* New York: Collier Books.
Tyson, Neil deGrasse, and Donald Goldsmith. 2004. *Origins: Fourteen Billion Years
    of Cosmic Evolution.* New York: W. W. Norton & Company.
Van Huyssteen, J. Wentzel. 2003. *Encyclopedia of Science and Religion.* New York:
    Macmillan Reference.
Van Till, Howard. 1986. *The Fourth Day: What the Bible and the Heavens are Telling
    Us about Creation.* Grand Rapids, MI: William B. Eerdmans Publishing Com-
    pany.

Vilenkin, Alex. 2006. *Many Worlds in One: The Search for Other Universes.* New York: Hill and Wang.

Villard, Ray. 2003. "Order Out of Chaos." *Astronomy* 32(11): 38–43.

Wallace, B. Alan, ed. 2003. *Buddhism and Science: Breaking New Ground.* New York: Columbia University Press.

Ward, Keith. 1996. *God, Chance and Necessity.* Oxford: Oneworld.

———. 1990. *Divine Action.* London: Collins.

Ward, Peter D., and Donald Brownlee. 2000. *Rare Earth: Why Complex Life Is Uncommon in the Universe.* New York: Copernicus-Springer-Verlag.

Webb, Stephen. 2004. *Out of This World: Colliding Universes, Branes, Strings, and Other Wild Ideas of Modern Physics.* New York: Copernicus Books.

Weber, Renee. 1986. *Dialogues with Scientists and Sages: The Search for Unity.* New York: Routledge and Kegan Paul.

Weinberg, Steven. 1977, *The First Three Minutes: A Modern View of the Origin of the Universe.* New York: Basic Books, Inc. (updated edition 1988).

Whitehead, Alfred North. 1962. *Science and the Modern World: Lowell Lectures, 1925.* New York: New American Library Mentor Book.

Wilber, Ken, ed. 1982. *The Holographic Paradigm and Other Paradoxes: Exploring the Leading Edge of Science.* Boston: New Science Library.

———. (ed.). 2001. *Quantum Questions: Mystical Writings of the World's Greatest Physicists.* Boston: Shambhala.

Witham, Larry. 2003. *By Design: Science and the Search for God.* San Francisco: Encounter Books.

———. 2005. *The Measure of God: Our Century-Long Struggle to Reconcile Science and Religion.* San Francisco: HarperSanFrancisco.

Wildman, Wesley J. 1996. "The Quest for Harmony: An Interpretation of Contemporary theology and Science." Pp. 41–60 in W. Wildman and M. Richardson, eds., *Religion and Science: History, Method, Dialogue.* New York: Routledge.

Wood, Kurt. 1993. "The Scientific Exegesis of the Qur'an." *Perspectives on Science and Christian Belief : Journal of the American Scientific Affiliation* 45(2): 90–94.

Worthing, Mark William. 1996. *God, Creation and Contemporary Physics.* Minneapolis, MN: Fortress Press.

Young, Louise. 1986. *The Unfinished Universe: A Radical New View that the Universe Is Perfecting Itself.* New York: Simon and Schuster.

Zajonc, Arthur, ed. 2004. *The New Physics and Cosmology: Dialogues with the Dalai Lama.* Oxford: Oxford University Press.

Zukav, Gary. 1980. *The Dancing Wu Li Masters: An Overview of the New Physics.* New York: Bantam Books.

Index

About the Author

KATE GRAYSON BOISVERT is an instructor in astronomy at Los Medanos College and a consultant at the Center for Theology and Natural Sciences.